Building the Future

Innovation in design, materials and construction

The Institution of Structural Engineers Informal Study Group 'Model Analysis as a Design Tool'

The Group, which was formed in February 1977, operates under the auspices of the Institution of Structural Engineers and presently membership stands at over 600 covering 40 different countries. Members come from a wide range of backgrounds including; research, design, engineering and contracting organisations, universities, government departments and local authorities, and utility companies.

The primary objective of the Group is to create opportunities for members of the Institution and the profession to exchange information on the use of testing and model analysis to solve design problems. The scope of the Group encompasses the whole spectrum of structural engineering applications including; conventional structures, bridges, foundations, pressure vessels, offshore, harbour and coastal structures, etc. It is intended to cover structures made of a wide range of materials and subjected to different loading conditions.

The Group's activities comprise the publication of a quarterly newsletter, organising bi-annual international seminars, visits to test centres in the UK and Europe, sponsoring specialist lectures and holding an annual competition for student dissertations on the application of physical modelling and testing in design.

Further information about the Group may be obtained from the Convenor, Dr F. K. Garas, Taywood Engineering Ltd, Taywood House, 345 Ruislip Road, Southall, Middlesex UB1 2QX, UK.

Building the Future

Innovation in design, materials and construction

Proceedings of the International Seminar held by the Institution of Structural Engineers and the Building Research Establishment, and organized by the Institution of Structural Engineers Informal Study Group 'Model Analysis as a Design Tool', in collaboration with the British Cement Association and Taywood Engineering

Brighton, UK
April 19-21, 1993

Edited by

F. K. GARAS
Taywood Engineering Ltd

G. S. T. ARMER
Building Research Establishment

J. L. CLARKE
Sir William Halcrow & Partners

E & FN SPON
An Imprint of Chapman & Hall
London · Glasgow · Weinheim · New York · Tokyo · Melbourne · Madras

**Published by E & FN Spon, and imprint of Chapman & Hall,
2-6 Boundary Row, London SE1 8HN**

Chapman & Hall, 2-6 Boundary Row, London SE1 8HN, UK

Blackie Academic & Professional, Wester Cleddens Road,
Bishopbriggs, Glasgow G64 2NZ, UK

Chapman & Hall Inc., One Penn Plaza, 41st Floor, New York NY10119, USA

Chapman & Hall Japan, Thomson Publishing Japan, Hirakawacho
Nemoto Building, 6F, 1-7-11 Hirakawa-cho, Chiyoda-ku, Tokyo 102, Japan

Chapman & Hall Australia, Thomas Nelson Australia, 102 Dodds Street,
South Melbourne, Victoria 3205, Australia

Chapman & Hall India, R. Seshadri, 32 Second Main Road, CIT East,
Madras 600 035, India

First edition 1994

Printed in Great Britain by Cambridge University Press

ISBN 0 419 18380 9

A catalogue record for this book is available from the British Library

Library of Congress Catalog Card Number: available

Scientific Committee

Dr G. Somerville (Chairman)	British Cement Association, UK
Dr Ing Klaus Brandes	BAM, Berlin, Germany
Dr J. W. Dougill	Institution of Structural Engineers, UK
Dr K. J. Eaton	Steel Construction Institute, UK
Dr F. K. Garas	Taywood Engineering Limited, UK
Mr Haig Gulvanessian	Building Research Establishment, UK
Prof. Y. Hasegawa	Waseda University, Tokyo, Japan
Prof. Peter Marti	ETH Hönggerberg, Zurich, Switzerland
Dr H. G. Russell	Construction Technology Laboratory, USA
M. Jean-Luc Salagnac	CSTB, France
Prof. T. P. Tassios	Athens University, Greece
Prof. K. S. Virdi	The City University, UK

Organising Committee

Dr F. K. Garas (Chairman)	Taywood Engineering Ltd, UK
Mr G. S. T. Armer	Building Research Establishment
Dr J. L. Clarke	Sir William Halcrow & Partners, formerly British Cement Association
Mr R. J. W. Milne	Institution of Structural Engineers
Mrs P. M. Rowley	Building Research Establishment
Mrs H. M. Stevenson	Seminar Secretary

Contents

Preface

Developments in the design of structures, improvements in existing materials and the introduction of new materials have historically gone hand-in-hand. The limited range of materials available for construction at any one time dictates the practical range of structural form and performance. It follows that current design methods, materials and construction techniques may not be suitable to meet the demands of particular applications. The introduction of a new material, or new practice may move the performance of the construction away from that which is reasonably well understood in the existing population of structures. By so doing, it may also lead to behaviour which cannot be adequately predicted by the current design methods. Developments such as these therefore bring risks as well as benefits. Innovations required to meet the clients' needs can also have a profound effect on the built environment. It is important that any deleterious effects are limited. Undoubtedly all these risks can be most satisfactorily limited by sufficiently high quality physical testing before any significant changes are introduced to construction.

This book is based on the proceedings of the International Seminar held in April 1993, entitled 'Building The Future' - Innovation in Design, Materials and Construction. The objective of the Seminar was to provide a forum for all those involved in innovation in construction to address the problems associated with the introduction of new materials, methods and processes.

Forty four papers are reproduced in this volume together with written discussions and summaries. The following subjects are covered:

General principles and philosophy
Timber
New structural materials
New concrete materials
New concrete techniques and masonry structures
Steel/concrete composite structures
Steel structures
New construction techniques

Together, these represent a state-of-the-art examination of the rôle of physical tests in the development of design methods for new materials, new construction techniques and new forms of construction including practical application of these new methods of construction of buildings, structures and civil engineering works.

F. K. Garas
G. S. T. Armer
J. L. Clarke

Acknowledgements

The editors gratefully acknowledge the work of the authors, session chairmen, the Scientific and Organising Committee and not least the delegates, which ensured the success of the Seminar.

Mrs Patricia Rowley and Mrs Celia Belbin have been responsible for the administration of most of our eight seminars and it is difficult to imagine how we could maintain the efficiency with which this job has been done without their professional expertise and dedication.

Our special thanks are given to Mrs Babs Roberts who prepared and formatted the manuscripts for the preprints and these proceedings. Mrs Julie Smith was responsible for work on the figures and photographs and we are grateful for her support.

PART ONE

GENERAL PRINCIPLES AND PHILOSOPHY

1 MACROMATERIALS

J.G. Parkhouse and H.R. Sepangi
University of Surrey, UK

Structural form is defined in this paper as any periodic shape, and any form (perforated, corrugated, tubular or latticed, for example) is shown to generate its own material from its parent material, just as foam is the vehicle for transforming a material like plastic into another material, foam plastic. It is shown that there are material properties existing at macroscopic scales which are relevant to structural performance, safety and economical design which as yet are not given the recognition we believe they deserve.

INTRODUCTION

It is well known that material strength is a property that is influenced by scale: tiny fibres of glass are much stronger per unit area than rods made of identical material. This can be explained by the defect patterns within the glass. What we suggest in this paper is novel: it is the more general idea that wherever there are patterns not only of defects but also of grains and, at the macroscopic scales, structural forms such as corrugation, perforation, tubularity and latticing, there is a different set of material properties for each scale and the sets associated with the macroscopic scales are particularly deserving of attention.

SHAPE AND FORM

Any shape can be modelled mathematically by a presence field, $s(x,y,z)$, which can take the value 1 or 0 depending on whether the point (x,y,z) is inside or outside the shape. Figure 1 shows a two dimensional shape: the points where $s = 1$ are covered by dark shading, while those points where $s = 0$ are covered by light shading. This shape, like every other, has a closed boundary corresponding to the black line on the figure. $s(x,y,z)$ may be defined by a common scalar function $f(x,y,z)$ by taking s as 1 wherever f is positive and 0 elsewhere. Shapes defined this way are not necessarily as simple topologically as the one illustrated as they may consist of unattached pieces and there may be holes, corresponding to islands and lakes of a typical land mass. The function f could correspond to height of ground above sea level and the boundary, i.e. the coastline, would be defined by $f(x,y) = 0$.

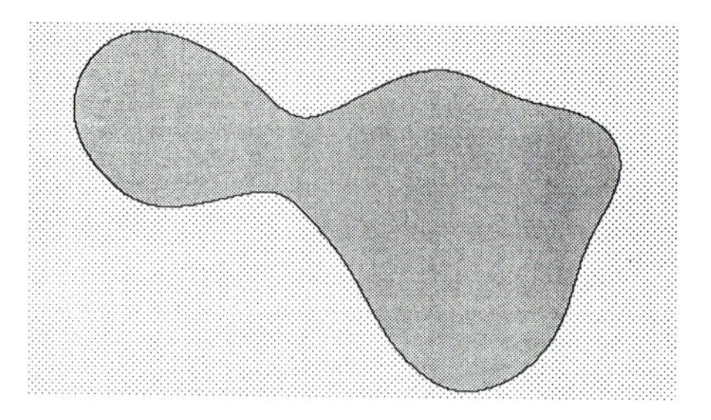

Figure 1. A shape defined by a boolean presence field, $s(x,y)$ taking 1 and 0 inside and outside.

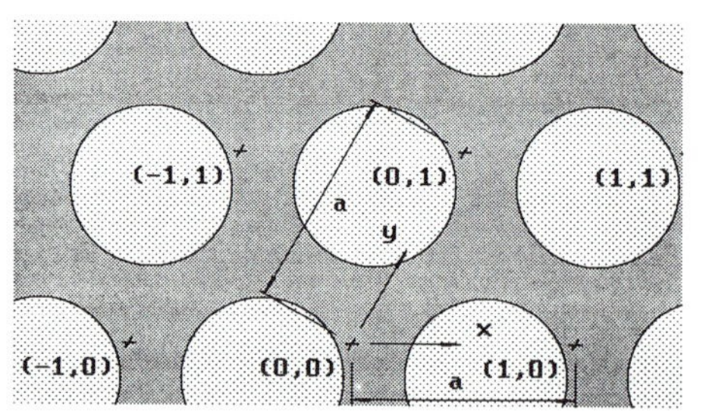

Figure 2. A periodic shape, of form, defined by a periodic presence, values field $s(x,y)$

When the function f is periodic in any or all of x,y or z the shape s that is generated by f is repeating, and we shall call any shape with this property a form. Figure 2 shows a perforated form, a shape which is repeated in two directions with a pitch of a. Pitch can range from the fraction of a nanometre of atomic lattices to the 20 m. member spacing of an offshore structure. The 'address' of any point within this form is its coordinates (x,y), only the periodicity of the form makes it appropriate to separate these coordinates into components (i,j) and (x',y') where (i,j) specifies the cell and (x',y') specifies the location within the cell. If the pitches in the x and y directions are a and b, then

$$x = ai + x'$$

and

$$y = bj + y'$$

Axes are not necessarily orthogonal: they are at 60 deg. to each other in the example of Figure 2 i and j must be integers and x' and y' must always lie between zero and a and b respectively. Varying i and j while keeping x' and y' fixed gives a set of *lattice points*. When a and b are both equal to 1, i and j correspond to the integer parts of x and y to the left of their decimal points, and x' and y' correspond to the non-integer parts. In other words form has just the same hierarchical nature as the arabic numbering system, which is able to describe any number using just ten digits: in a four digit number like the date 1066 the leftmost digit defines the millennium cell, the next the century cell and so on. Another sign of the hierarchical nature of forms is that there are two sorts of adjacent point in the x direction, $x + dx$ and $x + a$, one on the infinitesimal scale and the other, the adjacent lattice point, on a finite scale. Traditional continuum mechanics is founded upon the infinitesimal scale. With forms, finite scales and finite differences assume a new significance.

MATERIAL

A shape is given physical reality when it has material inside it, preferably material with noticeable properties. Less noticeable material, like air, is ideal for the outside if the shape is to be noticed. Mathematically, material is a property field, having values for properties like density, resistivity and stiffness at every point. For example suppose we have material of density $m(x,y)$ shown in Figure 3. Filling the shape $s(x,y)$ of Figure 1 with the material is the multiplication of $s(x,y)$ and $m(x,y)$ to give the structure shown in Figure 4. This confirms the statement made by Parkhouse[1] that

$$\text{shape} \times \text{material} = \text{a structure.}$$

Another assertion made in that paper was that

$$\text{form} \times \text{material} = \text{another material.}$$

When material is dispersed in a repeating pattern its presence is partial: a proportion of any volume considered is occupied by material, and provided the sample volume is chosen carefully this proportion has a value unique to the form. Volumes (or areas in 2-D) which satisfy this sample criterion are called cells, and two such 2-D cells are shown in Figure 5. In 2-D the boundary of one of these cells must be made up of four lines joining any set of adjacent lattice points and opposite lines should be identical to each other apart from a translation by a pitch length in the direction of the other axis. It is clear from the figure that every feature of the form appears exactly once in every unit cell, the complete boundary of one hole for example. We have defined sparsity, i, as the reciprocal of the partial presence factor, i.e.

$$i = \frac{\int_C \mathrm{d}A}{\int_C s \mathrm{d}A}$$

where C describes integration over a cell. In 3-D the integrations would be over volume. The sparsity, i, which Parkhouse has also called the dilution factor[2], is a

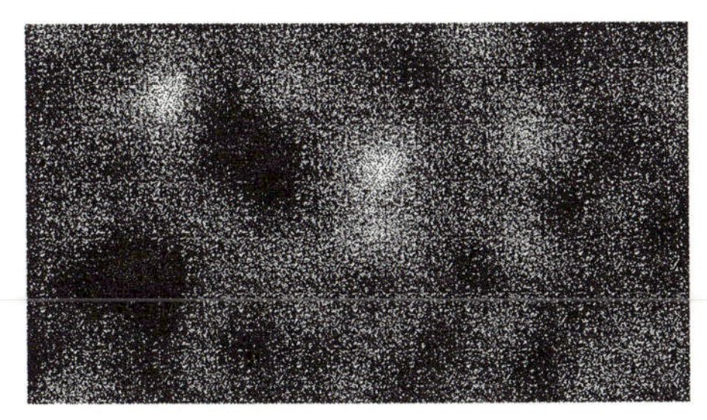

Figure 3. A material of varying intensity $m(x,y)$.

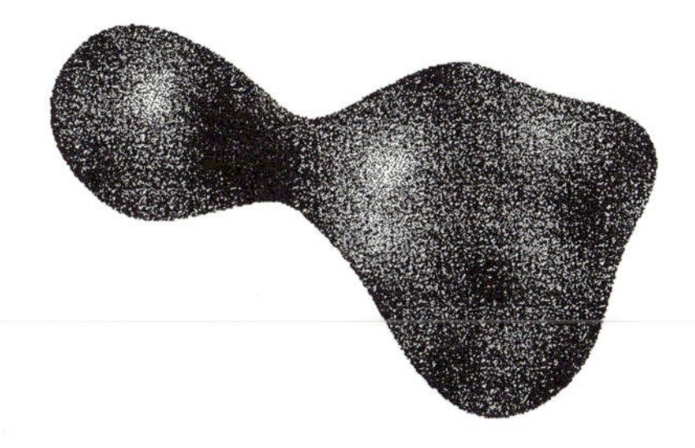

Figure 4. A structure composed of the shape of Fig.1 and the material of Fig.3.

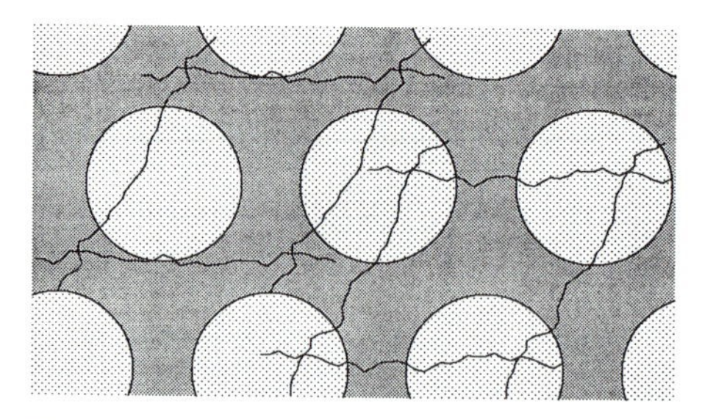

Figure 5. Two unit cells. Each contains just one of each feature of the form.

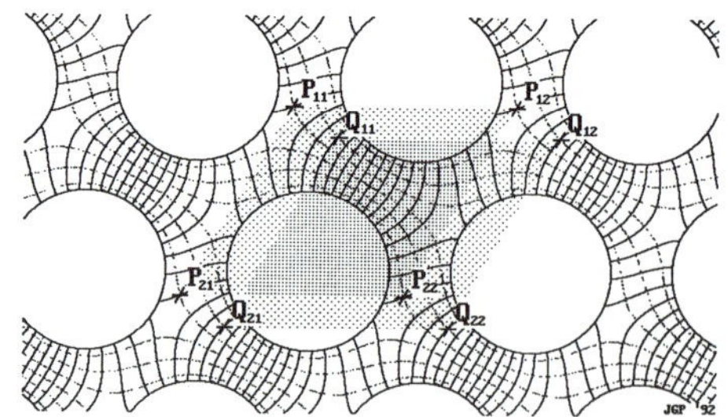

Figure 6. Equipotential solid lines and dotted flow lines through the form.

measure of how sparsely material is spread. It is a property which is independent of which particular point (x,y,z) within the form is being considered so it may be attributed to a uniform material solidly filling the whole volume. Suppose the perforated form is filled with a material of density ρ. Because of the voids in the form the average density will be ρ/i and this average density may be attributed to a uniform material solidly filling the whole area. The material of density ρ will be referred to as the *parent* material and the material of density $\rho' = \rho/i$ as the *equivalent* material. The parent material has been transformed by the form into a different material. Simply, **shape is a transformer of material.**

MATERIAL PROPERTIES

Where there is no density we conventionally agree that there is no material, which is why density may be regarded the most important material property. There are other material properties as simple as density: temperature and electric potential, for example. From these come more complicated properties like potential gradient, current density and resistivity. Resistivity, like density, is an important material property because of its constance: it is relatively unaffected by its environment and distinguishes materials from each other. This cannot be said of temperature. Figure 6 shows a flow pattern in a perforated plate: the dotted lines are currant flow lines and the solid lines are lines of equipotential. Resistivity has three components r_{xx}, r_{yy} and r_{xy} which may be estimated for the equivalent material from measurements of potential difference across adjacent lattice points and the summation of total current flow between adjacent lattice points. A close look at Figure 6 can reveal that both cells P and Q give exactly the same measurements for potential drops and summed flows, demonstrating again that the resistivities of the equivalent material solidly filling the plane are continuum properties of the same nature, but not the same values, as the resistivities of the parent material.

MACROSTRAIN

It should not now come as a surprise to discover that a new set of stresses, strains and stiffnesses arise as a consequence of a form. Figure 7 shows a latticed form deforming from one state defined by a,b and θ to another defined by a',b' and θ'. Note that the deformation pattern is exactly the same in each of the cases (b) and (d): the only difference is that different points on the same form have been chosen as reference points, and the point of the two examples is that the deformation states measured between adjacent lattice points are identical whichever reference point is chosen. The values of a,b and θ are precisely the same in both (a) and (c), and so are the values of a',b' and θ' in (b) and (d). The only difference between (b) and (d) is the rigid body rotations of the two lattices caused by the zero rotation imposed at the reference point. When θ is 90° and the strain is small, the deformation shown by the figures would correspond to engineering strains

$$\epsilon_{11}=\frac{a'-a}{a} \qquad \epsilon_{22}=\frac{b'-b}{b} \qquad \gamma_{12}=\theta-\theta'$$

These equivalent material strains are an entirely different set of strains from the parent material strains which also exist at every point. Since such strains in

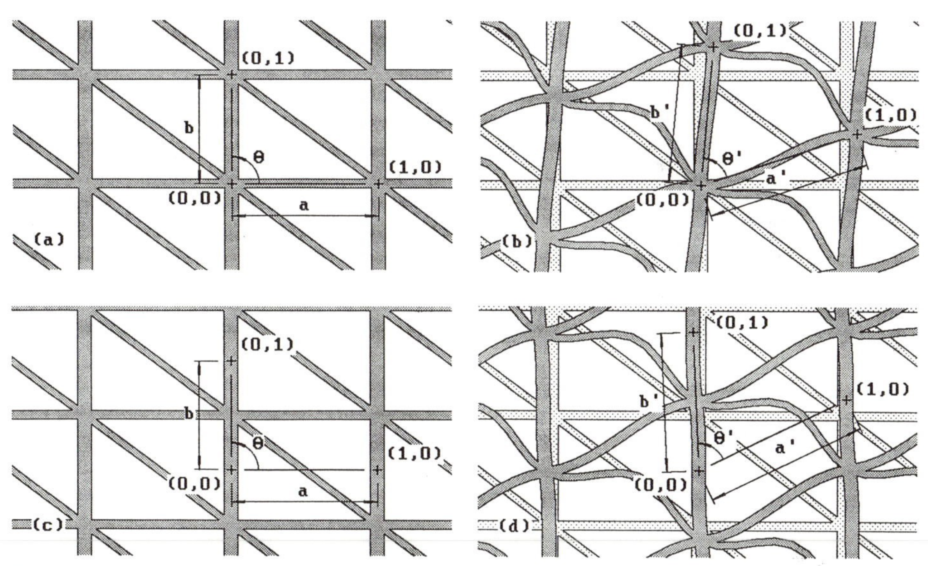

Figure 7. The latticed form on the left is shown on the right after it has undergone a deformation. The difference between the upper and lower illustrations is only the difference in choice of reference point, and this figure illustrates that whatever point is chosen for this purpose the measures of the undeformed state (a,b,θ) and the deformed state (a',b',θ') are the same.

engineering structures occur at rather large scales we shall refer to them as macrostrains. The terms macrostress and macrostiffness will be coined for the same purpose.

What we have shown is that material properties exist at scales other than the infinitesimal. The significance of this to engineering design and development would be difficult to over-estimate, but just now we must admit there are still theoretical difficulties to overcome: when a property like macrostrain is not uniform, the value at the point x could be defined in 1-D as that obtained by averaging either between x and $x + a$, or between $x - a/2$ and $x + a/2$, or between $x - a$ and x. This may be a trivial problem, simply requiring an arbitrary choice, but consider the difficulties of defining macroproperties where the form itself is not uniform, but has varying pitch, and worst of all where a form is discontinuous, as it is at its boundaries: such discontinuities of form are the rule rather than the exception in engineering structures so the measurement or computation of macroproperties over the most commonly encountered structures is as yet problematic.

A form that is periodic in one direction only is relatively free of these problems, its boundaries being confined to its ends. Figure 8 shows a beam, (a) in its undeformed state and (b) deformed. Its deformed state can be expressed in terms of the deformation of its neutral axis by the macrostrains ϵ, γ and κ. ϵ is the axial strain $(a' - a)/a$, γ is a shear strain equal to the angle indicated in the figure and $\kappa = 1/R$ is the curvature of the neutral axis. Which point on the neutral axis is chosen as reference point is again not going to affect the values of the macrostrains measured. What macrostresses, derived from axial force, shear force and bending moment, might be associated with these macrostrains, and could they be matched by a macromaterial stiffness? For incremental strains it turns out that there is always a *unique solid section*, which we call the *equivalent section*, which for a 3-D member is elliptical with its centroid at the neutral axis, which when filled with a *unique uniform material*, which we call the *equivalent material*, behaves for small increments identically to the original beam configuration. This requires the equivalent material to posses appropriate values of Young modulus, two shear moduli and a torsional modulus.

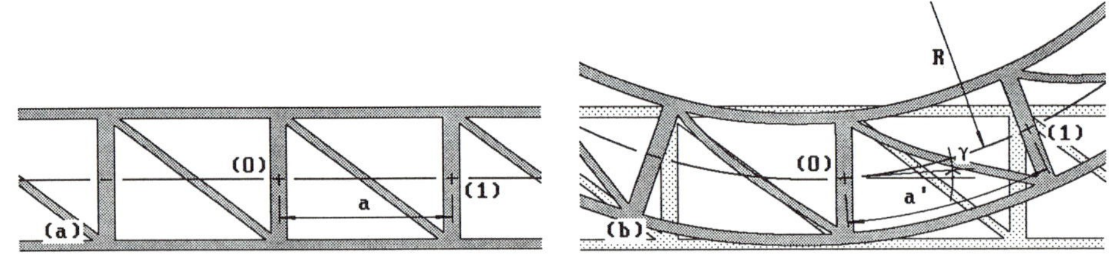

Figure 8. The deformed state of a form periodic in one direction only can be described by the axial strain $\epsilon = (a' - a)/a$, the shearing angle γ and the curvature $\kappa = 1/R$.

SPARSITY

Figure 9 shows three common engineering sections filled with the darker shaded parent material. When this parent material is of uniform stiffness, E, the equivalent solid sections are those illustrated behind, and these are shown filled with uniform equivalent material having such stiffness, E', that the axial and flexural stiffnesses of the equivalent construction are identical to those of the original sections of parent material. When the parent material is uniform it can be shown that the area of the equivalent section, A', must always be greater than or equal to the area of the original section, A. In the first two examples of Figure 9 the ratio A'/A is exactly 10, and we call this ratio sparsity: this definition for shapes is very similar in character to the earlier definition for forms. The elliptical section is that which has the same radii of gyration as the original one. Since both sections with their proper materials are identically stiff it must follow that E/E' must also be exactly 10, and the effect of both the I-section and the tube is to synthesize a material which is 10 times less stiff than its parent material. The densities of the two materials are also in this ratio. This is another illustration that **shape is a transformer of material**, and it also demonstrates that concave shapes like the I-section and the tube introduce sparsity by dispersing material away from itself. In this respect **structuring is a process of material dilution**[2]: latticing, stiffening and tubularity are all ways of making a little material go a long way. A steel I-section is structurally identical to a solid timber section: they serve the same purposes and it is no coincidence that the ratio of both stiffnesses and densities of steel/timber is about 10. Even a solid rectangular section like that shown to the right of Figure 9 has an equivalent elliptical section of greater area, which is why the solid ellipse is chosen: it is the least structured shape, always having a sparsity of one. A rectangle has a sparsity of $\surd(\pi/3)$.

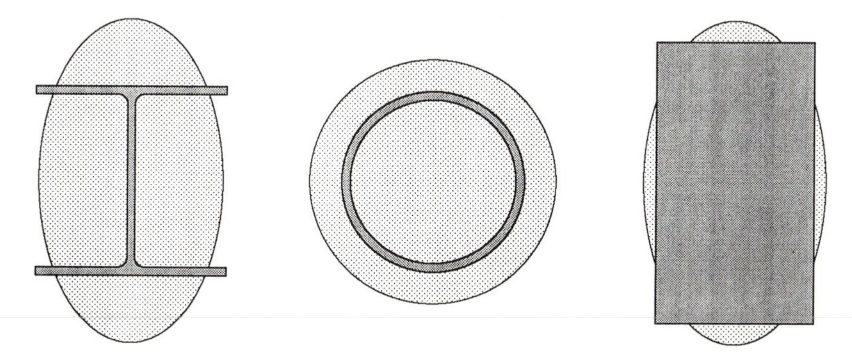

Figure 9. Three common sections in front of their 'equivalent' solid elliptical sections.

APPLICATION TO STRUCTURAL TESTING AND DESIGN

Hooke's Law, discovered in the mid 1600s, was of limited use in structural design until Cauchy expressed the *material* concepts of stress and strain in the mid 1800s when Hooke's Law was expressed as E = stress/strain, a formula that now pervades every structural calculation. We have yet to embrace this material reinterpretation at macroscales. Instead of leaving structure tests as force/displacement curves we could measure displacements of lattice points and estimate the summed loading between them, then compute macromaterial properties which properly reflect the performance of the parent material within the form of that particular structure. The importance of E = stress/strain is the universality of the materials like steel whose E we are constantly using: they are common to all manner of structures. And so it is with form: building frames, tubular steel lattices and prestressed concrete are examples of macromaterials that are used with minor variations over and over again. Their equivalent macromaterial properties like density, Young modulus, strength, toughness, durability and cost can be expected both to give a direct insight to their performance and to be usable directly in design, simplifying the process.

Figure 10 shows two four-chorded GRP lattices 1 m. tall buckling under axial compression, when were tested in 1979. Both had identical chords, but one was singly cross-braced and the other doubly so. The force/displacement curves for both structures are shown in the middle of the figure. Notice that doubling the bracing doubled the buckling load, but their failure modes were different. In the first the bracing did not buckle, only the chords, while in the second the bracing buckled as well, giving a much sharper peak to the load/displacement curve, a more brittle response. The chords had their centres on a rectangle 230 mm by 150 mm, so the radii of gyration of their sections were 115 mm and 75 mm. The ellipse with these radii of gyration has diameters 460 mm and 300 mm and an area of 108,000 mm^2. Dividing axial force by this area and dividing displacements by 1 m. transforms the force/displacement curves into macromaterial stress/strain curves.

The mass of the specimens is estimated as 650 gm and 800 gm. Dividing these masses by the volume described by the equivalent section multiplied by specimen length gives macromaterial densities of 6 and 7.5 kg/m^3, less than one thousandth that of steel. Such low densities are nothing new: the 51,800 tonnes of steelwork in the Forth Railway Bridge is spread within an envelope whose volume is 1,820,000 m^3, having a mean density of 28 kg/m^3. The strengths of the two macromaterials of 0.070 MPa and 0.128 MPa should be compared with about one thousandth of the strength of steel if we are to compare materials of the same density. For prismatic forms, like I- and box-section having a sparsity $i = A'/A$, the best we can expect from a parent material having density ρ, Young modulus E, and strength σ, is a macromaterial having density ρ/i, Young modulus E/i, and strength σ/i. When properties are so finely and uniformly reducible as they are in fluids we call it dilution, but this optimum solid material dilution is only achievable for prismatic members in their axial directions. Due to local buckling the strengths of macromaterials can be excessively diminished compared to their stiffness and density, and due to lateral members like bracing, stiffness can diminish more than the density.

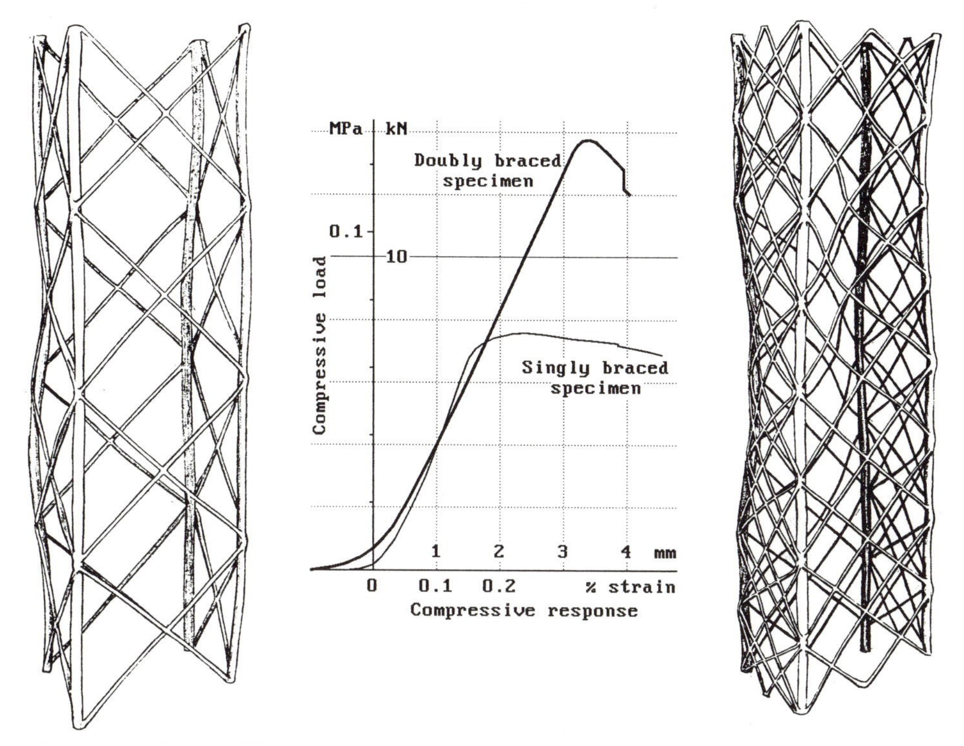

Figure 10. Two GRP lattices each 1 m. tall were tested in compression and are sketched in their buckled states. Their force/displacement curves are shown in the middle. The curve with the thicker line refers to the right hand doubly braced lattice. Doubling the bracing doubles the failure load but alters the failure mode from chord buckling to brace buckling and results in brittler behaviour at failure as indicated by the sharper load shedding of the thicker curve.

For some quantification of these effects refer to[2].

If principal properties like density, stiffness and strength can be transformed by such simple rules, what about toughness? Figure 10 shows that differences in bracing can radically alter the shape of a stress/strain curve. Macromaterials can show non-linear 'yielding' type responses that are nothing to do with ductile flow, and they can fail in true compression. In these respects they present engineering researchers with unexplored territory. Of particular interest is how important a yielding response of a macromaterial is to structural integrity: we know that a yielding response of a ductile parent material is essential in tensile applications, and that brittle parent materials like GRP can provide a yielding response in compression, as in Figure 10, or in the response of elastic Euler strut buckling. We also know that combined buckling and parent material plasticity can produce the

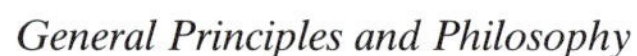

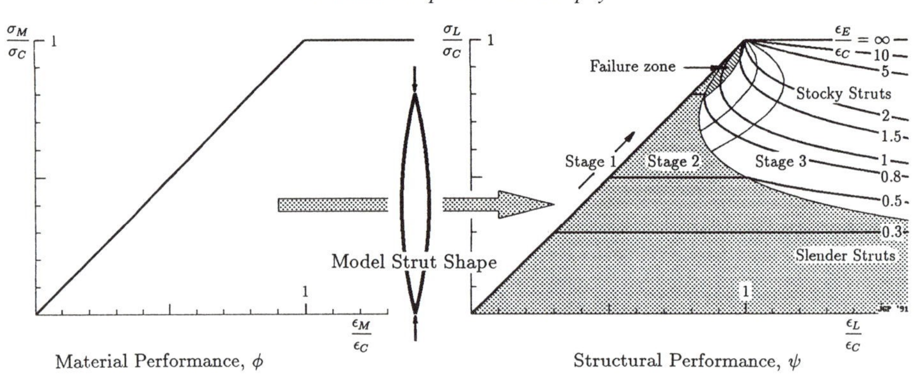

Figure 11. The elastoplastic material on the left is 'fed' into a *model structure*, the lenticular two-flanged structure in the middle of the figure. The response to compressive loading is shown on the right, each curve for a different slenderness.

most brittle responses, as shown in Figure 11, taken from[3]. The optimum strut, which reaches plasticity and elastic buckling at the same load has the most brittle imaginable response curve, like a breaking wave, the one marked '1' on the right of the figure. It seems clear that the shape of the stress/strain curve can be transformed both ways in compression, from brittle to yielding or from yielding to brittle, depending on the shape and form of the macrostructure. Certainly increased strength can be at the expense of increased brittleness, and indications are that very strong brittle parent materials are better for macromaterials working in compression than weaker ductile ones. This phenomenon needs further investigation as it may shed new light on the principles by which we should design.

THE WAY FORWARD

Presently material concepts for engineers are mostly confined to a laboratory scale, the scale defined by the width of our material test specimens. Materials' performance at these scales is widely available, and the materials industry is thriving on the development and manufacture of the burgeoning number of new materials coming to the market place. In contrast engineers have great difficulty knowing how to form new materials appropriately. We suggest this is because of a need for more useful shape concepts. The macromaterial concept that has been described is offered to meet this need. Ashby is recognising this need and in his recently published book[4] he not just provides a wealth of materials data in accessible form but introduces shape factors that for prismatic members effectively generate macromaterials.

The authors gratefully acknowledge sponsorship from BP International Ltd from 1986 to 1993 through their Venture Research initiative, and recently from the University of Surrey's Foundation Fund.

REFERENCES

1. Parkhouse J.G. 'Damage accumulation in structures.' Reliability Engineering **17** (1987) 97-109.
2. Parkhouse J.G. 'Structuring: a process of material dilution.' in Proc. 3rd Int. Conf. on Space Structures ed. H.Nooshin. Elsevier Applied Science Publishers (1984) 367-374.
3. Parkhouse J.G., Sepangi H.R. & Williams W.E. 'Structural simplicity through a lenticular pattern.' Int. J. Mech. Sci. In press (1992/3).
4. Ashby M.F. Materials Selection in Mechanical Design. Pergamon Press (1992).

2 DEVELOPMENT OF SKILL MODELS AS AN AID TO BUILDABILITY IN DESIGN

D.R. Moore
De Montfort University Leicester, UK

This paper deals with the development of Skill Models for use in assessing a design for buildability by rating the Task Difficulty inherent in its construction. Skill Concept Packages would be assembled to create Skill Models of the construction process. Assessment and modification of the design could be carried out as the design phase proceeds within a CAAD environment. This approach should aid in achieving good buildability whilst allowing for design creativity and innovation.

INTRODUCTION

The concept of buildability was seen within the construction industry as a tool for eradicating problems such as poor on-site performance, late completion and poor product quality. Such problems were generally seen to result from an incomplete understanding, by both constructors and designers, of the increasingly complex process of construction. Buildability was suggested as being a discipline which would enable the process to be understood, thereby eliminating problems. In practice, however, the utility of buildability was compromised. One compromising factor was the differences in the content of the definitions of buildability being put forward. Such differences were almost inevitable given the complexity of the problem being analysed. A second factor was the subjective nature of many aspects of buildability.

That buildability did not wholly succeeded is evidenced by the emergence of Quality Assurance (QA), which has its origins in the manufacturing industry. QA has, in turn, become compromised itself. Some companies have rejected QA as being unsuitable for the construction industry. This point was explicitly stated by a 1987 report on quality management systems within the industry.(1) Ten characteristics which caused construction work to differ from that contained within the quality standard BS 5750 were identified by the report. It has been argued that QA operational decisions which are made on inadequate information cause QA to be seen more as a promotional tool than a means of actually achieving quality.(2)

The industry is still attempting to overcome those problems which neither buildability nor QA were able to resolve. Further quality standards are being offered up as the solutions needed by the industry. It is suggested that these new standards will suffer the fate of their predecessors unless they are able to fill in the gaps in the knowledge of practitioners regarding the process of construction. Under such circumstances the improvement of the utility of buildability would be as equally valid as the development of a new concept or standard. This paper will put forward one possible technique for improving the utility of buildability as a means of dealing with design related problems in the industry.

DEFINING BUILDABILITY

Definitions of buildability vary in their precision. The definition suggested by CIRIA[3], for example, appears somewhat vague when compared to the totality of the definition of constructability (the American version of buildability)[4]. A more specific definition is that formulated by Illingworth[5], which examines the relationship between design and assembly: '**..the design and detailing which recognise the problems of the assembly process in achieving the desired result safely and at least cost to the client.**'. This will be the definition of buildability used throughout the remainder of this paper.

Both buildability and QA can be seen as part of total quality. Total quality has yet to be defined satisfactorily in construction specific terms, possibly because the industry took the QA route as being the relatively easy option. BS 4778:1987 Pt 1 defines total quality as:' **..the totality of features and characteristics of a product or service that bear on its ability to satisfy stated or implied needs.'** From this it can be seen that one aspect of total quality is problem prevention. The adoption of buildability could therefore be seen as a step towards problem prevention which did not achieve its full potential. Possible reasons for this failure can be found by examination of the philosophy of buildability.

PHILOSOPHY OF BUILDABILITY

Buildability is frequently discussed in terms of concepts which can be seen as being more akin to the organisation of manufacturing industry than that of the construction industry. Factory style construction processes can be perceived by designers as being constraints on design creativity and innovation. Buildability may not therefore be readily accepted by a building design profession whose main advertisement is the buildings produced from its designs. The elements of creativity and innovation are seen as essential in such circumstances. It is vital that designers feel able to accept and incorporate buildability in their designs without losing the freedom to be creative or innovative. The terminology of buildability does imply mass production rather than creative individualism. Repetition, rationalization and standardisation are all terms within the buildability philosophy. Simplification, in the sense of avoiding unnecessary technical difficulty, is also part of the buildability framework. By emphasising simplification rather than standardisation in design, buildability will be improved and creativity and innovation encouraged.

Buildability also requires that the designer should consider the design in terms of the practical consideration of both construction costs and technology.[6] Part of the practical consideration of any technology relates to assessment of the skills required for its successful implementation; an area requiring experience on the part of the assessor in order to minimise chaos.

CHAOS AND SKILLS

The buildability ethos requires the creation of an environment in which problems regarding the intended construction process can be identified. One source of problems is chaos. This is the state which results from one element of a system advancing more than the remaining elements. Chaos will exist until the elements lagging behind catch up. As advances in, for example, the production of construction materials occur some degree of chaos in the industry is inevitable. Only when the properties of the new materials are fully understood and utilised will chaos be reduced. Such system lags cause a variety of problems. One such problem is the assessment of the degree of skill required to allow new technologies within a design to be implemented successfully. If the design 'system' is accepted as working on the basis of fragment retrieval, in which fragments of existing knowledge are assembled to form a design, there is an obvious reliance on the quality of that knowledge. In this area the expertise and therefore the concepts of the designer and the constructor may be significantly different. Such differences can be referred to as 'concept differential'.

In the case of existing technology, experience has enabled some individuals to have an 'educated' concept of the skills required for particular tasks. Such individuals could be said to possess a 'concept package' which is similar to that of the operative carrying out a given task. This may result from the individual being involved in experiences not dissimilar to those of the operatives and therefore their assessments of skills required could both be classed as 'experience based'. This would result in a low concept differential between the two parties. For those without such a concept package assessment of skills required will be more difficult but not impossible. Such individuals can resort to 'vicarious experience based' assessments. One approach would be to distil the experience based knowledge of a number of individuals to form an average indicator of skills requirement. The various 'price-books' are examples of the vicarious approach. This approach will result in a high concept differential due to the implied nature of skill assessment. The argument that 'book knowledge' appears to miss the experimental aspect of expertise development was also suggested by Dreyfus and Dreyfus[7].

Peter Rogers[8] has recently stated that 90% of cladding units tested needed some form of modification. It was also stated that '**One of the major reasons..(is**

that)..each block obtains what is effectively a prototype.' In effect the cladding systems are 'tested' by the client during the post-construction phase. This situation may arise, at least in part, by the designers of the units not being constrained by a low concept differential with regard to the installation operatives. This is a problem solving environment rather than a problem prevention environment.

A problem prevention environment would seek to develop a means of assessing all levels of the design against existing concepts of skills prior to producing the components. In other words, it would eliminate the concept differential. This would be a third possible approach to skills assessment (neither experience nor vicarious based) and could be referred to as being 'actuality based'. The development of skill concept packages (SCoPs) in a form which would allow them to be applied during the design process would represent an 'actuality based' assessment system. This would aid both buildability and total quality through the adoption of a discipline requiring the attainment of a low concept differential in all aspects of a design.

CONCEPTS OF SKILL

Concepts of skill will vary with the subjective values of individuals. Consider the skill involved in bricklaying; is it simply the ability to lay individual bricks to form a particular pattern? Or is it the knowledge required to construct a particular part of a building using bricks? Depending upon the assumption made both the type and the level of skill required will be considerably different. Clarke[(9)] saw skill as being traditionally defined and also an elusive term, being historically specific only to changes in the social relations of building production. The development of skill could therefore be seen as a process whereby the labour resource attempts to 'catch up' with changes in the technology resource within the construction and design processes.

In such circumstances SCoPs developed as a means of modelling a skill would need to allow for the evolving of that skill within an industrial environment. This could require the inclusion of those subjective values which are clung to independent of experience, as identified by industrial psychologists and referred to as frustration. One particular aspect of frustration is fixation, a state in which actions will be repeated even after it has been demonstrated that they are not beneficial to the task in hand.[(10)] Without inclusion of such values SCoPs would run the risk of optimising rather than replicating a skill. Such values would be difficult to quantify as evidenced by Patemans[(11)] statement, in connection with Total Quality Management, that '**.. if you can not measure it do not include it..**'.

It is suggested that the inclusion of objective assessment criterion only would be a distortion of essentials of a skill. This would cause the resultant skill model to be devalued in the manner of buildability and QA. It is therefore important to determine a definition for skill. To this end existing areas of study can be examined.

The area dealing with the psychology of skill suggests one approach to the task of what may be recognised and modelled as skill. As there was with buildability, so there is also a difficulty in defining 'skill'.

DEFINING SKILL

The industrial concept of skill as being a result of specific training can be compared to the concept of skill used in psychology which is given by Welford(12) as being '**..all the factors which go to make up a competent, rapid and accurate performance.**' Control of such factors is by the brain which Craik (1943) was insistent must be thought of as a computer receiving many inputs which it then combines to produce unique, but lawful, outputs.

One possible description of a skilled construction performance deals with the manner in which inputs of data arrive at the brain-computer and are then processed. Data inputs arriving at different points in the process of, for example, laying bricks may need to be combined. This will require the bricklayer to exhibit short-term retention of the inputs in order to mentally record what has been done and compare this to what has to be done. To do this the brain-computer must be able to model the construction of the wall which is external to itself. The process of skill modelling also requires that SCoPs should be capable of being assembled so as to model the process of construction which is demanded by a design.

Craik suggested that the brain-computer or '**neural machinery**' is able to extrapolate from data which has been input and therefore test various results without actually taking the physical action(s) required to arrive at each result. This is in agreement with phase 3 of Krendal and McRuer's(12) skill development model and suggests that thought is a series of definable brain-computer operations. Krendal and McRuer proposed a model for the development of skill based on '**..successive organisations of perception..**'. There must therefore be a program for the practising of a task which the brain-computer has to learn and the level of skill of an operative will depend on his/her ability to learn and implement that program. When program implementation is to be assessed each factor within the task represented by a program will need to be considered.

MEASURING SKILL

The area of work measurement suggests that an operation can be decomposed into a series of predetermined time values, referred to as Predetermined Motion Time Systems (PMTS). However, construction tasks may be amongst some of the most complex tasks to decompose because they are seldom truly repetitive in the manner of, say, drilling specified holes in steel plates.

PMTS can be improved upon by considering the area of research concerning the effects on time taken for tasks involving simultaneous symmetrical and asymmetrical motions.(13) Such research has examined hand movements only, within a fixed station type task. For construction tasks the relationship of all body movements which are a response to carrying out the task needs to be quantified. The identification of a relationship between successive symmetrical and asymmetrical movements could be the basis for both the identification of tasks required by a given operation and the 'lawful' rules by which those tasks are compiled. In other words a skill model would seek to identify and implement the relevant program (composed of the required SCoPs) for a given operation.

Task difficulty assessment

In order to extract benefit from a Skill Model the SCoPs used must be capable of assessing the difficulty experienced in completing the tasks required by the modelled operation. This would be, in effect, the basis for the measurement of skill. It is obvious that some tasks are more difficult than others but it is not always so obvious what it is about a task that makes it difficult. To reduce this difficulty initially it is suggested that the assessment of task difficulty (TD) should be solely on the basis of objective criterion.

Attempts have been made to assess TD in various ways. One approach used by Moore(14) was to survey experienced roofing contractors regarding the difficulty experienced in completing given roof details. This investigation aggregated subjective assessments and in working to a consensus did indicate the recognition of varying TD within roofing tasks. This was not, however, a fully objective assessment but it did indicate a low concept differential regarding TD in an area in which respondents were experienced.

A further attempt at measuring TD is discussed by Raouf as being:
TD = log² 2D/C, where D = distance moved by the hand and C = lateral clearance of fit.(13) This method of calculating TD has seen further development but has not been explicitly included in PMTS. Assessment of TD in this manner indicates a relationship between the nature of the movement required and that which the body is capable of achieving. Preliminary research has shown that a relationship between body sections can be formulated in terms of the movement carried out at each section joint. Each joint has a natural limit on the movement which it is capable of undertaking and this can be measured. Certain joints can move in three axes whereas other joints will accept movement in only two axes. If a task can not be completed within the movement allowed on one joint then further movement must be carried out on successive joints until the task is completed. In extreme cases this further movement will take the form of perambulation to take the operative closer to the task or object under consideration.

By establishing the maximum movement possible at each joint it should be possible to establish a relationship between distance of reach and percentage of movement at each joint required to complete that reach. Such relationship based modelling suggests the possibility of reducing the size of knowledge base required to carry out skill modelling. It may be feasible to have one function statement for each degree of freedom on each joint, leaving the system free to select function statements from which to assemble the required movement(s) and thereby model the required task. The selection of function statements would be governed by a series of rules.

TD could then be assessed in terms of considering at what point in the movement range of one or more joints, in each possible axis, the task is being carried out. For example, a task being carried out at either of the extremes of a joints movement may be more difficult for the operative than one being carried out in the centre of the movement range. Consequently TD could be assessed as being higher. By summing the TD on each axis for each joint used an overall TD rating could be achieved.

SKILL MODELLING ENVIRONMENT

It is envisaged that skill modelling would be carried out within a CAAD environment and implemented at regular stages of the design process. This presents two particular problems. CAAD systems in general are memory hungry and the addition of a further sub-system would need to be done in a way which minimises additional memory requirement. CAAD systems also rely heavily on the calculating power of the system to carry out the mathematics involved in locating successive ordinates. In seeking to reduce memory requirement care should be taken not to place undue emphasis on the need for successive complex calculations. This returns us to the earlier point regarding the use of single function statements for each degree of freedom on each joint as perhaps being a suitable approach.

One approach to the provision of a skill modelling facility would be to develop a knowledge based system. An important first step is to determine which level of knowledge the system will need to attain; surface (heuristics, facts); domain/procedural, or deep (laws, principles). At present the technology of knowledge systems does not allow for the creation and use of deep knowledge. A simple rule based system will create surface knowledge and a more sophisticated hybrid system allows the creation of domain knowledge. The majority of design, planning and scheduling systems rely on domain or procedural modes.[15] If skill modelling is to be developed for use within a computer environment it would appear to be possible only if the required system can be shown not to rely upon the creation of deep knowledge. This is assuming that the present majority view of knowledge base systems is accepted. In such a case the precognitive behaviour typical of the third phase of learning a manual tracking task would have to be achievable by the skill modelling system without recourse to the creation of deep knowledge.

An alternative would be to consider the development of a self reproducing knowledge base existing within the design environment. An example is given by the development of cellular automata (CA's) which has suggested that they can be used to model various natural occurrences such as 'boid' movements, within the computer environment.[16] This ability to model nature could be of use within skill modelling of construction operations if the rules governing, for example, the laying of bricks are natural rules. If this is not the case then the rules resulting from use of CA's could be different to those actually used on site and would therefore be of little value.

CONCLUSIONS

It is important that buildability is returned to the design process agenda in order to place the emphasis on problem prevention within the industry rather than on problem solving. To achieve this buildability should also have a change of emphasis from standardisation to simplification, thereby allowing and encouraging design creativity and innovation. The emphasis on simplification should be achieved by adopting a design phase discipline which seeks to negate the influence of chaos within the construction system. Such a discipline could be provided by the process of skill modelling based on SCoPs which are actuality based and thereby reduce the chaos resulting from significant concept differential. Through the use of skill modelling the design team can be directed to technically difficult aspects of a design prior to construction taking place. Such aspects can then be further simplified on a 'what-if?' basis or replaced entirely. By placing the skill modelling system within a CAAD environment a design/re-design facility can be supplied in a convenient form. This may then cause buildability to be considered as an aid to design creativity and innovation rather than a restriction.

REFERENCES

1. McLellan A. 'BS5750 Battle' New Builder. New Builder Publications Ltd. (22 February 1990)
2. Tietz S.B. 'New materials and new technologies - their effect on building design and maintenance'. CIOB Technical Information Service Paper No. 88 (1988)
3. Construction Industry Research and Information Association 'Buildability: an assessment'. Special Publication No. 26 (1983)
4. Hon-Kueng, Lai 'Integrating total quality and buildability.' CIOB Technical Information Service No. 109. (1989)
5. Illingworth J.R. 'Buildability - tomorrow's needs?' Building Technology and Management. (February 1984)
6. Hon-Kueng, Lai. 'Integrating total quality and buildability.' CIOB Technical Information Service No. 109. (1989)
7. Dreyfus H.L. and Dreyfus S.E. *Mind over Machine: The power of human intuition in the era of the computer.* Basil Blackwell, Oxford. (1986)
8. Macalister T. 'When cladding fails to weather the test'. Guardian. (10 August 1992.)

9. Clarke. L. 'On the concepts of skill and training in the construction industry.' Proceedings 1983 Bartlett International Summer School. Geneva. (1984)
10. Brown J.A.C. *The Social Psychology of Industry*. Penguin Books Ltd. (1986).
11. Pateman J. 'A culture of customer satisfaction'. Chartered Builder. (May 1992).
12. Welford A.T. *Fundamentals of Skill.* Methuen & Co Ltd. (1968).
13. Raouf A., Tsuchiya K. & Morooka K. 'Effect of task difficulty and angle in a positioning task involving symmetrical and asymmetrical motions.' International Journal of Production Research. Vol 20. (Nov/Dec 1982.)
14. Moore D.R. 'Task difficulty and quality control'. (Dissertation) Nottingham Polytechnic. (1990).
15. Harmon P. & Sawyer B. *Creating Expert Systems*. John Wiley & Sons. (1990).
16. Levy S. *Artificial Life*. Jonathan Cape, London. (1992)

3 DURABILITY DESIGN: FORM, DETAILING AND MATERIALS

J.G.M. Wood
Structural Studies & Design Ltd. UK

Over the centuries builders slowly evolved structural forms and detailing which achieve durability for a range of traditional materials. The durability performance of modern construction materials is reviewed with examples of success and failure of steel, concrete, timber, polymers, and GRC. This shows that physical and chemical testing to quantify material characteristics and the determination of the microclimates in structures are essential for the achievement of durable buildings for the future, with existing and new materials. Durability design must evolve the form and detailing for structures to create suitable microclimates in which the rates of the deterioration of materials are consistent with the client's specification for the design life and maintenance regime.

INTRODUCTION

For the last thirty years structural analysis to reduce weight and first cost, has dominated the design effort in engineering offices. It has also dominated engineering education. The computer has enabled structural proportions to be minimised using strength criteria which generally presume no deterioration. Materials specification has similarly been concentrated on strength. The contractors and materials suppliers have made progress mainly in speeding construction, maximising early strength and minimising cost.

Now Marsham Street, the Palace of Building Regulations, has been classified as uneconomic to maintain because of a failure of durability. A substantial proportion of post-war construction is failing to meet its design life objectives, because of poor durability. This blight affects reinforced concrete, prestressing steel, new materials like GRC and plastics, galvanised wall ties and lintels in brickwork, soft wood window frames, steelwork, etc. Meanwhile UK structures of the 11th C to the 19th C remain durable with the prospect of centuries of future use.

It is time that explicit durability design was established as a major part of the design process in Engineering and Architecture. It is the form and detailing of structures and quality of materials which determine the reliability of structures, not the fine tuning of partial factors on loading and strength. However quantitative durability design will require substantially more physical and chemical testing to establish the characteristics of materials and to quantify the environmental processes which degrade. The modelling of degradation phenomenon can then enable deterioration to be predicted, as readily as strength, in the analysis of design options. This quantitative durability design can be used for major projects and for establishing more reliable simplified standards for routine design. The costs to society of premature deterioration of structures necessitate this change from hopeful empiricism to a rigorous scientific approach to durability.

TRADITIONAL CONSTRUCTION

The careful observation of the characteristics and behaviour of traditional design should be included in the education of all engineers and architects. This should continue as part of their 'Continuing Professional Development' as they walk around our cities and when they are on holiday or working abroad. The traditional styles of brickwork, stonework, metallic and timber construction evolved through a process of observing failures and remedying them in new construction, until a structural form and detailing consistent with the climate, available materials and use, was found. This was easy over the time scale of centuries with a tradition of settled craftsmanship passed from Master to Apprentice. However durability is only enjoyed by building owners who have given inspection and maintenance due priority.

While these observations can give us a qualitative understanding, it is only by testing to quantify the material characteristics and microclimates in and around the structure that we can the determine the long term sensitivity of rates of deterioration to changes in exposure. There are enough centuries old structures to enable us to judge the durability of traditional construction by comparison, without the need for extrapolation.

MODERN CONSTRUCTION

Many of the materials and forms of current construction have a short history. Their long term performance cannot be judged by superficial inspection. However the microclimate to which materials are exposed, and the early signs of deterioration in locally severe conditions, can often be spotted in the first year or two of the life of a structure. Within a decade too much modern construction is starting to show obvious signs of decay. So careful observation of the structures around us must be the first step in improving the form and detailing of new construction.

There are difficulties in recreating in the laboratory the range and complexity of the environmental conditions in the field. Accelerating the deterioration of 100 years of field performance into the year or so of the normal experimental programme or the minutes or hours of 'Rapid Testing' is not easy to validly achieve. Some tests merely mislead. Quantitative measurement of the range of performance in the field by chemical and petrographic testing of samples can show the start of trends after only one or two years (eg carbonation depth, chloride ingress). Details and materials showing poor and good trends can be identified to guide new design. This field study can be linked to laboratory testing and predictive modelling of deterioration processes. The data can provide a base for extrapolation to predict long term deterioration over decades. Long term exposure tests, like those on sulphate resistance, marine exposure, corrosion(1) and carbonation by BRE, provide an essential link in this process.

INNOVATIVE CONSTRUCTION AND MATERIALS

When new materials, structural forms, or unusual environments are encountered, the comprehensive evaluation of physical and chemical stability and the compatibility of ingredients in composite materials, should enable better predictions to be made than in the past. The early age field performance must be closely followed by testing, rather than waiting until the owners of structures sue for defects.

TECHNICAL LIFE

The objectives for design life are well discussed in broad principle in the papers to the 1990 IABSE Colloquium on 'The Design Life of Structures'(2). The new BS 7543:1992 'Guide to Durability of Buildings'(3) gives definitions and general guidelines. The difficulties arise with specific applications. The experience of durability failures over the last 30 years does not give confidence that compliance with British Standards will achieve 'fitness for purpose' for designers. It is a moot point whether the BS by themselves even meet the requirements of 'reasonable good practice' for normal design. The move to European Codes is further lowering the durability quality in some standards.

It is important that the design objectives for durability are agreed with the client in writing. The agreed balance between first cost, maintenance costs, 'technical' life, level of supervision of materials and construction, speed of construction etc should be set out for the designer, specifier and, if things go wrong, for the lawyers. Then a proper economy can be achieved for fast track construction of short life warehouses and sufficient rigour can used for major structures.

The concept of using a 'Technical Life' for design is summarised by Somerville(4). This is not the time to structural collapse, but the time for which the owner can maintain the economic use of the structure without undue maintenance cost. Replaceable parts may well have a shorter technical life than the main frame or foundations for which effective maintenance or replacement would be uneconomic. It is important to appreciate that serviceability failures, eg spalling of facades, water ingress, or poor appearance, can make continued use of buildings uneconomic. It is often secondary elements and details of connections, that trigger wider deterioration and govern the technical life of a structure.

MICROCLIMATES

The first lesson from traditional construction is the over-riding of importance of the control of water to achieving a stable microclimate in which deterioration is slowed or halted. The large overhanging roofs of wooden alpine barns and their cold and dry environment ensure that the construction material is kept within the range of humidities in which rot does not occur or is very slow. The detailing of stonework to shed water away from the face, with particular attention to the provision of drips and the avoidance of ponding surfaces by a slight slope, keeps stonework dry enough to minimise frost damage and reduce the effects of acid rain.

This attention to detail can be contrasted with the way in which poor detailing has aggravated the deterioration of concrete bridges from salt induced corrosion. The BS 5400: 1984, Pt 4, Table 13, 'Environment' classes only crudely classify conditions for chloride ingress in specifying concrete quality and cover. It equates seawater spray for concrete adjacent to the sea with surfaces directly affected by deicing salts. Road salts are spread as a solid, not the dilute 19,000ppm Cl^- of the sea. Where drainage is good and rain can wash off the surface salt, the specified 40 mm nominal cover with Grade 50 delays corrosion initiation for some decades. However where salt seeps through expansion joints and can accumulate on level bearing shelves or debris clogged drainage channels, it concentrates and the ingress of chloride and start of corrosion can be accelerated by a factor of ten.

For surfaces 'Protected by bridge deck waterproofing' BS 5400 permits, for Grade 50, the nominal cover to be reduced from 40 mm to 25 mm (ie 20 mm minimum) cover. The reality is that leakage of bridge waterproofing is widespread and the constraints on work on motorways can delay remedial works for years. The rate of ingress of chloride into concrete under leaking waterproofing is faster than that into a bare rain washed concrete deck with good falls. It is not good practice to reduce concrete cover under waterproofing. A 120 year design life for concrete bridges can only be achieved by improved materials combined with a radical review of the classification of 'environments' to accurately reflect the effects on chloride ingress of different details. Similar developments are needed in classifying exposure in relation to carbonation.

Temperature accelerates deterioration with a doubling of the rate of chloride ingress and of corrosion for every 10°C increase in temperature. The durability criteria in standards do not take this into account. Concrete cover for a marine structure which gives 60 years life in the North Sea, may only give 40 years in the Channel, 20 years in the South of France and 10 years in the Gulf. The effect of temperature on deterioration rates is so great that both climate and the variation of temperatures within structures need to be considered in durability design. This is desirable within the UK, but is essential for international work.

The cycling of temperature and humidity are as important in deterioration as the absolute values. Permafrost preserves, freeze thaw fragments. The stresses arising in materials and structural elements due to differential heating and cooling and wetting and drying are a prime cause of physical degradation. To quantify these effects the internal thermal and moisture gradients in structural elements and their rates of change need to be established, particularly for composite construction

EVALUATION AND TESTING OF MICROCLIMATES

The evolution of quantitative durability design will necessitate the physical testing of structures to quantify the microclimates combined with laboratory and site measurement of deterioration rates associated with known materials and environments.

Microclimates can be classified in terms of the cycles of temperature, relative humidity and concentration of aggressive ions at the surface or in materials. Figure 1 shows a simplified format for summarising this. For specific deterioration modes environments can be graded by deterioration rate of a reference material, eg corrosion loss of bare steel in marine conditions[5].

The variability of overall climate, the variations of microclimate around structures and the gradients within construction materials all need to be considered. Overall weather data are available in great detail from the Met Office. Developments in the calculation of heat and vapour flows into and out of buildings[6] to evaluate energy efficiency provide a basis for calculating variations in temperature and humidity gradients, through the day and the year, for walls and roofs. The monitoring of bridge structures for thermal differentials and movements[7] also provides good data and a basis for calculating thermal microclimates for the consideration of deterioration. Similarly flows of water vapour and liquid water and the rates of ionic diffusion of chlorides can be calculated where materials data is available. Thus there are established experimental and analytical techniques which can be applied to predict thermal

Concrete Temperature	Concrete Relative Humidity, RH %.						Concrete Saturated.			
							low	no	Cl	SO3
C - C	45	55	65	75	85	95	oxygen	oxygen	ppm	ppm
-20 - -10									Road	
-10 - 0									Salt	
0 - 10										
10 - 20										
20 - 30										
30 - 40										
40 - 50										

Open Car Park Entrance Ramp, Bare Concrete

Av. Duration	
(hatched)	>500 hr/yr
(dotted)	>50 hr/yr

Figure 1. Microclimate Check Chart.

and moisture conditions in structures, their change with time and the consequent rates of development of deterioration processes. The calculated effects need to be compared against site measurements and records of deterioration to validate and refine the whole procedure.

The grinding of surface layers off concrete to obtain mm by mm chemical analysis, the use of petrography and SEM with EDXA enable the early stages of deterioration and the rate of carbonation and chloride ingress to be more accurately detected(8). The variability of concretes makes it difficult to use this as a measure of relative microclimates between structures. But it is a powerful tool for evaluating the variation in chloride ingress and carbonation with the form of detailing and the local microclimates within structures. This data can be used to extrapolate future durability trends of existing structures to plan maintenance. It has also been used to set materials performance criteria for the design of major bridge and tunnel structures.

The heating and air conditioning of buildings can establish benign stable conditions for the occupants and the internal materials. However it can also, depending on the location of vapour barriers, create substantially increased humidities in the walls and roof slab which can enable corrosion to develop prematurely. Thus the environmental gradients within the structural materials must always be considered as well as surface microclimates.

Because of the greater stability of temperatures and humidities in construction materials it is both easier and more relevant to measure them rather than surface or adjacent air conditions. The measurement of temperatures is straightforward and computers now enable the mass of data to be more easily digested.

Alkali aggregate reaction damage can be substantially retarded(9) by reducing RH. A reduction from 98% RH to 92% RH can reduce the rate of deterioration by a factor of 5 to 10. To measure the internal RH in concrete wooden plugs, sealed into holes to equilibrate with the moisture level in the concrete for some weeks, have been used, Figure 2. The moisture content of the wooden plug is measured by a conventional electrical resistance meter or by oven drying. This has proved to be simpler, cheaper and more robust than electrical RH meters. It is much more sensitive and accurate in the 85% to 100 % RH range, which is important for both corrosion rate and AAR.

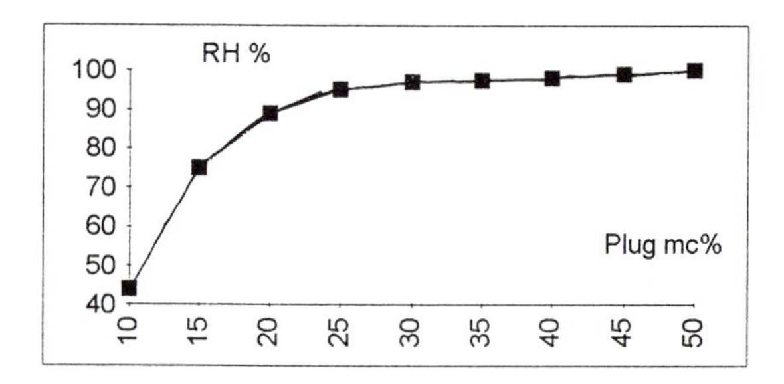

MC%	RH%
10	44
15	75
20	89
25	95
30	97
35	97.5
40	98
45	99
50	100

Figure 2. Relationship of moisture content of wooden plug to Equilibrium Relative Humidity in Concrete.

MATERIAL DETERIORATION CHARACTERISTICS

All construction materials deteriorate due to combinations of chemical, biological and physical processes. By identifying these processes and their rates under different environmental conditions and with variations in material quality, the environments in which deterioration rates will be sufficiently slow or cease can be identified. This can be tabulated in simplified format as in Figure 3, which can be compared with the envelope of temperatures and humidities for the microclimate of the structural detail in Figure 1. The form and detailing of structures must create these conditions and material and construction control quality must achieve the required quality.

For timber biological deterioration is sufficiently slowed by maintaining the equilibrium relative humidity (RH) consistently below about 75%. Some woods are tolerant of wetter conditions, (eg cedar and oak heartwood) and can be used externally or in saturated ground. Proprietary electrical resistance meters enable wood equilibrium RH and moisture content to be determined so that susceptibility of wood in buildings to deterioration can be easily determined by surveyors and builders. Copper salts and other wood preservatives can extend life in damp condition.

However damp wood accelerates the corrosion of metal fixing in timber and copper based preservatives further increase the corrosion rate in damp. So although the damp will not rot the timber the fixings will corrode through, so there is little real gain in durability! One notes that it is oak pins not bolts that join the frame of old Wealden Barns. The other destructive process for wood is the cracking from differential shrinkage and restrained shrinkage from wetting and drying. These characteristics of wood are well understood, and with good design and maintenance to achieve stable dry environments, or for oak piles stable saturated environments, it does last for centuries.

The corrosion deterioration rates for reinforced concrete also relate to temperature and RH but it is a two stage process of depassivation followed by corrosion. Depassivation can arise from carbonation or chloride ingress. In dry conditions carbonation develops rapidly, probably fastest at 50 to 60% RH in the sheltered and internal parts of the structures. Provided these stay dry the

Concrete Temperature		Concrete Relative Humidity, RH %.						Concrete Saturated.	
								low	no
C	C	45	55	65	75	85	95	oxygen	oxygen
-20 -	-10	0.00	0.00	0.00	0.00	0.01	0.03	**0.06**	0.00
-10 -	0	0.00	0.00	0.00	0.00	0.01	0.06	**0.13**	0.00
0 -	10	0.00	0.00	0.00	0.01	0.03	0.13	**0.25**	0.00
10 -	20	0.00	0.00	0.00	0.01	0.05	0.25	**0.50**	0.00
20 -	30	0.00	0.00	0.00	0.02	0.10	0.50	**1.00**	0.00
30 -	40	0.00	0.00	0.00	0.04	0.20	1.00	**2.00**	0.00
40 -	50	0.00	0.00	0.00	0.08	0.40	2.00	**4.00**	0.00

(shaded)	Pitting Corrosion

Reinforcement Corrosion Rates: mm/year. Concrete with ~1% Cl/opc, uncarbonated.

Figure 3. Deterioration Rate Chart.

reinforcement in the carbonated concrete will not corrode. At consistently high moisture levels carbonation is slow in the waterclogged pores of the concrete, so the reinforcement may not be depassivated for centuries with good quality concrete. 80-90% RH is low enough for carbonation to develop and still damp enough for corrosion. Even more severe is the cycle of drying to accelerate carbonation and wetting to accelerate corrosion which is a characteristic of English climate, of internal concrete which is periodically washed down or becomes wetted by leakage or where the vapour pressure gradients from the heated interior to the cold exterior produce high humidities in the winter months. The BS 8110 environment classes do not reliably represent the relative rates of carbonation and corrosion in buildings.

The most unsatisfactory feature of current concrete durability clauses is the reliance on 28 day strength as a measure of concrete durability. The five fold reduction in carbonation rates from reduced water cement ratios, achievable at little cost with plasticisers, is not properly exploited as specified strengths are set too low. The resistance of pfa and slag mixes to chloride ingress, which is ten times better than similar OPC mixes, is discounted. A shift to selecting mixes on the basis of long term bulk diffusion testing to determine resistance to chloride ingress, as has been done for major marine tunnels[(8)], would make a major improvement to the durability of bridge and marine structures.

MATERIAL INCOMPATIBILITIES

The most destructive processes in nature arise from strains induced by physical effects of thermal and shrinkage differential movements and the expansion from frost and salt crystallization. These processes have reduced the Alps to aggregate and we must not underestimate their disruptive effects on buildings. The swelling following chemical changes from sulphates, AAR or polymerisation is similarly disruptive. The values of coefficient of thermal expansion and Young's Modulus (E) are well characterised. The moisture movements are less well publicised but are frequently more destructive than thermal differentials. Figure 4 sets out some of the physical constants for a range of construction materials.

Material.	**Young's Modulus** E kN/mm2	**Thermal Expansion** microstrain/C	**Moisture Movement** 100% --> 50% RH microstrain	**Alkali Sensitivity**
Steel	200	12	0	Passivates
Cement	14	18	3000	
Mortar (1:3), w/c = 0.5.	20	11	1200	
Concrete (1:2:4) w/c = 0.5.	30	12	400	
Glass	70	7	0	Reactive
Aggregate, Typical Granite	45	7	0	
Polypropylene	1	120	0	
Nylon	3	90	1500	

Figure 4. Some Typical Properties of Concrete, Mortar and Reinforcement.

Many materials failures in construction arise when materials with different movement characteristics are fixed together without adequate provision for differential movement or where their bond strength is not sufficient to resist the differential stresses that arise. This principle was well know 2,000 years ago.

> Matthew 9. v16. "*No man putteth a piece of new cloth unto an old garment, for that which is put in to fill it up taketh from the garment, and the rent is made worse.*"

Recently the theological ignorance of the concrete repair industry has led to the introduction and now abandonment of epoxy patch repairs and the use of high cement content mortars which expand and shrink and detach themselves in the long term from the concrete.

The difficulties with GRC materials arose from a combination of large moisture movements due to their high cementitious content and a tensile strength, and more importantly tensile strain limit which falls with time as the alkalis weaken the glass fibre. Much has been done to reduce this by enhancing the resistance of the glass fibre, coating panels to reduce the wetting and drying cycles and avoiding shapes in which temperature and moisture gradients arise between the face and ribs of the material. Fixings which locate, but do not restrain, have also helped. It is a classic case of a material where failure to carry out a durability design review to evaluate the incompatibilities within the material and relative to the building frame has blighted the commercial development of a material with great potential.

The facing of concrete or steel structures with bricks, granite or glass creates similar problems of differential movement between the rapidly heating and cooling, wetting and drying face, and the more stable interior. The durability and replaceability of the sealants which must accommodate the resultant movements in the joints between the cladding elements can have a major influence on overall durability. The sensitivity of the durability of the fixings and materials behind the cladding to water ingress as seals deteriorate must be explicitly considered in design.

The incompatibility of dissimilar metals creating bimetallic corrosion should never be forgotten, but often is.

CONCLUSIONS

Design to achieve structures with substantially improved durability can be readily achieved. However this will require the form and detailing of the structures to be developed to create microclimates in which the decay processes are slowed or halted, as well as improving the materials. Reinforced concrete structures, away from road and sea salt can be designed and constructed to achieve 1000 years before corrosion starts. This will cost only marginally more than recent structures for which 60 years was hoped for, but only 20 years achieved. Most established construction materials can have their reliable durability similarly enhanced. For new products rigorous testing and extrapolation should enable us to break the cycle of 'innovation, trial, decay, try something different' that bedevils developments in construction materials.

REFERENCES

1. Treadaway K.W.J. et al. 'Durability of corrosion resisting steels in concrete,' Proc. Instn. Civ. Engrs, Pt 1, 86, Apr., pp 305-331 (1989).
2. Somerville G. (Editor) *The Design Life of Structures*, Blackie, London. (1992).
3. BS 7543:1992, *Guide to Durability of Buildings*, BSI London (1992).
4. Somerville G. 'Some reflections on design life', pp 257-264, Somerville G. (Editor), *The Design Life of Structures*, Blackie, London, (1992).
5. Billington C.J. and Guy R.G. 'The life of steel structures in the marine environment', Proc. Int. Seminar 'The Life of Structures' Brighton BRE/IStructE, (1989).
6. CIBS Guide Vol.A. Chartered Institute of Building Services, London.
7. Ho D. and Liu C.H. 'Temperature distribution in concrete bridge decks', Proc. Instn. Civ. Engrs, Pt 2, 91, Sept., pp 451-476 (1991).
8. Wood J.G.M., Wilson J.R. & Leek D.S. 'Improved testing for chloride ingress resistance of concretes and relation of results to calculated behaviour', 3rd Int. Conf. Deterioration and Repair of Reinforced Concrete, Bahrain Soc.of Engrs/ CIRIA, (1989).
9. Wood J.G.M. & Johnson R.A. 'The appraisal and maintenance of structures with alkali silica reaction.' The Structural Engineer Vol 71, No 2, pp 19-23, 19 Jan. (1993).

4 SOME ASPECTS OF THE POSSIBILITIES AND PERSPECTIVES IN NON-DESTRUCTIVE TESTING IN CIVIL ENGINEERING

P. Cosmulescu and P. Strateanu
Polytechnic Institute, Romania

The paper deals with some of the main problems of the non-destructive testing methods currently in use in civil engineering. The technical and economic importance of non-destructive testing is analysed concerning many responsibilities of the modern technique with a view to assure: increased productivity and profits, increased serviceability, safety. In order to do this, basic research and development must be carried out and accelerated. In this we will meet the needs of the world's advancing technology.

INTRODUCTION

General considerations

The rapid development in the use of non-destructive testing (NDT) methods in recent years has taken place in answer to the many requirements of modern production techniques. Scientists and engineers in many countries have made a great contribution to the rapid growth and application of non-destructive testing.

The term 'non-destructive testing' is often considered to be concerned only with the detection and location of flaws. However a great variety of non-destructive tests are in worldwide use in order to detect variations in the physical and mechanical characteristics of materials.

The problem of product quality involves two distinct aspects:

- the first aspect generally refers to the conception, calculations, design and choice of suitable materials, technology of manufacture, etc. which will assure adequate quality of the product
- the second aspect refers to determining and certifying the quality of the product. This problem interests everyone involved in industry, including management, control engineers and all those participating in the manufacturing processes.

Success will be assured if these methods and techniques can be used to determine the behaviour of materials in their anticipated uses. But there is no universal non-destructive technique that is applicable to all situations.

Those unfamiliar with the basic nature of non-destructive tests expect them to provide magic solutions to production problems.

We maintain that it is not an 'aberration' to affirm that when choosing between two products, to be used for the same purpose, the one with the higher quality cost will be cheaper in the long run.

It has long been recognized that perfection is unattainable and that any attempt to achieve perfection in production is unrealistic and costly. Sound management is not interested in perfection, but in optimum quality level.

Motivations

A great variety of non-destructive tests[1] are used in civil engineering construction. Many types of manufactured materials such as cement, concrete, ceramic, bitumen, glass and sheet metal, whose quality is accurately controlled, are produced, with the aid of non-destructive tests. Such tests are made in the investigation of soil foundations, road construction, steel or concrete structures, and so on.

Engineers, designers and most quality inspectors tend to exaggerate requirements, while technologists tend to underestimate them

The designer works with known loads in mind and on the supposition that the materials and workmanship are near perfect. Then, knowing that the load may accidentally be exceeded and that materials and workmanship are never perfect, he applies a high safety factor.

In addition, there are today many new materials and technological solutions whose performance characteristics are not completely known. These create greater problems, many of which have been solved through the use of suitable non-destructive tests.

This paper discusses the wide application of non-destructive ultrasonic and atomic testing, and the circumstances in civil engineering under which their use if preferable.

TYPICAL ULTRASONIC AND ATOMIC NON-DESTRUCTIVE TESTING METHODS USED IN CIVIL ENGINEERING

Introduction

Application of non-destructive tests in industry can usually be classified under two heads:

- on the basis of the industry in which they are used
- on the basis of the physical principles of the methods.

Both classifications[1] have their merits.

From the point of view of this paper, a classification based on applications appears most useful, and we have accordingly adopted this.

Ultrasonic methods
The importance of use of ultrasonics for non-destructive testing of material is widely recognized.

The propagation of ultrasonic waves in a material is related to the elastic properties of the material and the homogeneity of its structure. Any investigation of this relationship involves observation of the intensity and direction of the waves, and measurements of the time taken for the waves to pass through the material. Successful commercial instruments are now available for this purpose. Striking a specimen and listening for the characteristic 'ring' has been used as a means of detecting flaws.

Specific application of ultrasonic testing methods in civil engineering include weld inspection and concrete inspection. The ultrasonic techniques used to test welds are generally well known.

The ultrasonic tests for concrete have ben developed more recently. Non-destructive testing of concrete by ultrasonic methods are summarised in Figure 1. Ultrasonic pulse methods(3) are the most used in testing concrete in situ and in the laboratory.

Resonance methods for testing concrete (see Figure 2) are less used. They are generally applied in laboratory investigation. The most common form of these testing methods is still by handheld probes with the interpretation of indication of defects being entrusted to an operator.

Atomic methods
Methods based on the radiations emitted by radioactive substances which include a radioactive source in combination with a radioactive detector are referred to as radioisotope methods. They are usually used to measure a variety of physical characteristics in the laboratory and in situ. The advantages of radioisotope instruments are given below:

- the measurements are performed non-destructively
- they are versatile and require simple safety devices
- the penetrating nature of high-energy gamma radiation enable measurements to be made through the walls of concrete or steel
- electrical power supply is not required
- they are compact, portable instruments, little space is required for installation. This has permitted many applications such as geotechnical prospecting at great depths in boreholes, weld radiography of the pipe, etc.

Figure 3 is a schematic diagram showing typical applications of atomic methods in civil engineering. X and ϒ radiography has been included under the radiography with penetrating radiations because both are electromagnetic radiations.

The chief merits of gamma-ray sources are small size, the high penetration of the radiation compared with industrial X-ray sources in common use, the relatively low cost compared with X-ray units, the independence from electricity

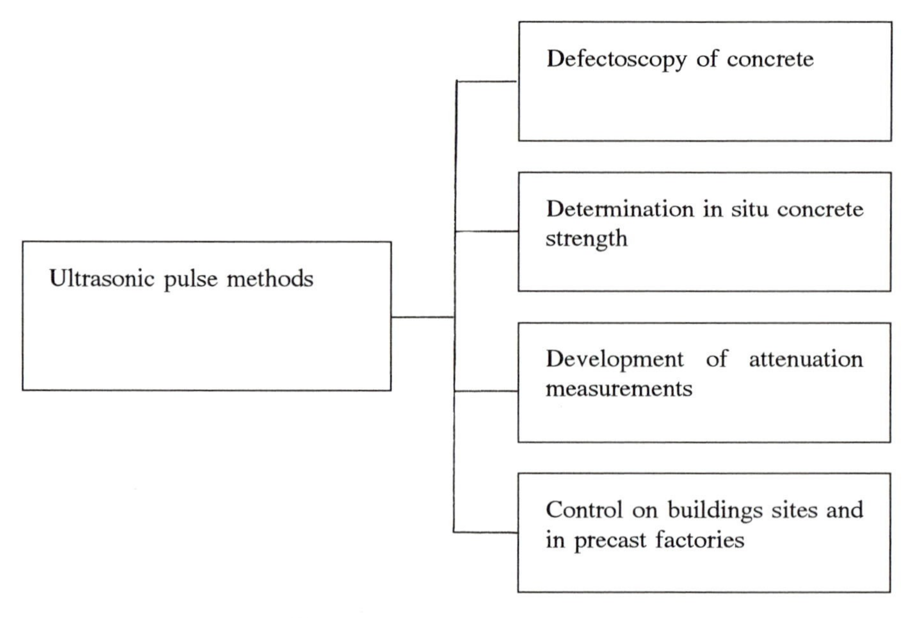

Figure 1. Ultrasonic pulse methods

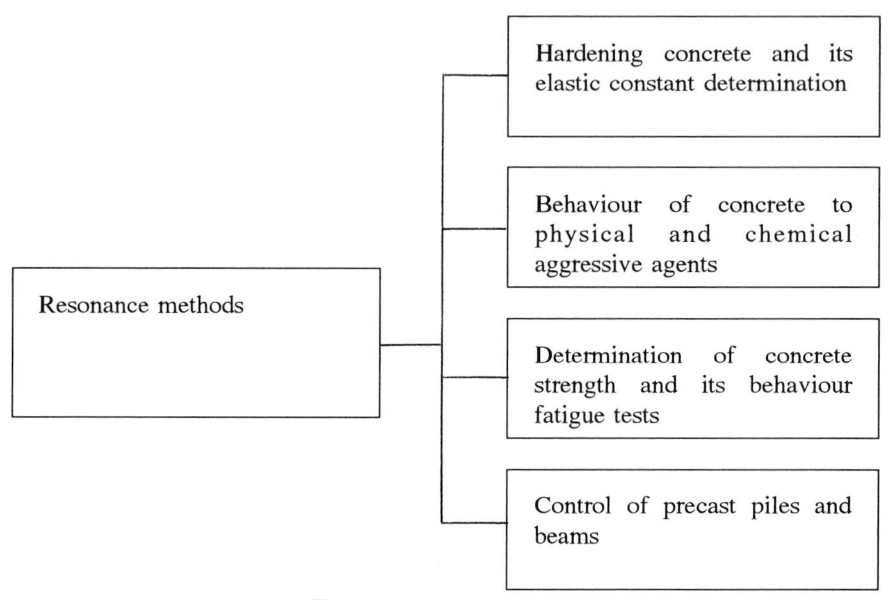

Figure 2. Resonance methods

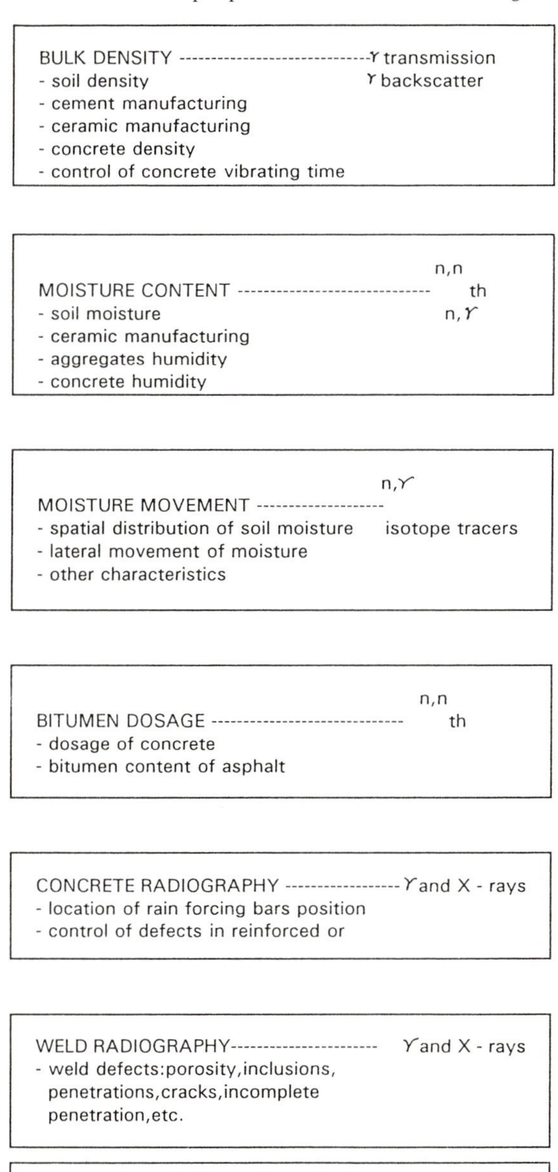

Figure 3. Atomic methods

and water supply and the lower image contrast which permits a larger range of material thickness to be recorded on one exposure film.

The amortization of radioisotope instruments is evaluated in terms of the operating time during which the savings resulting from the installation are equal to the initial cost of the instrument. This period is usually[1] between six months and two years, but in particular applications it can be a great deal less.

FACTORS AND CRITERIA FOR USE IN NON-DESTRUCTIVE METHODS

Technical factors

The principal technical factors to be considered when determining the best test method and technique for a particular problem are the following:

- the basic principles of the problem to be solved and the performance of the various methods of non-destructive testing
- correct application and interpretation of the methods
- techniques which have been used
- the kind of results to be expected
- the advantages, disadvantages and limitations of each test

It should be noted that no single method is satisfactory; on the contrary, the non-destructive methods supplement each other. Some research workers and engineers are tempted to overestimate the possibilities of the particular method in which he specializes, or which he frequently applies.

Economic factors

The economic factors criteria to be applied when using non-destructive instruments vary greatly according to the particular applications and the specific financial circumstances. They include:

- aid in establishing a better product design
- optimum quality level
- increased productivity
- increased profits

The non-destructive test can reduce manufacturing costs if it highlights undesirable characteristics at an early stage. For example, it finds defects in and locates the position of the reinforcing bars of concrete structures, it finds weld defects, etc. It thus saves time and money that would otherwise be spent in further processing or assembly.

When deciding whether it is profitable to adopt the non-destructive test, the main problem is to determine the inspection cost. Cost criteria are: what is inspected and also, the tolerance and the degree of perfection required. The interrelation of the above factors is shown in Figure 4.

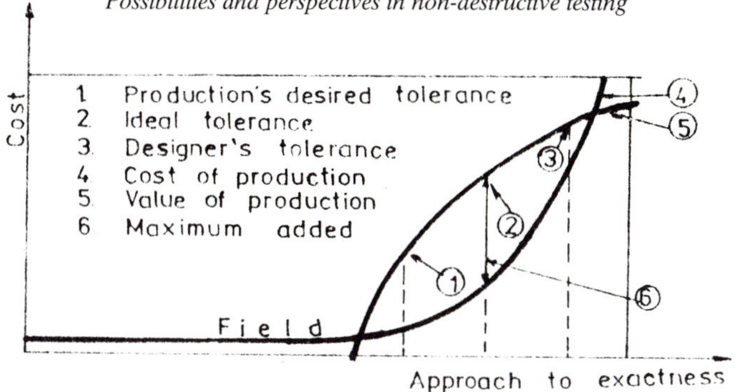

Figure 4.

Obviously, the cost of a product rises rapidly as quality comes closer and closer to perfection. Also, quite obviously, the value of a product is zero until a minimum degree of quality is attained.

The non-destructive test can affect profits in a variety of direct and indirect ways because there are many industrial operations designed to improve materials, but which sometimes spoil them instead.

The fabricating of steel products involved many kinds and necessary stages of inspection. These are shown by the schematic diagram in Figure 5 (stars represent control operations).

Ultrasonic or atomic non-destructive tests can be used for some of these inspection operations, depending upon the desired quality of the end product. Every control engineer attempts to save operations while at the same time making sure that the final product meets acceptable standards. Not all operations can be saved.

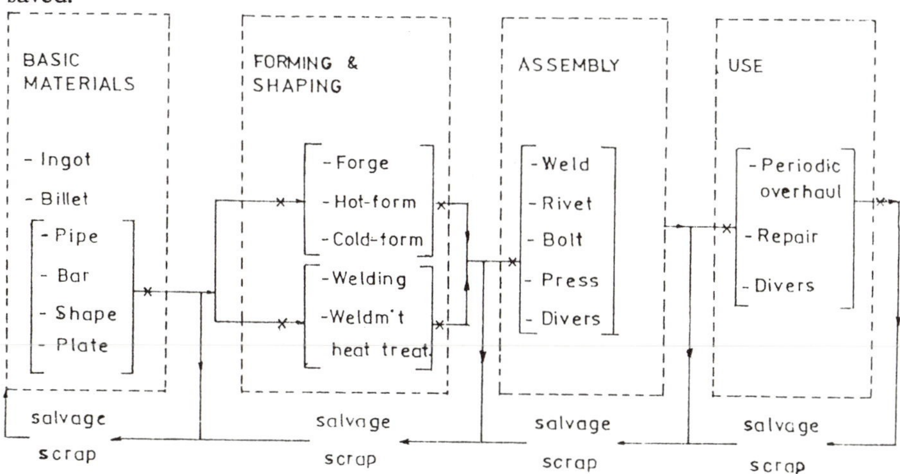

Figure 5.

The economic justification for partial inspection is based more or less upon the cost of the non-destructive tests. This can depend on:

- automation of test method
- quantity of parts to be inspected
- tolerance permitted in interpretation of test results
- sensitivity required from the test method
- cost of the test personnel required

More and better-trained personnel are required in the field of non-destructive testing.

CONCLUSION

We have used the term 'civil engineering' in this context in the wide sense of the disciplinary domain. Civil engineering can be considered as that field of engineering concerned with the planning, design and construction for environmental control, development of natural resources, buildings, transportation facilities and other structure required for the health, welfare, safety, employment and pleasure of mankind.

Because of the large number of non-destructive test methods, and the great diversity of problems to which they are applied, it is not possible to deal here with all the applications in the field of civil engineering. However, the following general observations can be made.

In ultrasonic testing and in non-destructive testing, generally, the ultimate outcome could be the development of semi-automatic equipment to replace the manual equipment. Some useful beginnings have been made in the establishment of spectroscopy, in particular, in association with computer processing of signal data to extract information on the nature of a defect which otherwise would not be obtainable.

Improvement is necessary to standardize the approach to no -destructive testing, with regard to test procedures, the interpretation of the results, and standards of acceptance and rejection.

In future, it may be possible to speak of a new disciplinary subject, namely 'defectology', to include two principal aspects:

- non-destructive testing
- quantitative defects measurement, ie quantitative relationships to be established regarding the types, sizes and positions of the flaws and their influence on the mechanical and technological characteristics of the products of civil engineering.

REFERENCES

1. Cameron J.F. and Clayton C.G. *Radioisotopes instruments*. Pergamon Press (1971)
2. Cosmulescu P.P. 'The analysis of the dynamics of moisture and bulk density in ceramic manufacturing by non-destructive testing'. Rev. Construction Materials. Romania 3 vol.2 (1972)
3. Jones R. and Facoaru I. 'Non-destructive testing of concrete'. Ed Tehn, Romania (1971)

5 VALIDATING INNOVATION: A GRAPHIC AID FOR DESIGNERS

R.W. Birmingham and A.L. Marshall
The Universities of Newcastle upon Tyne and Sunderland, UK

The reasons designers of engineering structures may be reluctant to take advantage of the opportunities presented by developments in material science and production technology is discussed. A chart based system is introduced that enables the designer to rapidly and economically explore alternatives at an early stage of design. The example of a low cost beam is used to demonstrate the construction and use of these charts. Future work to extend the system to complex structures, and implement it on PC based software is outlined.

INTRODUCTION

Designers in the field of structural engineering are by nature conservative. Confidence is placed in the tried and tested, while new ideas are treated with great suspicion. This is understandable for the risks associated with the new are hard to both identify and quantify(1).

Innovations that do occur tend to be forced by new requirements. The arrival of fresh challenges demands new solutions. This route to innovation has been termed as 'demand pull'(2), indicating that the impetus for change comes from the market place, which drags creative solutions from the designers and technologists. Innovation in this case is an inevitable response to more exacting and demanding problems. But there is nothing inevitable about an alternative route to innovation. Discoveries in science and technology can create unexpected opportunities for improvement in both the design and manufacture of engineering artefacts. This route can be termed as 'science push'(2), indicating that here the impetus for change comes from the development of new technology, before the arrival of a market need.

Although change is possible under these circumstances, it is not inevitable. If the market is not forcing change, then why disturb the status quo? The investment in conventional technology (equipment, training, expertise) makes any innovation an expensive proposition, and vested interests mitigate against even examining new options(3).

Experience with particular methods, materials, and forms, progressively builds a degree of expertise in a specific area for every designer. When searching for solutions to problems the designer draws on that expertise, while tending to disregard alternatives using different technologies. For a given problem, the civil engineer may develop designs using reinforced concrete, while the structural engineer may produce alternatives in steel. Neither probably will explore possible solutions using materials in each others domain, or alternative materials such as glass reinforced plastic (GRP).

When new or alternative technology is outside the designer's area of experience, and while his or her own expertise can produce satisfactory results, then there is little motivation to develop the additional skills required to explore innovative possibilities.

If the designer could rapidly compare the probable outcome of using alternative materials or structural forms, the decision to explore or ignore those options would be soundly based. If he or she could perceive the design trend due to some innovative change, in the same way that an expert in that field could intuitively anticipate the likely optimal solution, a rational decision could be made to either invest in a more detailed analysis, or to continue with the established technology.

At the Engineering Design Centre based in the Universities of Newcastle upon Tyne and Sunderland, a graphic chart based system is being developed which is intended to give the designer of engineering structures just this information. Using the system a rapid indication of the potential of many alternatives, including those outside their own area of expertise, can be obtained.

THE INTEGRATED APPROACH TO DESIGN ALTERNATIVES

The analysis of a structural design problem leads to three areas which can be examined independently: material, form, and loading condition(4). Conventional procedures successfully optimise each of these in isolation. Difficulties arise when trying to combine the optimum solution from each area, as production constraints often prevent the optimum form being made from the optimum material(5). Empirically derived data (manufacturing 'know how') has to be combined with the technical analysis to produce a practical design. This is the designers expertise.

Procedures developed have shown that charts of the geometric properties of structures, juxtaposed with charts of the physical properties of materials, enable the alternative combinations of material and form to be examined in an integrated way(5). In addition it is found that by plotting a database of manufactured structural components on such composite charts, the manufacturing constraints associated with each material are presented explicitly. These charts allow a designer to make valid comparisons of optimal solutions, including those from areas outside their own experience.

The generic form of the composite chart is a cross, Figure 1, with each of the four axis representing one variable. These are either a material property (such as Young's modulus, yield stress, or the density), or a geometric property (such as the second moment of area, section modulus, or cross-sectional area). In the four quadrants of the charts the attributes of alternative solutions are plotted. In general two quadrants contain the physical characteristics of the structure, and the

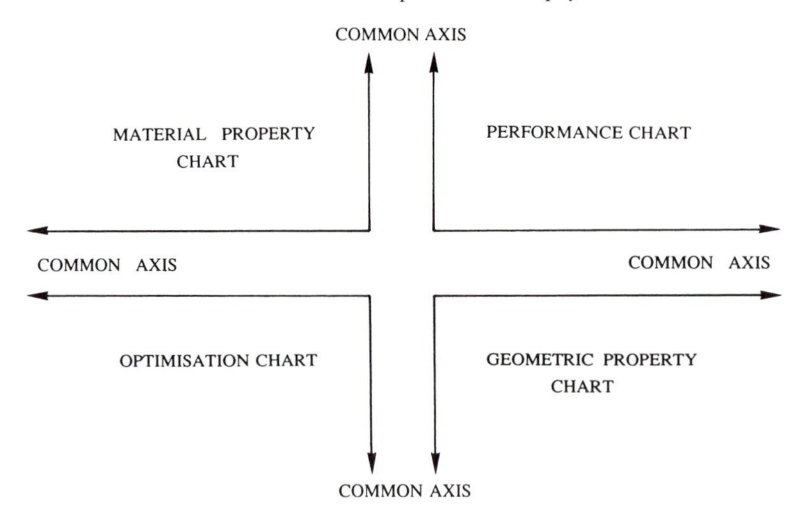

Figure 1. The generic form of the composite charts for comparing alternative combinations of material and form

other two indicate the potential performance of these alternatives in terms of load, weight, cost, or other measure of performance.

Each chart models one type of loading condition and is based on the fundamental equations of structural analysis. For the beam problem this model provides an elegant solution.

ISOLATING MATERIAL SELECTION

In the design of engineering structures the primary problem is to establish the optimum combination of material and form to support the required load. The loading condition can stress the structure in several ways, such as tension, compression, torsion, shear, or flexure. If the loading condition and mode of failure have been established for an element, it has been shown that the material selection procedure can be isolated from the problem of the element's geometry(4). This can be demonstrated with the example of a beam element with yield as the performance criterion, i.e. an element subject to bending for which permanent distortion must not occur. Elasticity theory(6) shows that the onset of yield occurs when the bending moment, M, is given by:

$$M = Z \quad \sigma$$

where: M = bending moment
Z = Section modulus, I over y
σ = Yield stress.

From this it can be seen that the performance of the element is the product of two components which can be optimised independently. These are the yield stress, a material property, and the section modulus, a geometric property.

With the material property separated from geometry, material selection can be made by a graphical system[4] which is clear, concise and simple to use. In this particular case the performance property is the yield strength, but for other problems different properties are used. If stiffness is required, Young's modulus is the property limiting performance. The property to be optimised would be density in the case of minimum mass design, or density times cost (in pounds per kilogramme) for minimum cost.

On a graph with axes of the performance property against the optimising property all materials can be plotted. An example is shown in Figure 2, where the values of the yield stress of selected materials are plotted against their density. As can be seen, similar materials clump together into islands, the spaces between indicating that no materials exist with just such a combination of properties.

ISOLATING THE SELECTION OF ELEMENTAL FORM

The separability of the material and geometric properties allow equally the material and the geometry of an element to be examined and optimised in isolation. For geometric properties the method employed can mirror the property charts used to plot the characteristics of materials, only here the properties of different structural forms are used. Once again the type of element has to be established (beam, tie rod, strut, etc.), and the mode of failure to be examined specified. Identification can then be made of the geometric property which limits performance.

As already stated, for a beam element which must not yield, the section modulus (Z) is the relevant geometric property controlling performance. If

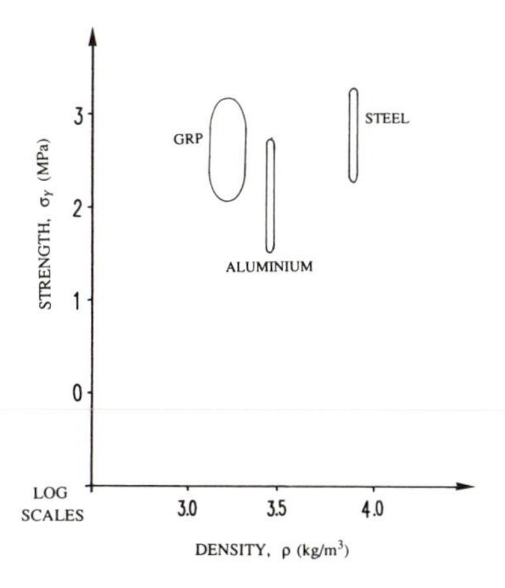

Figure 2. A material property chart

deflection were taken as the mode of failure, then the second moment of area of the cross-section (I) would be used as the performance property. To construct a geometric property chart the performance property forms one axis, the other being the measure of the property to be optimised. If weight is the criterion, then minimum volume is required. This can be expressed as volume per unit length, which for a prismatic beam is cross-sectional area (A). Other properties may be used as the criterion such as surface area (if maintenance or corrosion are important factors), depth of beam or enclosed volume (in space limited applications) or of course cost.

Plotting the relevant geometric values such as Z and A from a sample of actual beams will create a scatter diagram describing current manufacturing practice. When this is done using manufacturers published data [(7)(8)(9)] it is found that the data points do group together, implicitly forming a long thin 'island' for each material. This indicates the limits for single process manufacture of I beams, as shown for three materials in Figure 3. Such a chart need not confine itself to standard I beams. It could include castellated beams or entirely different structures altogether, such as tubes, or even space frames.

In Figure 3 the overlapping groups are for steel, aluminium, and GRP. Three materials are plotted on a chart with geometric axes. This chart too could be called a material property chart, but it is not plotting the physical properties of each material, it is plotting their manufacturing properties in the geometric domain. In this case the steel sections are hot rolled, the aluminium extruded, and the GRP pultruded. Three different manufacturing methods, but each is the appropriate process to produce equivalent sections from their respective materials.

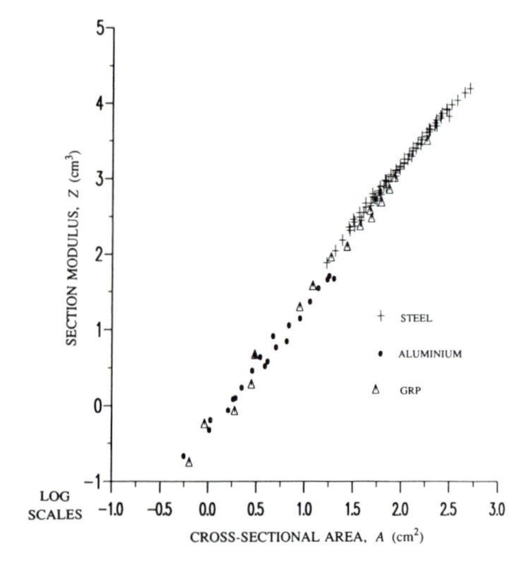

Figure 3. A geometric property chart

LINKING THE PHYSICAL AND GEOMETRIC DOMAINS

In the above procedures the physical properties have been modelled in the material domain, and the manufacturing properties modelled in the geometric domain. This was possible because the two domains were seen to be separable. Linking these domains can be achieved by recombining the separated elements. To continue with the beam example, linkage is achieved by returning to the original equation of analysis:

$$M = Z\,\sigma$$

A graph can be drawn with axes of Z against σ, and contours drawn indicating constant values of bending moment. This chart plots the potential performance of the beam. For a given loading condition the maximum value of M can be calculated, and the relevant contour identified. When the objective is to optimise a property which is influenced by both the material and geometric properties of the structure, another chart can be produced. For a minimum mass beam this chart will have axes of density against A, and contours of constant mass per unit length.

The most widely used criterion of selection is of course cost. The cost of the beam can be presented in a variety of ways. One simple approach is to convert the mass function used in the previous example, into a cost function. This can be achieved by introducing a variable indicating the cost of each beam in pounds per kilogramme. If this cost variable is combined with the geometric variable of sectional area to form one axis of the optimising chart, and the material property of density used as the other axis, it can be seen that the mass function used in the first example now becomes a cost function measuring pounds per metre length, and contours of constant cost can be drawn. The value of the cost variable could be set at the price, ex-works, of each section, although more sophisticated measures of cost could also be used.

The four types of chart discussed above can model a variety of loading conditions. The four necessary charts are listed here. The specific axes that are used in the example of a minimum cost, maximum strength beam are given in each case bracketed:

1. Material property chart, indicating physical properties of materials. (Example axes: ρ against σ.)
2. Geometric property chart, indicating manufacturing properties of materials. (Example axes: Z against A times cost.)
3. Performance chart, with contours of equal performance. (Example axes: σ against Z.)
4. Optimising chart, with contours of equal value of the objective function to be optimised. (Example axes: A times cost against ρ.)

Each of these charts has both axes common with two other charts, enabling a complete design problem to be presented in one drawing. The example of a minimum cost beam, with yield taken as the mode of failure is shown in Figure 4. The four charts are drawn back to back, one in each quadrant.

A common axis links neighbouring charts, allowing a progression to be made (in either direction) around all four. In Figure 4 two sets of arrowed lines link across the four quadrants of the composite chart illustrating the procedure used to select a minimum cost beam. Given a maximum bending moment of 1MNm, indicated in the top right quadrant by a solid line, links are made from this line to possible options in both the adjoining charts. Note that from any point on this line links are made to the same material in the materials property chart, top left, and the geometric property chart, bottom right. From these two points, lines converge into the bottom left quadrant, their intersection indicating the cost per unit length of a suitable beam. In Figure 4 the cross chart links are shown for steel and GRP, and indicate that in this case steel can produce a lower cost beam.

The links that connect the information in all four quadrants of the chart allow them to be used in several ways. The starting point can be in any one of the four quadrants, and conclusions drawn from the others. In Figure 4 the route

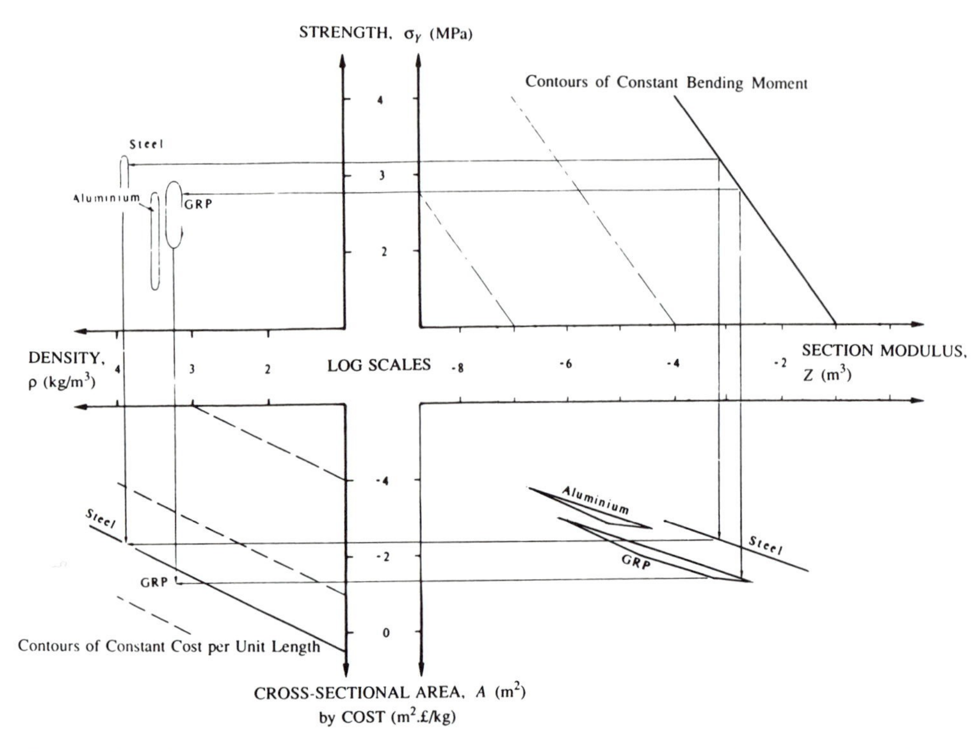

Figure 4. A composite chart for selecting minimum cost beam elements when yield is the failure mode. Steel is shown to be less expensive than GRP to support a 1MNm bending moment.

would be reversed if a maximum cost were specified at bottom left, with the maximum bending moment found top right.

FUTURE RESEARCH

This paper introduces procedures that integrate the selection of material and form in the design of engineering structures. Although only a small sample of data has been presented, the examples do demonstrate that a graphic presentation is possible which allows direct comparison of alternative concepts. Existing techniques do enable such comparisons to be made, but these rely on iterative analysis of specific cases which require considerable design effort for each alternative proposal.

Each composite chart is modelling a specific load transfer mechanism (tension, compression, torsion, shear, or flexure) and a specific failure mode (deflection, yield, or collapse), combined with a particular object function (cost or mass). For these cases there are theoretically 30 charts. A few of the combinations are trivial, and their resulting charts very simple, but in most cases the charts take the form indicated in Figure 1. In some cases the analysis is more complex and the resulting charts are inevitably more involved. These lack the elegance of the standard chart form, but will become more useable when implemented on computer software. In a very few instances the theoretical case has no physical meaning. In these cases there is no identifiable equation of analysis, and no chart can be drawn.

Further research to develop the system is planned in two directions. One element of the further work proposes to extend these methods to include more complex structural forms than the prismatic elements so far examined. These would include trusses, space frames, grillages, and sandwich forms. To enter such structures onto the charts the effective values of the relevant variables have to be found.

It is also planned to implement these procedures in the form of a computer based design guidance system, from which any case can be called at will. In order to achieve this, the theoretical background to all expected cases is required, and the form of each composite chart established. In addition the full range of variables must be known, in order to allow the design of a supporting database which will be populated with material properties, and manufactured structural forms.

CONCLUSION

New possibilities arising out of technical developments present the designer with a dilemma: their potential can only be identified after considerable design analysis, but this effort can only be justified if the potential is already known. In addition when the new opportunities are in technical fields with which the designer has little or no previous experience, conventional design analysis may be misleading unless information on the limitations imposed by manufacturing technology is researched and included.

This paper demonstrates how a chart based system can be used to explore the design of engineering structures. The example used here is confined to commercially available I sections, but indicates how more complex structural forms could also be included. When the system is computer based it is evident that more complex problems could be tackled. Such a chart based design guidance system, backed by a computer mounted database of materials and structural components, would allow designers to make a quick appraisal of alternative concepts. Only if the potential improvement on the established concepts is shown to be substantial need they consider investing in a more detailed exploration of new technology, and the acquisition of the necessary expertise to undertake a complete design study.

REFERENCES

1. Littler D.A. 'The management of industrial innovation.' Proceedings of Seminar. Liverpool University, (1977).
2. Harrington K. 'Concrete as a fabrication for simple hulls'. PhD Thesis. Sunderland Polytechnic, (1987).
3. Abernathy W.J. 'Mapping the winds of creative destruction', in Tushman, M. C. (Editor), *Readings in the Management of Innovation*. Ballinger Publishing Co., (1988).
4. Ashby M.F. 'Materials selection in concept design'. Materials Science and Technology, pp. 517-25, (1989).
5. Birmingham R.W. and Wilcox J.A.D. 'Charting the links between material selection and structural form in engineering design'. Journal of Engineering Design, (in press for 1993).
6. Southwell R.V. *An Introduction to the Theory of Elasticity*. Clarendon Press Oxford, (1936).
7. 'Structural Steels', Ref No GS 5025 12 6/91, British Steel, (1991).
8. 'Alcan Buyers Guide', Alcan Metal Centres Ltd.
9. 'Design Manual: Engineered Composite Profiles', Fibreforce Composites, (1988).

6 A DIRECT COUPLING OF EXPERIMENT AND ANALYSIS: A HYBRID COMPLETE-MATRIX FORMULATION

K. Brandes
Federal Institute for Materials Research and Testing, Germany

Hybrid method within this context means the methodical coupling of the experimental as well as the analytical treatment in only one procedure for solving mechanical (physical) problems. Based on the least squares method, a principle has been defined capable to create a comprehensive regression analysis similar to the Complete Matrix formulation in Finite Element Analysis.

INTRODUCTION

According to the development in computer-oriented methods of structural analysis with new numerical methods, eg the finite-element method and the boundary element method, on the one hand and on the other hand the advances in experimental analysis, the requirement arises for a more sophisticated technique of mathematical treatment of recorded data including the incorporation of mechanical (physical) models.

During the last decades, several hybrid methods have been presented which are suitable in the respective context[(1)(2)(3)]. However, on the basis of the least squares method as defined by Gauss nearly two hundred years ago[(4)], a principle can be defined being the fundament of a generally applicable method which offers a systematized technique comparable to the finite element method. Looking in the fundamental book[(4)] by Gauss that created the method of least squares 1809, surprisingly one can find an advice to combine statistical and physical methods.

It should be mentioned that the basis of the presented method is not a real novelty in the field of mathematical sciences, because a similar technique is known in Applied Statistics[(5)(6)] and Econometrics[(7)]. However, the application to problems of engineering mechanics in a systematic manner is not common practice and shall be introduced by a few examples.

STATEMENT OF THE METHOD IN MATHEMATICAL TERMS

In all cases of hybrid analytical-experimental analysis, both problems have to be stated

- the problem of regression analysis[8] and
- the mechanical (physical) problem in terms of a mechanical model.

Subsequently, with the notation of Draper and Smith[8], the following set of equations arises:

$$(Y\text{-}Xb)^T(Y\text{-}Xb) \rightarrow \text{Minimum} \qquad (1)$$

$$d - Cb = 0 \qquad (2)$$

Using the method of Lagrange's undetermined multipliers λ_i [6][8], the formulas (1) and (2) can be transferred to the Lagrangian function

$$(Y\text{-}Xb)^T(Y\text{-}Xb) + \lambda(d\text{-}Cb) \rightarrow \text{Minimum} \qquad (3)$$

Equation (1) leads to the normal equation of regression analysis with the vector b of parameters b_i. Equation (2) describes the mechanical model in terms of the parameters b_i (restrictions or constraint equations). Complete differentiation of the Lagrangian function (3) with respect to b and λ, a set of linear equations for b and λ is formed:

$$\left[\begin{array}{c|c} X^TX & C^T \\ \hline C & 0 \end{array}\right] \left\{\begin{array}{c} b \\ \hline \lambda \end{array}\right\} - \left\{\begin{array}{c} X^TY \\ \hline d \end{array}\right\} = \{0\} \qquad (4)$$

The solution of the unconstrained least squares problem is

$$\hat{b} = (X^TX)^{-1}\,X^TY \qquad (5)$$

The relevance of this type of formulation becomes more clearly in connection with the examples which concern problems with a series of normal equations the parameters of which are different and only linked together by the constraint conditions.

The difference of the solutions of b or $\hat{b}$ respectively gives a measure for the adequateness of the mechanical (physical) model which is well known in econometrics as Lagrange Multiplier Test[6].

OVALIZATION OF A CYLINDRICAL TANK AS A RESULT OF FOUNDATION SETTLEMENT

Very often, tanks for oil storage become oval during construction when they are filled with water for lifting the top part for mounting. In most cases, the settlement of the foundation causes the ovalization. After a case of damage, we measured the displacements of a tank of a diameter of 54 m and a height of 16 m[9], Figure 1. The vertical displacements at the fundament could be measured very accurately (0.1 mm deviation) but the changes of diameter at the top could only estimated roughly (5cm deviation).

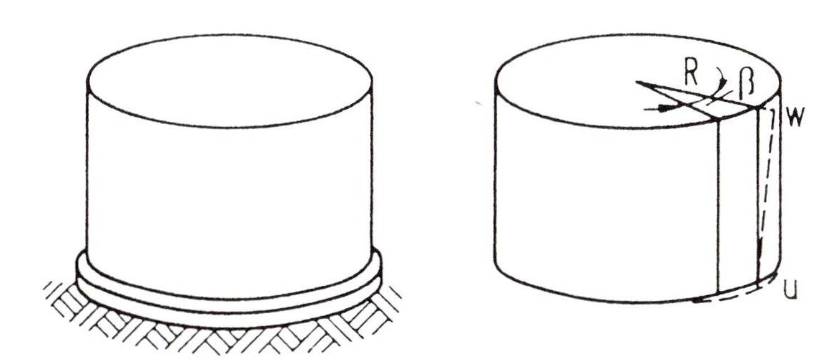

Figure 1. Cylindrical oil tank. Definition of displacements u (vertical in axis - direction x) and w (radial).

For analyzing the deformation of the tank, in particular the correlation of the vertical displacements at the foot circle to the radial displacements of the tank wall at the top, an approach basing on the theory of thin shells has to be applied.

The stress function F

$$F = \sum_n (F^c_n \alpha \cos n\beta + F^s_n \alpha \sin n\beta) \qquad (6)$$

is a solution of the governing partial differential equation of F (t = wall thickness)[9]

$$\Delta\Delta\Delta\Delta F + \frac{1-\upsilon^2}{c^2} - \frac{\partial 4F}{\partial \alpha^4} = 0 \qquad (7)$$

$$c^2 = t^2/(12 R^2) \qquad (8)$$
$$\alpha = x/R$$

The displacements are[9] (Figure 2)

$$\text{radial} \quad w = \sum_n (w^c_n \alpha \cos n\beta + w^s_n \alpha \sin n\beta) \tag{9}$$

$$\text{vertical } u = \sum_n (u^c_n \cos n\beta + u^s_n \sin n\beta) \tag{10}$$

$$u^c_n = -w^c_n/n^2 \; ; \; u^s_n = -w^s_n/n^2 \tag{11}$$

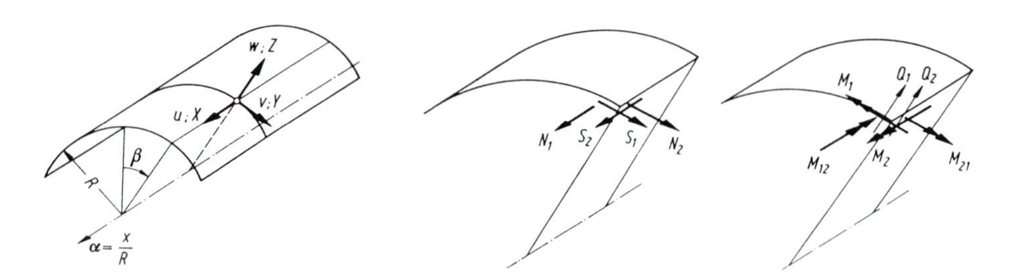

Figure 2. Definitions of mechanical quantities of the theory of thin shells[9].

At the foot circle, the vertical displacement u has been measured at 16 equidistant points. The radial displacement at the top circle (x = 16 m) could only be estimated at 6 diameters and only as a deviation of the diameter from the nominal value. According to the measuring method at the top, only the portions of the even numbered trigonometric functions could be evaluated (n = 2, 4, ...).

The equations for the two circles of measurement are:

at the foot circle (u_M: measured values at points β_M):

$$X^T X u - X^T u_M = 0 \tag{12a}$$

at the top circle (w_M: measured changes of diameter):

$$Z^T Z w - Z^T w_M = 0 \tag{12b}$$

By adding the constraint equations, the following system of equations arises (n = 1, 2, 3, 4)

The result is dominated by the equ. (12a) because the scatter of the measure of w_M is too large. For n = 4, the result, Figure 3, corresponds very closely to the measured data.

While the presented example is a boundary value problem which is solved by applying a boundary method, in the next section, a domain method is used for the solution[10]; a regression finite-element method.

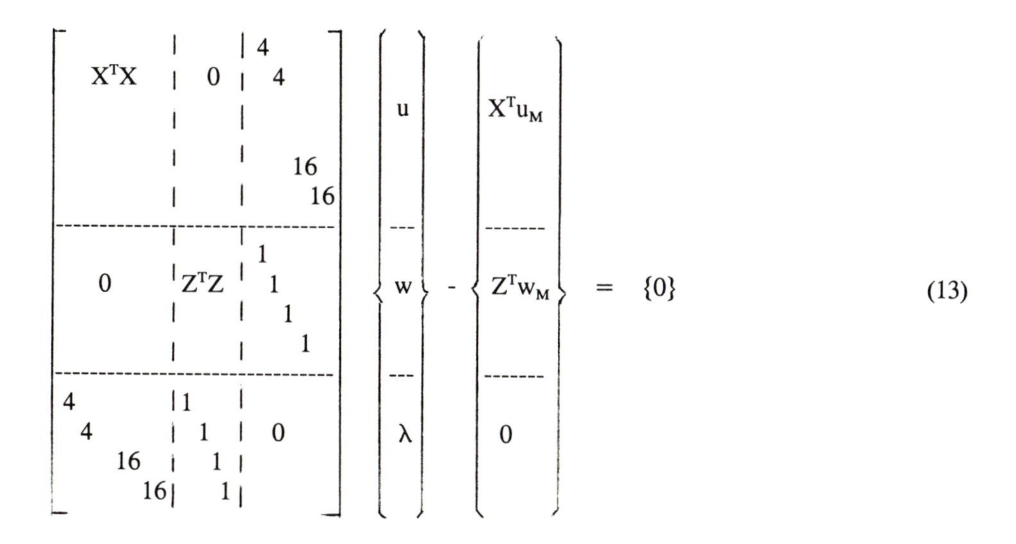

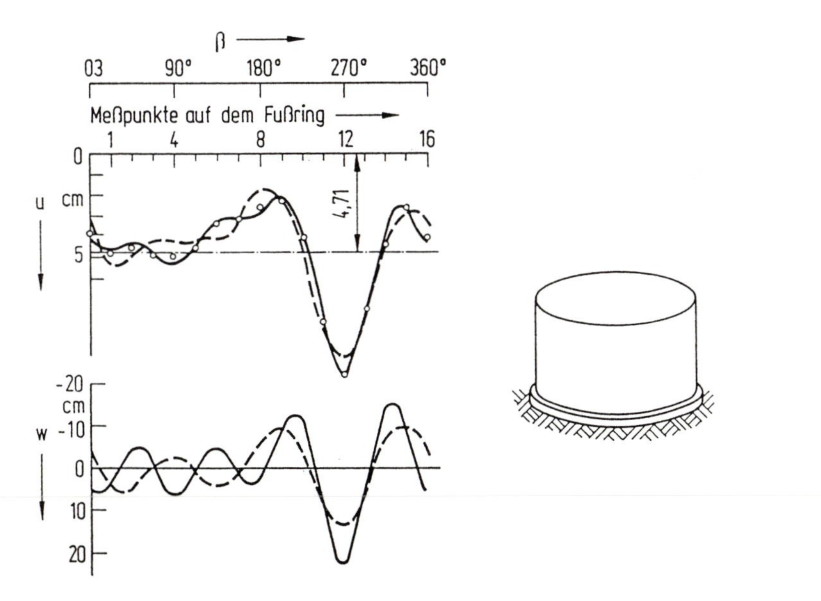

Figure 3. Results of the hybrid analysis for n = 4 (full line) and n = 3 (dotted line)

RESIDUAL STRESSES IN A ROLLED I-GIRDER

Caused by the cooling down process of rolled steel girders, residual stresses exist being of interest in some cases. The values of these stresses are evaluated by cutting the girder into pieces after applying strain gauges at several points[11].

The difference of the measured values performed before and after the cutting are the strains generated by the residual stresses.

For many girders, the residual stresses have been evaluated within a research program for one of them the results are presented in Figure 4. The distribution of residual stresses over the cross-section is needed for further investigations. In the case under consideration, even polynomials are applied to represent the normal stress distribution in mathematical form,

$$\text{flange: } \sigma_{fl} = a_o + a_1 \; \bar{y}^2 \tag{14}$$

$$\text{web: } \sigma_w = b_0 + b_1 \; \bar{z}^2 \tag{15}$$

$$\bar{y} = 2y/b, \quad \bar{z} = 2z/h$$

h: height of girder
b: width of flanges (see Figure 4)

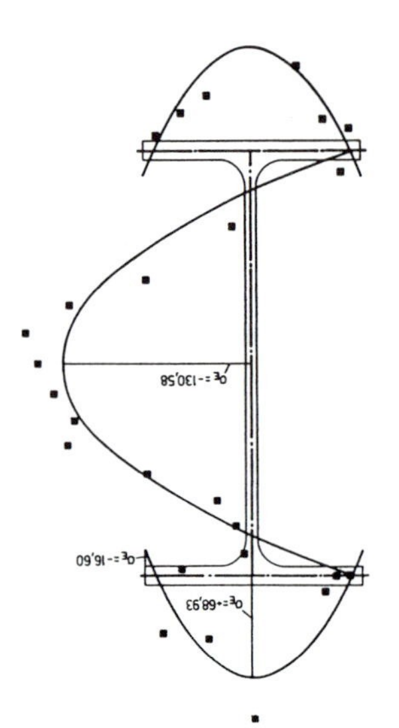

Figure 4.
Residual stresses in a IPE 200 rolled steel girder. Measured values and regression curves.

There are two constraints in the case of even polynomials,

- the stresses at the connecting points of web and flange have to be equal
- the sum of stresses as integrated on the cross-section must be zero.

In terms of the parameters:

$$a_0 - b_0 - b_1 = 0 \tag{16}$$

$$2 \int_{fl} t \, \sigma_{fl} \, dy + \int_{w} s \, \sigma_w \, dz = 0$$

t : thickness of flange
s : thickness of web

For the girder in question (IPE 200) that means

$$1700 \, a_0 + 567 \, a_1 + 1072 \, b_0 + 357 \, b_1 = 0 \tag{17}$$

The system of linear equations which includes the normal equations of the different elements of the cross-section and the two constraint equations contains the parameters a_0, a_1, b_0, b_1 and two Lagrange parameters λ_1 and λ_2:

$$\left[\begin{array}{cccc|cc} X^TX & & & & 1 & 1700 \\ & & & & 0 & 567 \\ & & Z^TZ & & -1 & 1072 \\ & & & & -1 & 357 \\ \hline 1 & 0 & -1 & -1 & 0 & \\ 1700 & 567 & 1072 & 357 & & \end{array}\right] \left\{\begin{array}{c} a_0 \\ a_1 \\ b_0 \\ b_1 \\ \hline \lambda_1 \\ \lambda_2 \end{array}\right\} - \left\{\begin{array}{c} X^T\sigma_M^{fl} \\ \\ Z^T\sigma_M^{w} \\ \\ \hline 0 \\ 0 \end{array}\right\} = \{0\} \tag{18}$$

Solution: $a_0 = 69$; $a_1 = -84$; $b_0 = -131$; $b_1 = 200$ N/mm^2

$$\lambda_1 = -76; \quad \lambda_2 = 0.012$$

For the unconstrained problem, the solution is (see equation (5)):

$$\hat{a}_0 = 51 \quad \hat{a}_1 = -53 \quad \hat{b}_0 = -140 \quad \hat{b}_1 = 249 \text{ N/mm}^2$$

The pattern of the matrix equation is similar to that of the finite-element method which is a 'domain method' of solution of boundary value problems as well as the presented hybrid regression method.

GENERALIZATION

According to the concurrent methods for solving boundary value problems of engineering mechanics approximately - the finite-element method and the boundary element method - a systematics can be presented concerning least squares principle for applied regression analysis - that means a hybrid analytical experimental technique.
For clarification, an example as given by Zienkiewicz in his well known textbook(12) is used and transposed to applied regression analysis, Figure 5,(13).

It is assumed that measured values of a quantity are available at many points of the different elements of the structure (Figure 5). For the description of the distribution of the values, shape functions are introduced similar to those of the finite-element method. The solution can be found by applying the least squares method taking in this concern the place of the energy principles of the finite-element method. The interconnection of the element stiffness matrizes can be performed in different ways. The 'stiffness' matrix [K] of the entire system in Figure 5 is put together to a matrix of minimal band width. However, with respect to a special type of incidence matrix, the matrix equation (19) is formed which is of maximum band width. In terms of matrix transforms, there are different ways of reaching a more compact pattern of equations.

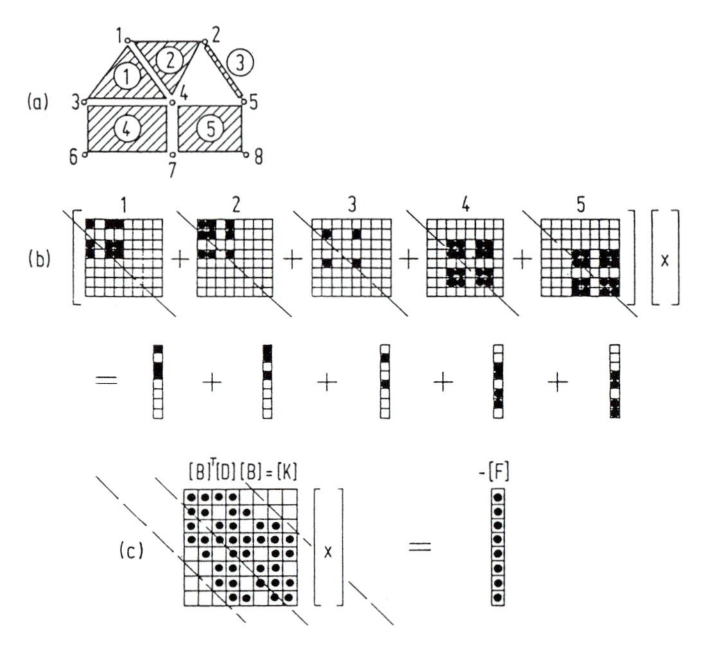

Figure 5. Structural systems to be topographically represented in equ. (19)(11).

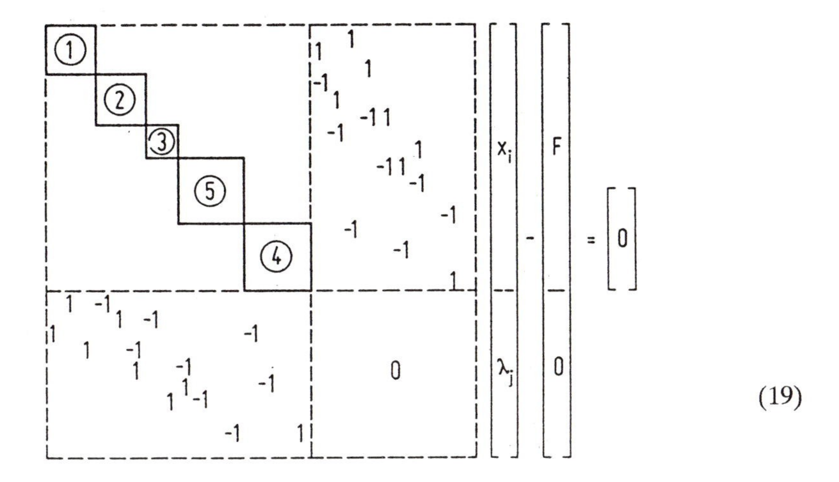

(19)

However, this type of formula has the advantage of being easy to survey. In the finite element method a similar pattern has been developed recently that is called Complete Matrix Method[13].

In applying the presented method, the question is answered together how has to be the weighting of the errors of the observations of different physical quantities the values of which shall be analysed in a regression analysis: The weighting is determined by the mechanical (physical) model which specifies the links between the parameters - the constraints.

CONCLUSION

Starting from the least-squares principle which was created about two hundred years ago and taking up a hint by Gauss[4], a hybrid method is presented that points out a systematic pattern similar to that of the finite-element method (FEM) or the boundary element method.

REFERENCES

1. Laermann K.-H. 'Recent developments and further aspects of experimental stress analysis in the Federal Republic of Germany and Western Europe'. Experimental Mechanics 21 (1981), 49-57.
2. Barishpolsky B.M. 'A combined experimental and numerical method for the solution of generalized elasticity problems'. Experimental Mechanics 20 (1980), 345-349.

3. Brandes K. 'Anwendung hybrider experimentell-analytischer Methoden auf bautechnische Probleme'. VDI-Bericht Nr. 731 (1989), 367-381.
4. Gauss R.F. 'Theorie der Bewegung der Himmelskörper, welche in Kegelschnitten die Sonne umlaufen'. Ins Deutsche übertragen von C. Haase, Hannover, Carl Meyer Verlag 1865 (Lateinische Originalausgabe 1809).
5. Wilks S.S.: 'Mathematical Statistics'. Princeton University Press 1950.
6. 'Encyclopedia of Statistical Sciences'. John Wiley & Sons, New York, 1983, Vol. 4, 456 ff.
7. Chipman J.S. and Rao M.M. 'The treatment of linear restrictions in regression analysis'. Econometrica, Vol. 32 (1964), 198-209.
8. Draper N.R. and Smith H. 'Applied regression analysis'. Second edition. John Wiley & Sons, New York, 1981.
9. Brandes K. 'Verformungen großer Tankbehälter durch unter schiedliche Fundamentsetzung während der Montage'. Bauingenieur 55 (1980), 307-312.
10. Brandes K. 'Zur Systematik numerischer Näherungsverfahren in der Statik und Dynamik der Konstruktionen'. Nuclear Engineering and Design 18 (1972), 469-485.
11. Lindner J. und Kurth W. 'Ermittlung der Eigenspannungen in Walzträgern IPE 200 vor und nach dem Spannungsarmglühen'. Bericht Nr. VR 2010, Technische Universität Berlin, Institut für Baukonstruktionen und Festigkeit, Dez. 1979.
12. Zienkiewicz O.C. 'The finite element method in engineering science'. Second edition. McGraw-Hilt, 1971.
13. Ebel H. 'Statische Berechnung von ebenen Stabtragwerken mit dem Gesamtmatrixverfahren'. Bauingenieur 64 (1989), 267-295.

7 EXPERT SYSTEM: DESIGN OF TANK STRUCTURES (TASDE)

A. Stachowicz
Cracow University of Technology, Poland

This paper describes an expert system for tank structure design /TASTDE/. The system was worked out and is used at Cracow University of Technology. On the basis of this system the following aspects were analysed:
1) the possible areas of application of this type of systems;
2) the particular character of expert systems applied to this design;
3) the possibility of aiding the phase of creative design by looking for new solutions in reference to civil engineering structures.

INTRODUCTION

The possibilities created by computers seem not to have been fully exploited in the design of civil engineering structures. The main interest is still focused on the employment of computers in the static and dynamic analysis of structures, the visualization of results or in the automatization of simple operations.

Not to depreciate the meaning of these fields of application, it is worth drawing attention to the fact that the design phase which is crucial for solution quality (i.e. the phase in which one looks for a conception of the solution and new proposals connected with construction and technology) is still carried out without the significant aid of computers.

In other fields of design, computers are applied much more widely. As an example one may point to architectural designs not to mention the design of cars, ships or airplanes. The reasons for this situation come not only from the particular character of the design of civil engineering structures but also from laws which govern the software market. Putting aside considerations whether changes in this situation will come about in the near future, one may say that this is one of the directions that design can take, a direction which would result in the development of the form of civil engineering structures.

DESIGNATION

S - system, design system, S_D - designing system
Ss - designed system (system model of structure)
US - undersystem (or subsystem)

X, Y - input and output set: $x \in X$, $y \in Y$
M, V - objects: decisive and valuation
P: X x M -> Y - function of a process
K: M x Y -> V - functions of a process description
vx, vy, vm - values of: input, output, decision
Q - a set of permissible solution
f - mapping, R - relation, o - composition
DB, KB - data and knowledge base of system
KBS, KBD - knowledge base of structures and design
RK - a representation of knowledge
BMi, BM - base model and set of models
OMi, OM - opening model and set of models
W - set of variables of state; $w \in W$; $w(\alpha)$ - logical values of a variable
α - rules, F - facts

GENERAL NATURE OF TASTDE

The aim of this system is to work out, in dialogue with the user (the designer of a structure), the best possible structural system under given conditions (resulting from given assumptions). The criteria for the selection are set up by the user. Suggestions proposed by the system are based on the criterion of the material consuming index or the building cost index (if the current coefficients have been given by the user). These decisions are supplemented by the set of compromises if the user chooses a few criteria.

The process function:

$$P: X \times Y \rightarrow Y \tag{1}$$

we may present in the following way:

$$P = P_1 \circ P_2 \circ P_3 \circ P_4 \tag{2}$$

where:

$P_1: X_1 \times M \rightarrow Y_1$ - proposal of structural system
$P_2: X_2 \times M \rightarrow Y_2$ - static calculations and dimensioning
$P_3: X_3 \times M \rightarrow Y_3$ - construction and drafting
$P_4: X_4 \times M \rightarrow Y_4$ - solution description

$$X_i = X \cup Y_{i-1} \; (i<1) \quad Y = \bigcup_{i=1}^{4} Y_i$$

In consequence the system is defined by:

$$(x_i,y_i) \in S \Leftrightarrow P_i\,(x_i,m^*) \in Q_i \wedge \exists m^* \in M$$

$$\text{such that} \quad K_i[m^*,P_i(x_i,m^*)] = \inf_{m \in M} K[m,p_i(x_i,m)] \tag{3}$$

The general system structure is presented in Figure 1. The structure of subsystems is of course also intricate and unlike the constant general system structure is dynamic. The current structure is set up on the basis of the values of input and process factors by the control module of a subsystem. An example of the subsystem structure US2 for the cylindrical tank of capacity $V=15{,}000\ m^3$ is presented in Figure 2.

Figure 3 shows the structure of a proper space of knowledge data base used by the control module.

The system realises the elements of structure design process in the space of system knowledge. This knowledge consists of :

- knowledge of designed structures and their relationships with environment KBS
- knowledge of design (and calculations) KBD
- strategy of effective searching for a solution.

The first two elements create the inner domain of the system. The general principles of the structure defining KBS, KBO are based on the system description of a structure and are presented in[9] and their application to this system in[10][12]. The above-mentioned description is based on the description of physical structure which may be found in[1].

The knowledge which belongs to the inner domain of subsystems USi (i=2,3,4) is algorithmatized and written as a set of procedures though excluding control modules. These modules use simple rules of deduction in order to analyse R(x,y,w,m) relationships. The knowledge representation for these modules is written in the following form:

$$KRi = \langle X,Y,W,M;Ri\rangle \quad (i = 2,3,4) \tag{4}$$

where:

M - decisions (made by the system or designer)

$$R_i(x,y,w,m) = \bigcap_{j=1}^{Ki} R_j\,(x,y,w,m)$$

W - variable of state

The knowledge which belongs to the inner domain of the subsystem US is only partly algorithmatized. There are mainly ‘non-number’ sets and ‘non-function’ relations in knowledge representation. The functions of US_1 are as follows:

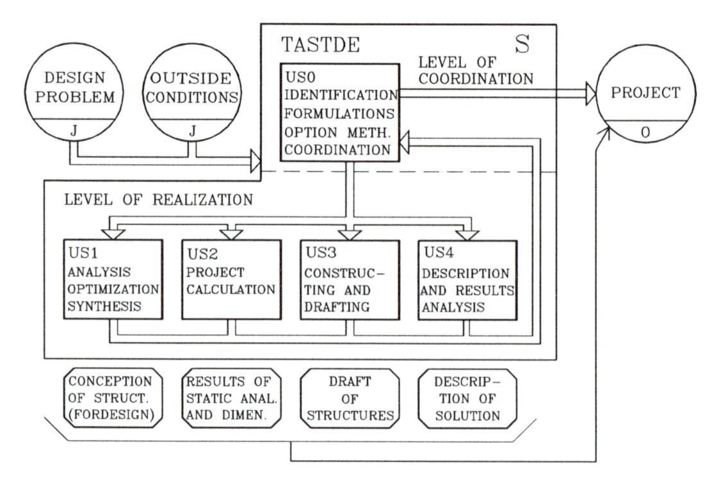

Figure 1. General scheme of TASTDE

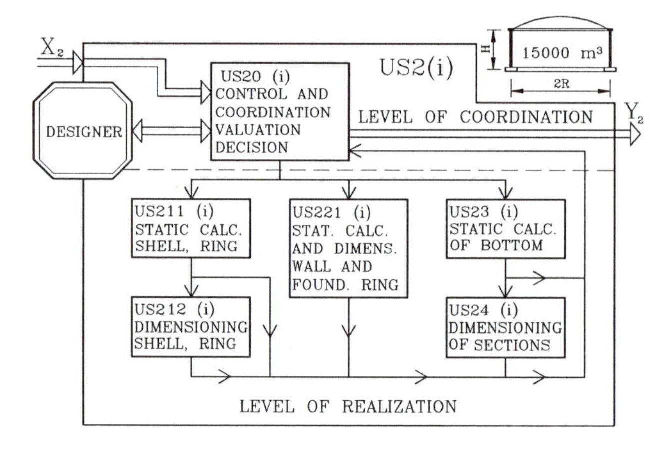

Figure 2. Scheme of system for cylindrical tank structures (V=1500m^3) design

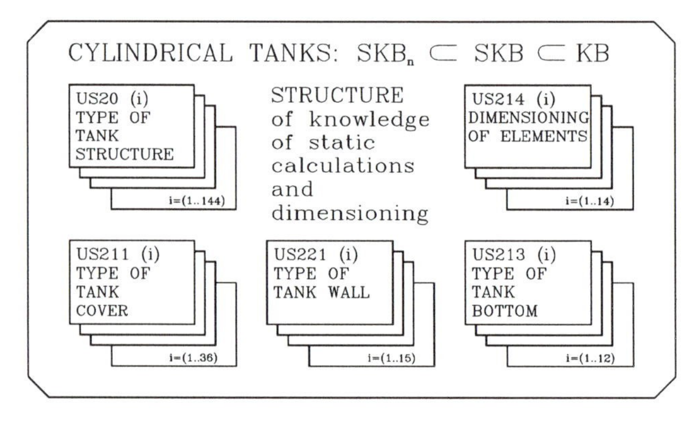

Figure 3. Knowledge base of structures: an example for cylindrical tanks

i) structure determination on the basis of :

$$R_{11}(x,y,w,m) = \bigcap_{j=1}^{K11} R_j(x,y,w,m) \tag{5}$$

which defines the mapping

$$f_1 : \bar{S}_D \rightarrow \overline{\overline{SS}}_{S1} \subset \bar{S}_S \tag{6}$$

where:

$\bar{S}_S$ - set of all structures definable on the basis of system knowledge of structures

$\overline{\overline{SS}}_S$ - subset of real structures taking into consideration the constraints

ii) analysis of the set $\overline{\overline{SS}}_S$ (selection or extension) on the basis of :

$$R_{11}(y,w,m) = \bigcap_{j=K11+1}^{K12} R_j(y,w,m) \tag{7}$$

$$f_2 : \overline{\overline{SS}}_{S1} \rightarrow \overline{\overline{SS}}_{S2} \subset \overline{\overline{SS}}_{21} \vee \overline{\overline{SS}}_{22} \supset \overline{\overline{SS}}_{21} \tag{8}$$

iii) valuation $s = \in \overline{\overline{SS}}_{22}$ assuming $s \doteq BM_i \in BM$

$$f: BM \rightarrow OM$$

$$< x,y,m > \in OM_i \Leftrightarrow P_i(x^*,m^*) \in Q_i \wedge \exists m^* \in M_i$$

$$\text{such that } K_i[m^*,P_i(x_i,m^*)] = \inf_{m \in M} K[m,P_i(x_i,m)] \tag{9}$$

$$\vee K_i[m^a, P_i(x_i,m^a)] = S_p(in,r) \in O(\inf K)$$

where:

m^a - designer decision contradicts system decision

Sp - sphere of radius r

O -surroundings of inf K

iv) comparative valuation $s \in \overline{\overline{SS}}$ and choice s^*

$$(OM_i,m) = OM_i^* \Leftrightarrow OM_i \in OM \wedge \exists m^* \in M$$
$$\text{such that } K[m^*,OM_i] = \inf_{m \in M} K[m,OM_i] \qquad (10)$$
$$\text{and also } K_i[m^a,OM_i] = S_p(\inf K,r) \in O\ (\inf K)$$

The general structure US_1 is shown in Figure 4. The current structure (US_{11},US_{12}) is defined by the module US_{10} on the basis of the relation R(KBD) which defines a mapping

$$f_k:\overline{\overline{SS}}_S \rightarrow KBS \subset KB \qquad (11)$$

EXPERT-SYSTEMS AND DESIGN OF CIVIL ENGINEERING STRUCTURE

i) In every design problem which is of a general enough nature, we have:

$$X = X_1 \cup X_2 \qquad (12)$$

where:

$$X_1 = \{x: vx = w(\alpha_i)\} = X_2/X$$
$$X_2 = \{x: \exists f: vx \rightarrow R\}$$

X_1 defines the construction form and its quality relationships with the environment whereas X_2 defines the parameters of the structure or relationships with the environment. Whenever the application of computers is beyond the analysis of the constant topology and unique quality relationship with the environment, the representation of knowledge in form must be taken into account: $RW = <\alpha,F>$

ii) application of system description of a structure (S_S), a design method (S) and in consequence the function[9]:

$$f = f_1 \circ f_2: S_S \rightarrow S_D$$
$$f_1: S_S \rightarrow S,\ f_2: S \rightarrow S_D \qquad (13)$$

enables the effective defining the structure of the system knowledge data base S_D and its representation.

iii) taking into account (12) every S_D system has a multilevel structure and the decomposition of lower levels[2] is also possible:

$$P(x,y,w,m) = P_o(x_1,x_2,y,w,m_o) \circ n\ P_i(v_i,x_1,x_{2i},y_i,w_i,m_i)$$
$$\inf_{m \in M} K[x,y,w,m] = \inf_{m_o \in M_0} \{ \min_{m_i \in M_i} \sum_{i=1}^{n} K_i(v_i x_1,x_{2i},y_i,w_i,m_i)\} \quad (14)$$

It enables one to obtain the effective solutions of complex problems of the analysis and the valuation of the structure (for vx_i) and also effective logic structure RW.

iv) Expert-system, which strategy of solving design problems is worked out by a group of experts, ensures:
- correct solution (consistence with the state of art and the standards)
- optimization (taking into account the bounds and criteria assumed by users) of design solution possible to be defined in system knowledge data base (KBS). So expert-systems may be elements of the improvement of design quality.

v) Searching new (S_D $\notin$ solutions it may be used to:
- aiming (KBS presentation)
- verification of decisions made by the designer (comparison with $m^* \in M$)
- inspiration (generation of extensions $\overline{\overline{SS}}_S$ on formal structure)

vi) It is possible to use expert-systems more widely for searching new constructions by:
- generalization of the design strategy incorporated into system (on the basis of theory of design)
- extension in knowledge representation beyond a monotonic conclusion (and classical logic of first order[6]) which ensures effective work with incomplete information (extension beyond KBS). The trials of using non-monotonic reasoning[3][5] and alleged logic[7] are the purposes of the work carried out at Cracow University of Technology under the leadership of the author.

vii) Given that:

a) civil engineering design theory can be separated from engineering design theory[11];

b) theory $KR \subset KBD$ can be considered independently of KBS (defining $KR \subset KBS$); it would appear possible to work out universal systems of coordination and control for the design of civil engineering structures. The functioning of the system is, of course dependent upon co-operation with KBS, though the choice is unlimited provided that it is consistent with the structure Figure 5. The effective creation of KBS seems to be possible for a specified class of structure only because only then can the mapping be defined (12). The range of effective application of a system can be defined from the relation:

where:

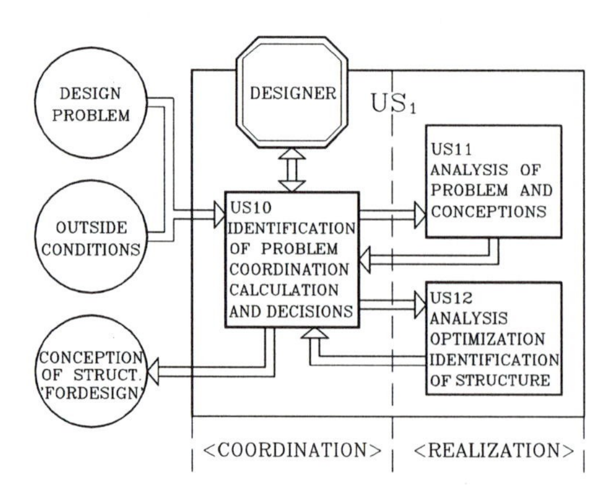

Figure 4. Scheme of subsystem US_1

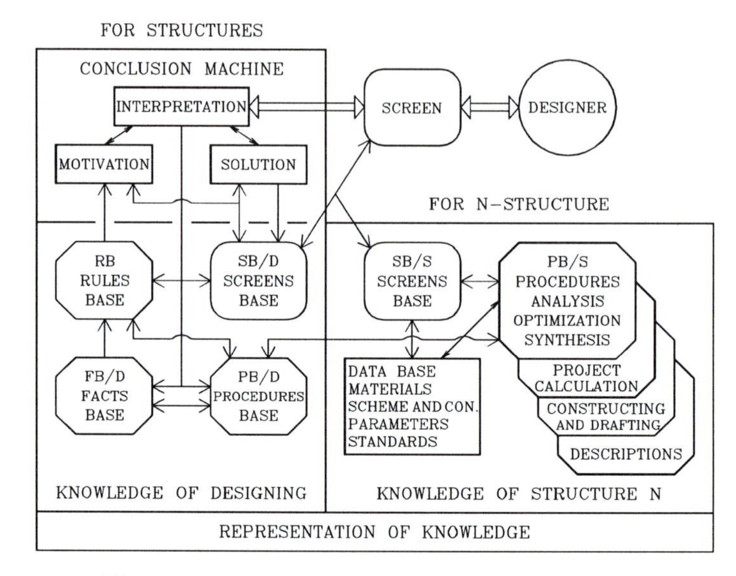

Figure 5. Representation of system knowledge

$$\Delta KBS = \overline{KBS} \,/\, \bigcup_{i=1}^{n} KBS_i \tag{15}$$

$\overline{KBS}$ - the set of all possible structures

KBS_i - the type structure defined in the system
Only when

$$\Delta KBS = 0 \tag{16}$$

can a system be said to have universal applications.

REFERENCES

1. Hiroyuki Yoshikawa. 'General design theory and CAD System' - Proc. Mech. Com. in CAD/CAM, Tokyo 80, p. 80-96. (1980)
2. Jendo S. and Stachowicz A. 'Multilevel optimization in civil engineering', ZAMM - 66, p. 199-200. (1988)
3. McDermott D. and Dogle J. 'Non-monotonic logic I', Artificial Intelligence vol. 13, pp. 41-72. (1980)
4. Mesarovic M.D. and Vasuhiko Tokahara. General System Theory - Ac. Press, N.York, (1975)
5. Lukasiewicz W. 'Formalization of knowledge and ignorance, introduction to non-monotonic reasoning', CC-AJ, vol.3 No.1-2, pp.7-31 (1986)
6. Moore R.C. 'The role of logic in artificial intelligence', Tech. Note 355, SRJ International (1984)
7. Reiter R. 'Non-monothonic reasoning', Ann. Rev. Comput. Sci. vol.2, pp.195-204 (1987)
8. Stachowicz A. 'On applying of systems CAD in formulation of design problems for building structures' - Proc. VIII Int. Symp. CAD/CAM, Zagreb 1986, pp.178-186 (1986)
9. Stachowicz A. 'Optimum structural design and CAD as a problem of system theory', Cracow Uniw. of Tech. Monogr. 59, pp.137-157, (1987)
10. Stachowicz A. 'CAD-System: concrete tanks design' in polish), Proc. III Pol. Conf.: 'Design of concrete tanks. Cracow 1990, pp. 153-158 (1990)
11. Stachowicz A. and Ziobron W. 'Water Tanks, Static Analizis And Optimal Formfinding' in polish), Arkady, Warszawa (1986)
12. Stachowicz A. and Sidwa M. 'Expert System: tank structure design', Proc. V Int. Civil Engin. Conf., Kosice pp.140-144 (1992)

8 NON-LINEAR BEHAVIOUR OF GRILLAGE SLAB 'ORTHO'

K.S. Virdi
P. Ragupathy
City University, UK
P. Hajek
Czech Technical University, Prague

Grillage slabs could, in some cases, be considered as open grid structures with appropriately substituted properties for the component beam elements. In a nonlinear analysis, adequate modelling of the behaviour of materials becomes necessary. Thus, tensile cracking and compressive crushing of concrete, and yielding of steel reinforcement need to be included in the analysis. This paper describes a comparison between test results and numerical results obtained from a newly developed method of nonlinear analysis of reinforced concrete structures with rigid and semi-rigid connections. The paper examines a simplified approach for including the effect of interaction between torsion and flexure.

INTRODUCTION

A rigorous method of nonlinear analysis of isolated restrained beam-columns has previously been developed by Virdi(1)(2). The method can be used for an ultimate load analysis of columns of a variety of cross-section, including material and geometric nonlinearities, following a variety of load paths, as well as allowing variation of cross-section along the length of the column. The method has been extensively verified by tests on composite and reinforced concrete columns.

The above method of analysis of isolated columns has recently been extended to include the problem of 3-dimensional frames with or without side-sway. The analysis takes proper account of the behaviour of flexible connections, such as those encountered in precast frames. A computer program *SWANSA* has been written, based on the new method of analysis. The method offers significant advantages over 3-d finite element analysis, with which it compares in accuracy of results(3). In this paper, the program has been used to study the behaviour of a newly developed grillage slab system *Ortho*. The aim of the study was to determine whether testing, which can be expensive and requires time, could be supplemented by a rigorous method of analysis, in examining all aspects of the behaviour of a new structural system.

SUMMARY OF THE NEW NUMERICAL METHOD

Most structures show a reduction in stiffness with increase in applied loading. This results in instability, and eventually to collapse. The behaviour of nonlinear structures can be studied by following the load-deflection response, much as it is done in experiments. The analysis begins with an initially low level of applied external loading. Iterations for obtaining a solution for the deflected equilibrium state take place in three principal phases: at a section to determine moment thrust curvature relations, along the length of the member to determine the member deflected shape, and at nodal points to ensure equilibrium and compatibility through any flexible or rigid joints. The external forces are increased in steps until, for a given load factor, an equilibrium deflected shape cannot be found. Such a load is taken as the ultimate load of the frame.

Torsion

In the present implementation, it is assumed that members behave linearly for torsion. While failure in flexure can be described by any theoretical or experimental stress-strain curve, it is further assumed that torsional moment and shear forces do not affect the flexural moment capacity of the cross section. This approach is reasonable when analysing in-situ or precast reinforced concrete frames.

Moment Thrust Curvature Relations

Virdi's approach(4) for calculating moment thrust curvature relations using Gauss quadrature formulae has been retained. The advantage in using the above method is that any cross-sectional shape with mixed material properties can be modeled using a number of quadrilaterals. Arbitrary stress strain relations can be used, and strain softening does not pose any computational problem.

Equilibrium Deflected Shape of Beam Column

Consider a space frame as shown in Figure 1. A typical beam-column A-B can be identified, between global nodes i and j. Assuming a deflected shape for the space frame, the end deformations of the individual beam-column can be evaluated by transforming the global displacements to member displacements by using a standard transformation matrix. The method is based on the observation that there exists a combination of member end forces that would satisfy the assumed end deformations, if the member can stay in equilibrium with these forces. The beam-column theory given by Virdi(1)(2) has been used to calculate the equilibrium deflected shape for an individual member. The method is based on starting with an assumed deflected shape and modifying the deflections and end forces iteratively. Second-order Newton-Raphson technique is used to speed up convergence.

Frame Equilibrium

A numerically computed nonlinear stiffness matrix is used to solve the space frame problem. Initial geometrical imperfections in the beam-columns are permissible. By providing suitable moment rotation relations for the joints at the ends of beam-columns, the effect of flexible joints can be included in the analysis. This approach of numerically computed stiffness matrices allows a convenient method for the analysis of frames with partially rigid joints. If the end forces of all the members at nodal points are in equilibrium, then the assumed deflected shape is considered to be the equilibrium deflected shape for the frame. If not, an iterative technique is used to modify the deflections, using the out of balance forces at the nodes. Rigid joints can be analysed by assuming very high values for joint stiffness. Similarly, pinned connections are assumed to have zero stiffness. Rigid jointed frames, and grillages, can therefore be easily modelled. The theoretical background of the technique is described in detail in Ref (3). Only a brief outline of the method is given here.

NUMERICALLY COMPUTED STIFFNESS MATRIX

Considering a beam-column as shown in Figure 2, the end forces $\{p\}$ can be assumed to be functions of end deformations $\{\delta\}$.

$$\{p\} = \{f(\delta)\}$$

These end forces are transferred to global axes by multiplying by a transformation matrix. The member forces in global coordinates become:

$$\{P\} = \{P(D)\}$$

where $\{D\}$ are the global deformations. In the context of iterative procedures, if the external nodal forces were represented by $\{F\}$, the unbalanced nodal forces $\{R\}$ in the global axes system may be written as:

$$\{P\} - \{F\} = \{R(D_n)\}$$

where $\{D_n\}$ are the global deformations at the nth iteration. The aim of iteration is to reduce the unbalanced forces $\{R\}$ to an acceptably small value.

The generalised Newton-Raphson method for iterative solution suggests that an improved set of deformations $\{D_{n+1}\}$ is given by

$$\{D_{n+1}-D_n\} = -[R'(D_n)]^{-1} \ \{R(D_n)\}$$

By substituting $\{R\}$ in terms of $\{P\}$ and $\{F\}$

$$[\{P_n'\}-\{F_n'\}]\{D_{n+1}-D_n\} = -\{P_n\}+\{F_n\}$$

since $\{F\}$ is constant, $\{F'\}=0$.Thus,

$$[K_n']\{D_{n+1}-D_n\} = \{F_n\}-\{P_n\}$$

where $[K_n']$ is an assembly of member incremental stiffness matrices. The

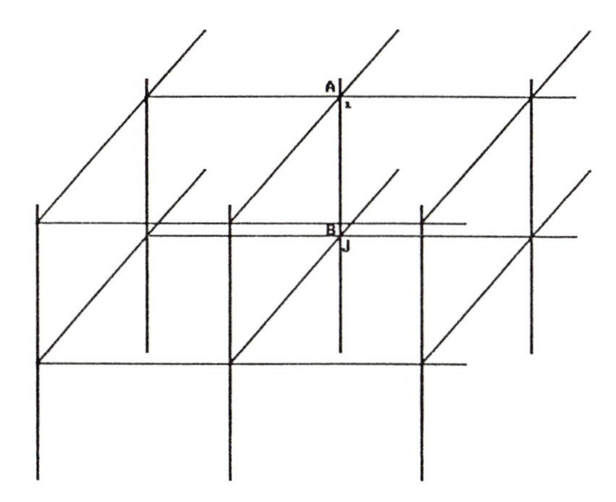

Figure 1. Typical member in a space frame

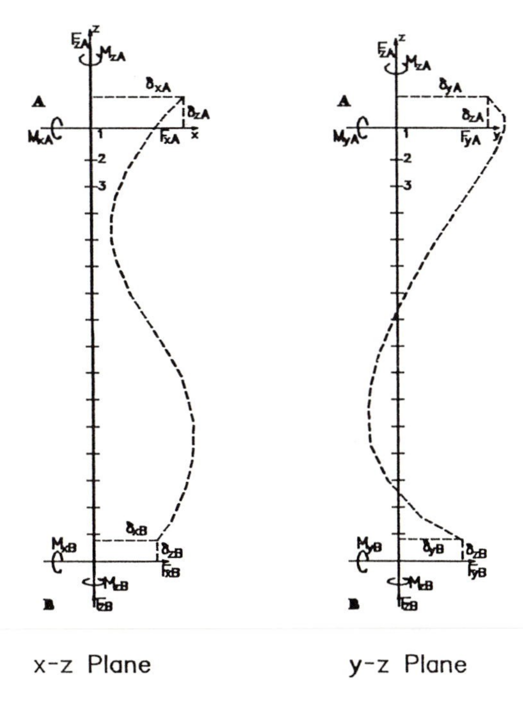

Figure 2. Sign convention for member actions and deformations

elements of the member incremental stiffness matrix are given by:

$$K_{ij}=\frac{dP(\delta_i)}{(d\delta_j)}$$

In the general case, the 'stiffness' matrix would be a full matrix except for the components corresponding to torsion. In line with assumptions described above, the column and row representing the torsional force and torsional rotation in the above matrix are made zero except for the derivative of torsional force with twisting rotation.

The steps necessary to calculate the individual components of the above matrix are now discussed briefly. Figure 2 also shows the notation used for forces and end deformations. The displacement measured along the axis of the member is relative to End B, located according to the relative displacement of End A respect to End B.

STEPS FOR EVALUATING THE NUMERICAL STIFFNESS MATRIX

The beam-column equilibrium position is calculated by keeping end displacements Θ_{xA}, Θ_{yA}, Θ_{xB}, Θ_{yB} and $(\delta_{zA}$-$\delta_{zB})$ (initially assumed zero) constant. The end moments M_{xA}, M_{yA}, M_{xB} and M_{yB} and end forces F_{zA}, F_{xA}, F_{yA}, F_{xB} and F_{yB} are obtained from the above calculation.

The beam-column equilibrium position is then calculated for incremented values of the stress resultants at the two ends in turn. For example, with the axial load $F_{zA}+dF_{zA}$, and all the end rotations being kept constant, the axial deformation δ'_{zA}, end moments M'_{xA}, M'_{yA}, M'_{xB} and M'_{yB} and end transverse forces F'_{xA}, F'_{yA}, F'_{xB} and F'_{yB} are thus obtained. The process is repeated with other components of end deformations. The above step by step procedure enables the member stiffness matrix to be evaluated by a numerical procedure, rather than in a closed form. The key advantage is that this procedures requires no limitation on constitutive relations of the materials or any limitation on the geometric shape and constitution of the cross-section.

The element incremental matrices and the end force matrices are transformed to global axes using standard transformation matrices. After assembly by considering the equilibrium at nodes, the following equations are obtained:

$$\{F\} - \{P_n\} = [K]\{dD\}$$

where, $\{F\}$ is the vector of external forces on the nodes
$\{P_n\}$ is the vector of out of balance forces in the nodes
$\{dD\}$ is the vector of increments in nodal displacements
and, $[K]$ is the overall incremental stiffness matrix

The above simultaneous equations are solved for increments in nodal displacements, and the frame deformations are modified. The new values, in turn, are used in the beam-column sub-program to calculate the out of balance forces and the new incremental member stiffness matrix. The process is repeated until the results converge.

For the next load increment, the above converged values for deformation are extrapolated and used as the initial deformations. The load is increased in steps, until the nodal displacements fail to converge. The maximum load for which convergence is obtained, is taken as the ultimate load.

COMPUTER PROGRAM *SWANSA*

A computer program, labelled *SWANSA* (SWay And No-Sway Analysis), has been developed based on the above method for linear or non-linear analysis of 3-dimensional precast concrete sway and nonsway frames. Joints can be rigid, pinned or flexible (semi-rigid). The program can also compute the ultimate load of single beam-columns, continuous beam-columns, or grillages. The output includes deflections, moments, shear forces, axial forces, strains, and torsion at all the member stations and at global nodes.

The program *SWANSA*, and its preprocessor, have been written in *FORTRAN77*, and the two can be installed on any machine with the *UNIX* or *OS/2* operating systems. A post processor for *SWANSA* has also been developed for use with the drafting system *MICROSTATION* from *INTERGRAPH*. The post processor has facility to view and produce hard copy of deflected shapes, bending moment diagrams and shear force diagrams.

The program *SWANSA* has been validated against experimental and computational results published from several sources(5)(7), as well as tests on eight column-beam subframes at City University, full details of which are also given in Ref(3).

EXPERIMENTAL BEHAVIOUR OF FLOORING SYSTEM *ORTHO*

A new type of hollow brick element *ORTHO*, suitable for ceramic reinforced concrete floor slabs has been developed at the Czech Technical University. An accompanying paper(8), gives details of the system, as wells of test on a full scale floor panel. Results from the test are now compared with computations from program *SWANSA*. The aim of the study is to further validate the program *SWANSA*, in its application to grillage slabs. It has already been indicated that in program *SWANSA*, interaction between torsion and flexure has been simplified. It is intended that the program can be used for a more extensive parametric study to demonstrate the range of applicability of the *ORTHO* flooring system.

ANALYSIS USING PROGRAM *SWANSA*

In order to study the influence of torsion, a number of cases have been run, assuming full shear modulus (1.00G), as well as two reduced values (0.25G and 0.10G). The continuous floor is idealised as a grillage of T-beams, the flange width being equal to the rib spacing. The basic data for the analysis is as described in another paper[8] in the proceedings. In order to simulate the uplifting of the corner points, which were not held down in the test, it was assumed in the analysis that 25% of the edge length in each corner was free to deflect, and only the middle 50% of each edge was held down. Results are compared with experimental values for deflections as well as strains at selected points within the floor.

Load Deflection Response

Figure 3 shows the load deflection response at the centre (Point 9) of the floor. The References to points are to those shown in Figure 5 of Ref (8). It will be noted that for all three values of shear modulus, the deflections are overestimated for low values of the applied loading. For 1.00G and 0.25G, good correlation is obtained for loads upto about 80% of the failure load. Above that value, results with a low value of shear modulus (0.10G), appears to predict the trend better. It may be concluded that, if the analysis could model the progressive deterioration in the shear modulus, the correlation between theoretical and experimental results would be excellent.

Figure 4 shows the comparison for deflections at the four corner points. The test results and the computed results show considerable differences. It must be pointed out that part of the discrepancy arises from the manner in which the edge support has been modelled. The results will be significantly altered, if a greater parts of the edges were assumed to have been held down. Thus, the discrepancy occurs not only due to inadequate modelling of the shear modulus, but also due to the uncertainty regarding the contact length. Figures 5 and 6 show the deflections for points 14 and 15 respectively. These points are on adjacent edges. In view of the orthotropy of the system, some differences in the response are only to be expected. Thus in the test, Point 14 moves upwards, while Point 15 deflects downwards. In the computed results, both points move upwards. It should be pointed out that, in both cases the deflections are very small, and the trend is exaggerated in the graphs because of the scale. A perspective of the deflected shape obtained from the analysis is shown in Figure 8.

Figure 7 shows a comparison between the measured and calculated strains. It would appear that the computed strains are consistently lower than the measured strains. This could also be attributed to the higher value of the torsional rigidity assumed in the calculations.

In all cases, lowering of the shear modulus results in the computed results approaching the experimental results. Clearly further work needs to be done in the modelling of the interaction between shear and flexure, in the nonlinear range.

CONCLUSIONS

A new method of analysis of reinforced concrete space frames has been described. The method takes into account material and geometric non-linearities,

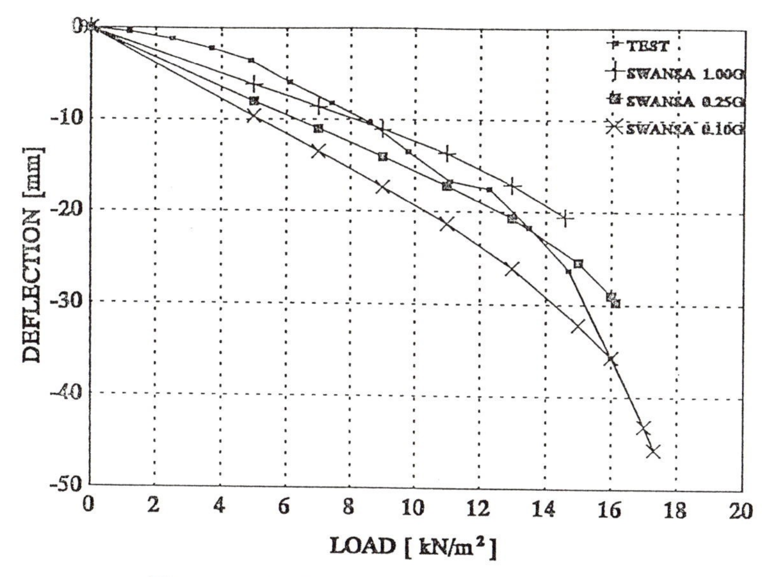

Figure 3. Deflections at Central Point 9

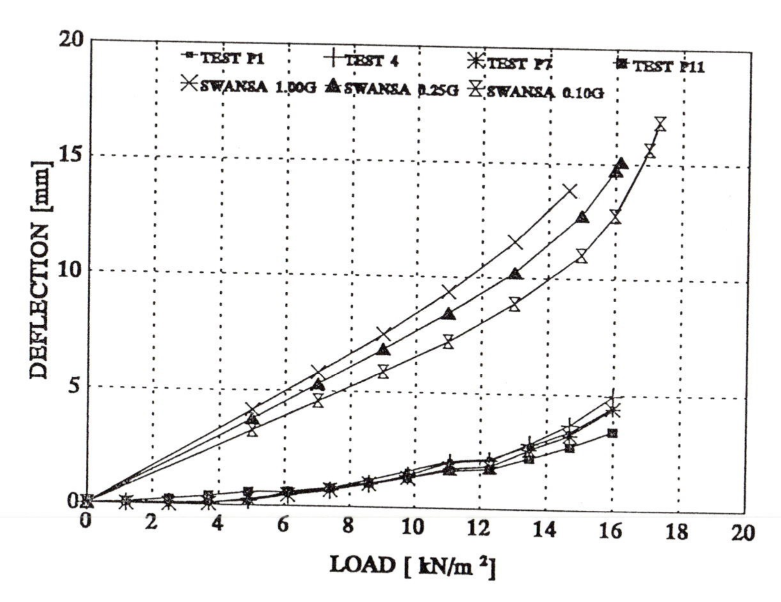

Figure 4. Deflections at corner points

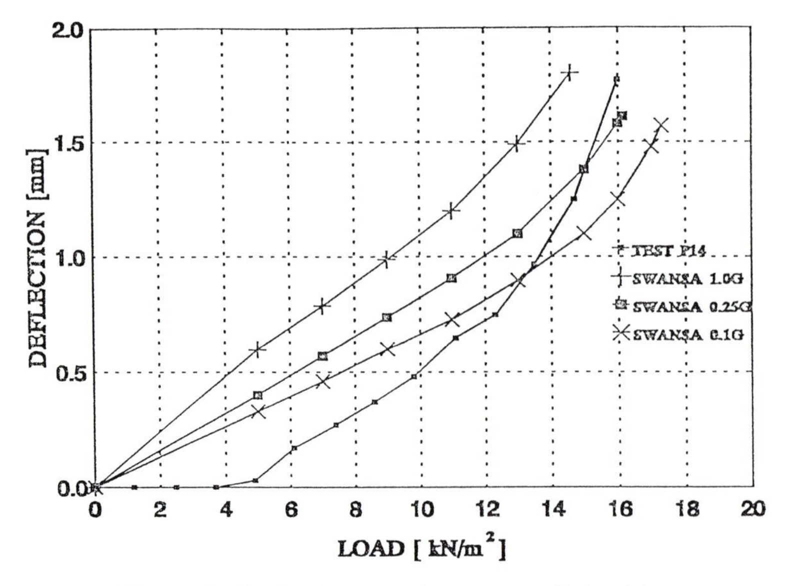

Figure 5. Deflections at edge support Point 14

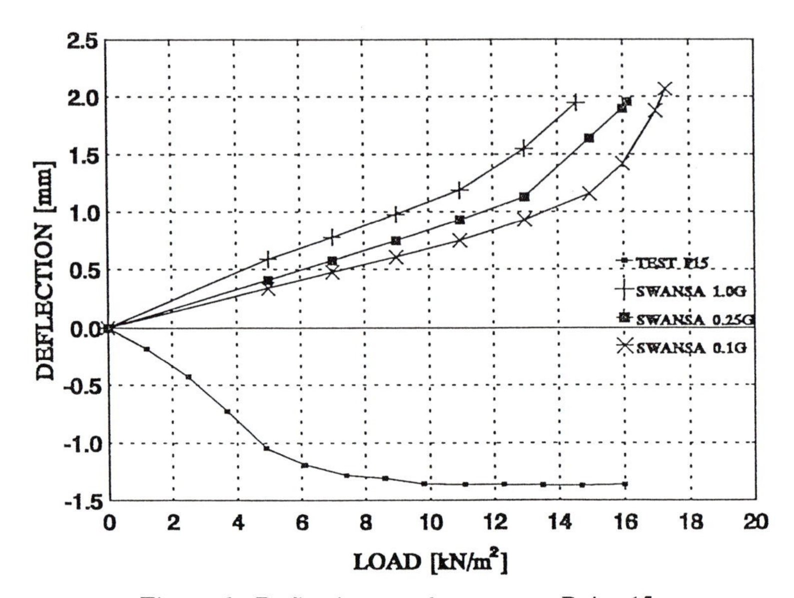

Figure 6. Deflection at edge support Point 15

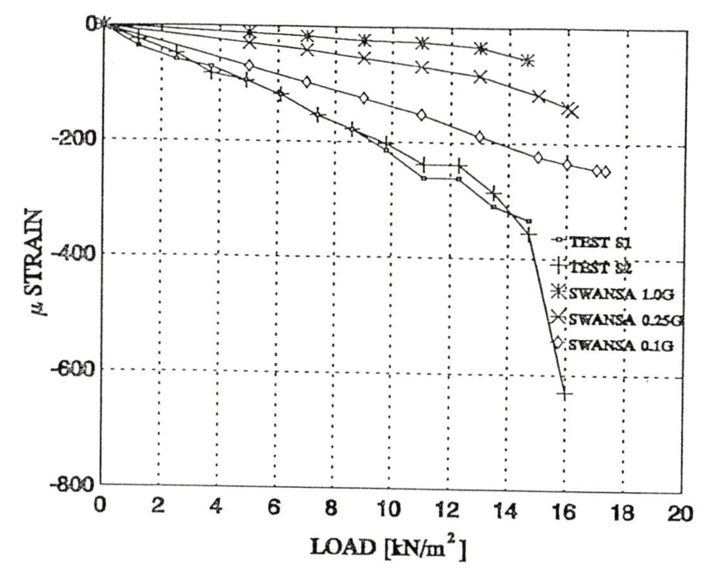

Figure 7. Strains at centre of the floor

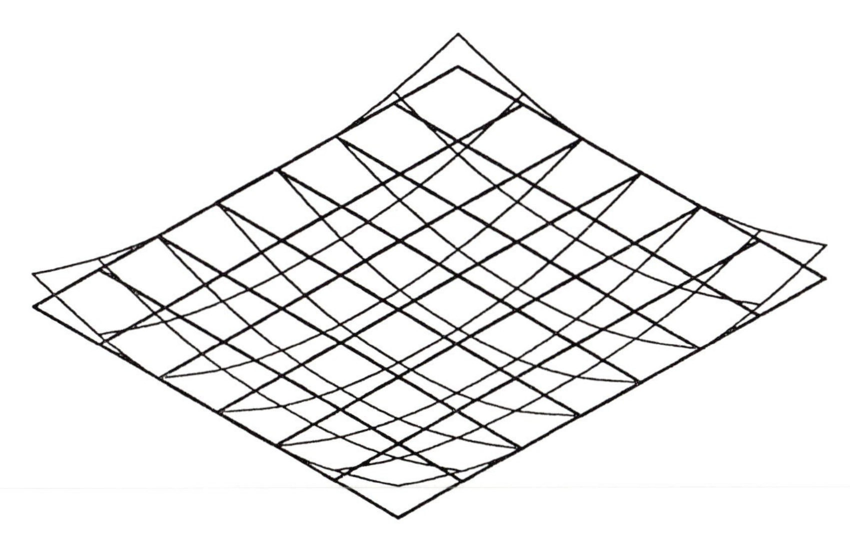

Figure 8. Perspective view of the computed deflected shape of the floor

as well as the semi-rigid nature of connections used by most suppliers of such frames. The method has been incorporated in a computer program *SWANSA*, and has previously been verified against experimental and analytical results. In this paper, the program has been used to study the behaviour of floors made from a new type of flooring system *ORTHO*. It is concluded that good correlation has been obtained for the failure load and for the central deflections of the test floor. However, discrepancy has been observed for deflections of the corners and the strains at the centre. The results show a need for improvement in the modelling of interaction between flexure and shear in a rigorous analysis of this kind.

REFERENCES

1. Virdi K.S. and Dowling P.J. 'The ultimate strength of composite columns in biaxial bending'. Proceedings. The Institution of Civil Engineers Part 2. Vol 58, pp251-272 (March 1973)
2. Virdi K.S. and Dowling P.J. 'The ultimate strength of biaxially restrained columns'. Proceedings. The Institution of Civil Engineers Part 2. Vol 61, pp41-58 (March 1976)
3. Ragupathy P. 'Nonlinear behaviour of precast concrete frames with flexible joints'. PhD Thesis, Structures Research Centre, Department of Civil Engineering, City University, London, (to be submitted in 1993)
4. Virdi K.S. 'A new technique for the moment thrust curvature calculations for columns in biaxial bending'. Sixth Australasian Conference on Mechanics of Structures and Materials, Christchurch, New Zealand (August, 1977)
5. Seniwongse M. 'The deformation of reinforced concrete beams and frames up to failure'. The Structural Engineer, Vol 57B, No 4, (December 1979)
6. Franklin H.A. 'Nonlinear analysis of reinforced concrete frames and panels'. PhD Thesis, University of California, Berkeley (1970)
7. Ernst G.C. et al. 'Basic reinforced concrete frame performance under vertical and lateral loads'. ACI Journal, p261 (April 1973)
8. Hajek P. 'A new type of hollow brick element 'ORTHO' for grillage and ribbed slabs'. Building the Futures: Innovation in Design, Materials, and Construction, The Role of Physical Testing. Joint IStructE/BRE International Seminar, Brighton (April 1993)

CHAIRMAN'S REMARKS

Dr G Somerville, BCA, UK

'Building the Future' is the theme of this Seminar. In the first Session, we have covered 'General Principles and Philosophy' and it is important that we get these right, to avoid repeating past mistakes and to be clearer on what exactly we are trying to achieve, in balancing adequate technical performance against economy (preferably measured in life-cycle terms.

An essential part of that is the introduction of innovation into practice in a sensible and responsible manner, while integrating it into the current wisdom and practices. Unfortunately, we have no structured framework for doing that. It could be said that much research is second hand, in having to pass through several pairs of hands before application; inevitably, much is lost in this translation - time, understanding, efficiency. It is often said, truthfully, that the construction industry is fragmented, and this is particularly true in its attitude and use of R&D.

The industry operates in a commercial mode, with varying and variable standards. Moreover, client requirements, in terms of performance criteria, do change - especially true at the present time. To cope with that, in introducing innovation, it is necessary to consider all the essential elements in the package; these are:

- performance criteria
- design loads
- design (predictive) models
- margins/factors of safety
- material specifications
- minimum standards of execution (workmanship)
- a maintenance strategy

All these elements interact; all have to be considered equally, if the innovation is to be successful. In practice, this is more likely to occur if the innovation is brought through as a joint venture, while harnessing the appropriate skills and know-how.

In this session, we have had 8 interesting papers which address some of these issues. In particular - for 'success' overall, we need to:

- underpin generically, as typified by Mr Parkhouse
- tackle the difficult gaps, as typified by Mr Moore
- ask questions about performance criteria, as typified by Dr Wood
- develop systems of checks and balances, validation and calibration, as typified by the other papers
- balance the different elements, while being mindful of the overall environment in which we operate (commercial, political, social, technical).

PART TWO

TIMBER

9 APPLICATION OF MODERN WOOD BASED COMPOSITES IN CONSTRUCTION

I. Smith and Y.H. Chui
University of New Brunswick, Canada

Manufacturing technology has enabled engineered wood composites that can be tailored to specific construction applications. Properties of parallel strand lumber and prefabricated wood I-joists are discussed in relation to design. Attention is drawn to need to consider serviceability explicitly to avoid problems associated with, for example, creep deformation and vibrational serviceability. Discussion focuses on where hypothetical limits on use of wood based composites in construction lie.

INTRODUCTION

Products such as lumber, glued laminated timber and plywood are well established as construction materials for large and small buildings, bridges and utility structures. The manufacturing technology for such products is relatively simple, but tends to require high grade sawlogs if, for example, high strength lumber or plywood veneer are being recovered. Need for high grade sawlogs with reasonable diameters conflicts with what is available as a commodity. Past exploitation and modern forest practices dictate that our primary source of wood fibre for construction is conifer species grown in managed forests(1). As has been shown in numerous studies, optimal production of bio-mass requires that trees be harvested on fairly short rotations, ie. 50 to 60 years for temperate softwoods grown in Canada or Scandinavia, 20 to 30 years for Southern pines grown in USA and Radiata pine grown in Chile, Such material tends to have relatively low density, high levels of grain distortions and high knot content. All these features militate against cost effective traditional construction products with desirable physical and mechanical characteristics, eg.(2)(3). Lumber, glued laminated timber and plywood continue to be available, but are now supplemented by newer wood based composite products that overcome many limitations.

Parallel strand lumber (PSL) is an example of a modern engineered wood based composite material. PSL is manufactured by gluing strips of veneer together so that fibres are 'all' oriented along the axis of a member. This material is a proprietary product produced under the supervision of a quality control agency,(4). Table 1 illustrates typical factored resistances for selected products used in beams in Canada. These values assume dry service conditions and 'standard' load duration,(5). It can be seen that PSL has much higher factored resistances than lumber and higher factored resistances than glued laminated timber, when compared on a depth-to-depth basis. Parallel strand lumber is available in section sizes up to 178 x 606mm. Lengths of members are restricted by transportation rather than manufacturing considerations.

Table 1. Normalised Factored Resistances of Selected Beam Products (on a per mm of thickness basis)

Section(1) (mm)	Bending Resistance M_r (Nm/mm)	Shear Resistance V_r (N/mm)	Flexural Rigidity EI (10^9Nmm^2/mm)
a) Spruce-Pine-Fir lumber (select structural grade):			
38x286	202	172	20.5
140x241	142	121	9.9
b) Douglas Fir Glued Laminated Timber (20f-E grade):			
80x266	271	319	19.5
80x304	355	365	29.0
130x608	1423	729	232
c) Parallel Strand Lumber:			
45x286	376	467	26.4
89x241	276	398	16.2
89x606	1573	1001	256

(1) The smaller dimension is the thickness, and the larger the depth.

Prefabricated wood I-joists are composite products manufactured by several North American companies as proprietary products under supervision of a quality control agency,[6]. Flanges are of machine stress rated lumber or laminated veneer lumber, and webs of plywood or Oriented Strand Board. The sections are glue assembled as a continuous strip with special flange to web details. Members can be cut to 'any length'. Typical uses are as floor or roof joists over single or multiple spans, and as secondary framing members. North American consumption of wood I-joists has increased substantially over the last few years, and was estimated at 36 million linear meters in 1987,[7]. These products are a beginning in the battle to place wood on an equal basis to materials such as steel with respect to ability to manufacture structurally efficient cross section geometries without waste. Often wood I-joists are light enough to be hoisted into place without mechanical equipment, leading to significant cost benefits on the job site.

The next two sections of this paper illustrate special considerations during development and use of modern wood based structural composites.

RELATIVE CREEP AND CREEP RUPTURE

Bledsoe and Knudson reported results of tests on two experimental parallel strand lumber products with apparently the same properties based on short-term laboratory tests,[(8)]. Products A and B were identical with respect to manufacture, except that different pressing strategies were used. Bending creep tests were performed and performances of the PSL products compared with that of a control of 2400f-E grade machine stress rated Douglas fir lumber. For a 38 x 89 section average relative creep values were 0.3 and 0.5 for Products A and B respectively, after 600 hours (25 days). The sawn lumber showed about the same response as Product A. These results correspond to a fairly high stress level of 0.6 times the characteristic 5 percentile strength. Also after 600 hours loading, creep rupture effects resulted in failure rates of: 6 percent for Product A, 36 percent for product B, and 15 percent for sawn lumber. These results graphically illustrate need to account for creep related behaviour of wood based composites at development and design phases in structural applications.

Because PSL products have relatively low variability in strength (c.o.v. approximately 0.1) compared with machine stress rated lumber (c.o.v. approximately 0.2); it can be deduced that Product A had slightly inferior relative creep characteristics but about the same creep rupture behaviour as the sawn lumber.

The concept of allowing for creep deflexion in wood based composites is not new. In, for example, in the now redundant British Standard Code of Practice CP 112 'The Structural use of Timber' a method is given for adjusting elastic moduli of oil tempered hardboard,[(9)]. The effect of creep deformation is included in proportion to the mix of load components classified as short-, medium- and long-term durations, and according to the ratio of the applied to the permissible stress in each of those duration categories. An upper bound on the adjustment of elastic moduli is that 50 year values are 1/4.52 times 1 minute values (relative creep 3.52).

It appears likely that Eurocode 5 'Common unified rules for timber structures' will recommend a much simpler approach in estimation of creep deflexion,[(10)]. Table 2 presents proposed relative creep factors for lumber and fastenings, and glued laminated timber. According to Larsen[(10)] relative creep values for service classes 1 and 2 will be about 50 percent higher for plywood, and 100 percent higher for particleboard and fibreboard. It seems reasonable to suppose relative creep of products such as PSL will eventually be classed along with lumber or glued laminated timber. Other new wood based composites may well exhibit much higher levels of relative creep, eg. Oriented Strand Board.

Table 2. Provision relative creep factors for Eurocode V,[10].

Load duration class	Service Class[1] 1 and 2		3
	Lumber and Fastenings	Glulam	All[2]
Permanent and long term	0.50	0.30	1.00
Medium-term	0.25	0.15	0.50
Short-term and Instantaneous	0	0	0.25

(1) Equilibrium moisture contents in most softwood lumber:
Class 1 - not exceeding 12 percent, Class 2 - 18-20 percent, Class 3 - may be higher that 18 to 20 percent.

(2) All signifies lumber, fastenings and glued laminated timber (glulam).

VIBRATIONAL SERVICEABILITY OF FLOORS

Annoying vibrations in buildings are felt by humans as a result of transient loadings that contain frequency components which coincide with natural frequencies of whole or parts of building systems. Both natural frequencies and levels of vibration amplitudes are dependent upon the mass of a structure. Light-weight structures tend to produce responses with 'high' frequency content when excited dynamically. Recent moves by code committees to incorporate proper vibration criteria in assessment of serviceability reflect increased user dissatisfaction, which stems from optimising designs based on static load response,[11]. This has become common with the advent of many high performance 'engineered' composites, including, but not exclusively, those of reconstituted wood.

Prefabricated wood I-joists are an example of products that are highly engineered, with low mass per unit length compared with traditional wood products for the same design span. Because of this, floors built with prefabricated wood I-joists are potentially problematic,[7]. Studies were recently conducted by the University of New Brunswick, Wood Science and Technology Centre (WSTC) on Wood I-joisted and conventional lumber joisted floors. Results are summarised and discussed here to illustrate comparative behaviours of systems.

Table 3 summarises characteristics of floors tested by WSTC. Edges of floors that contained ends of joists were simply supported and outer joists were free to vibrate. Other tests where all four edges were supported are reported elsewhere.

Table 3. Details of test floors

Floor	Span (m)	Width (m)	Joist		Flooring	
			Type(1)	Spacing	Type(2)	Thick-ness
1	5.9	3.413	241mm deep I-joists with 9.5mm OSB web and 38x89mm MSR lumber flanges	488mm	OSB	15.8mm
2	6.8	3.413	Ditto	488mm	OSB	15.8mm
3	4.673	5.852	241mm deep I-joists with 9.5mm plywood web and 45x38mm LVL flanges	488mm	CSP	15.8mm
4	6.502	7.924	356mm deep I-joists with 9.5mm plywood web and 38x89mm MSR lumber flanges	610mm	CSP	19.0mm
5	7.010	6.827	Ditto	488mm	CSP	15.8mm
6	4.06	4.2	38x235mm No.1 spruce-pine-fir lumber	600mm	OSB	18.5mm
7	3.18	3.6	38x184mm No.1 spruce-pine-fir lumber	600mm	OSB	18.5mm

(1)(2) OSB signifies Oriented Strand Board, LVL signifies laminated veneer lumber, MSR signifies machine stress rated, CSP signifies Canadian Softwood plywood.

Floors were first tested by a hammer impact approach, to determine their natural frequencies and viscous damping ratios, as unloaded systems. Impacts were also applied to floors using higher energy impacts resembling those encountered in service due to humans walking. These 'heel impact' tests produced transient responses for which frequency-weighted root-mean-square acceleration, A_r was measured. It has been found that A_r correlates well with human response. This latter type of tests was applied with and without imposed mass (see Table 4).

Table 4. Test results

Floor	Natural frequencies (Hz)					Mean viscous damping ratio (%)	A_r at quarter point (m/s^2)
	f_1	f_2	f_3	f_4	f_5		
1	13.9	15.2	18.7	21.4	25.2	2.6	2.411*
2	10.2	12.4	16.0	18.3	22.4	3.5	1.292*
3	15.2	19.4	22.2	24.2	26.3	2.3	0.417+
4	11.8	14.5	16.9	18.2	19.7	1.7	0.413+
5	12.5	15.2	17.1	19.8	27.6	1.1	0.412+
6	16.8	21.6	26.1	30.4	34.2	2.2	0.321+
7	21.0	28.2	33.0	39.3	43.5	2.6	0.396+

* with no imposed masses
+ with imposed masses equivalent to a uniform load 0.475 kN/m^2.

Primary test results are summarised in Table 4. The quarter point A_r values are response measurements at the intersection of quarter span and quarter width, with an impact near the centre floor position. Values are averaged over the four quadrants of the plan.

It is well established that people are sensitive to both vibration frequencies in the range 4 to 8 Hz (which corresponds to natural frequencies of internal human organs), and A_r values in the order of 0.45 m/s^2 or greater,[(12)]. Conclusions are drawn based on the above mentioned studies and computer simulations of free and forced vibration responses of a broader range of systems. It is concluded that unlike lumber joisted floors a large part of the discomfort caused by vibrations in wood I-joisted floors is due to the low frequency components of vibrations. This is a direct consequence of the long spans possible with such joist products. Long span wood I-joist floors using deep joists can have acceptable vibrational performance when designed on the basis of a static deflexion limitation, due to their relatively large mass i.e. relative to shallow I-joists. Floors with shallow I-joists are more prone to annoying high level vibrations if designed on static deflexion behaviour alone. In such cases proper dynamic response analyses are appropriate. Overall, it seems necessary to utilize vibration serviceability criteria and models that recognize the orthotropic nature of wood I-joisted floors if consistently reliable designs are to be produced.

INHERENT LIMITS

If one speculates about hypothetical limits on use of wood based composite materials in building construction, three issues are paramount: strength properties, stiffness properties and damping characteristics (material damping, inertial mass). Other issues such as a durability and fire performance can be dealt with effectively by chemical treatment or appropriate detailing, using existing knowledge.

Approximately speaking, paper represents an upper bound on strength of wood based composites. Pulp has a tensile strength of about 1 to 2 GPa, which is around half the value for man-made fibres such as Kevlar 49 and Carbon (standard). Thus currently available wood based composites only attain 3 to 5 percent of that potential, i.e. about the same as clear wood. Pulp has a modulus of elasticity in the range 10 to 100 GPa, which compares with about 130 GPa for Kevlar 49 and 235 GPa for Carbon (standard). Available wood based composites can attain a modulus elasticity of up to 15 GPa. Thus, it can be deduced that wood based composites cannot reach the levels of strength and stiffness properties attainable with composites containing man-made fibres. Wood fibre is of course much cheaper and more energy efficient to produce, so the contest is not lost. In a relative sense existing wood products exploit stiffness potential fairly well, but not strength potential. This is highly promising as, except for axially loaded ties, section geometry can be manipulated to overcome inherent lack of stiffness in the material, provided the material is formable. Attention should therefore continue to be directed at development of wood based products that can be cast during manufacture. In principle chemical treatment and adhesive technology can be applied to improve recovery of the strength potential of raw wood fibres. These latter approaches should also lead to reduced creep in wood composites.

The viscous damping ratio of wood is reasonably constant at 1 to 2 percent,[(13)]. Combined with the low mass per unit volume this means that structural systems of future wood based composites are likely to derive damping capabilities from sources other than inherent characteristics of those materials. Appropriate design strategies will include identification of construction geometries and detailing that eliminate troublesome (typically low order) vibration modes. Also increased use of damping devices such as tuned mass dampers is likely to avoid excessive motion due to transient loading from wind, earthquake or during normal use. Improved mechanical properties will almost certainly mean therefore that designers will have to address vibrational serviceability of systems built from wood based composites. Proprietary systems are those likely to dominate volume applications, as product optimisation will tend to be beyond the non-specialist designer.

CONCLUSIONS

Use of modern wood based composite materials in construction has expanded the scope of wood solutions. It has become necessary to consider issues such as creep and creep rupture behaviour, and vibrational serviceability at both product development and application stages. With appropriate measures such factors can be integrated or eliminated. Gazing into the future it is hypothesised that significant further advances are possible. The future is likely to belong to high performance proprietary wood based materials that can be moulded during manufacture of components. Design strategies will have to account for dynamic response characteristics of building systems and components, owing to the inherently low damping capabilities and low mass of engineered wood based composites. Advantages with respect to raw fibre cost and energy requirements in manufacture should ensure that tomorrow's wood solutions compete effectively with those of composites containing man made fibres.

REFERENCES

1. Murray C.H. 'A global perspective on wood supply to the construction industry'. Proceedings of International Timber Engineering Conference, Vol 1:1.3-1.19, TRADA, UK (1991)
2. Barret J.D., Kellogg, 'R.M. Strength and Stiffness on second-growth Douglas-fir dimension lumber' Forintek Canada Corp., Vancouver, BC (1984)
3. Senft J.F., Bendtsen B.A., Galligan W.L. 'Weak wood-fast grown trees make problem lumber' Journal of Forestry, 83 (8): 476-484. (1985).
4. McNatt J.D., Moody R.C. 'Structural composite lumber' Progressive Architecture, 12.90: 34-36 (1990).
5. Canadian Wood Council. Wood Design Manual, CWC Ottawa (1990).
6. American Society for Testing and Materials. Specification for establishing and monitoring structural capacities of prefabricated wood I-joists. ASTM D5055. (1990).
7. Leichti R.J., Falk, R.H., Laufenberg, T.L. 'Prefabricated wood I-joists: an industry overview', Forest Products Journal, 40 (3): 15-20 (1990)
8. Bledso J.M., Knudson C.R.M., 'Creep rupture testing: a necessary part of structural composite lumber development', Presented 24th International Particleboard/composite Materials Symposium, Pullman,WA (1990).
9. Chan W.W.L. 'Strength properties and structural use of tempered hardboard', Journal Institute of Wood Science, 8 (4-issue 46): 147-160 (1979).
10. Larsen H.J. 'An introduction to Eurocode 5', Proceedings International Timber Engineering Conference, Timber R & D Association, High Wycombe, Bucks, Vol 1:1.41-1.57 (1991).
11. Ohlsson S.V. 'Serviceability criteria, especially floor vibration criteria' Proceedings International Timber Engineering Conference, Timber R & D Association, High Wycombe, Bucks, Vol. 1:1.58 - 1.65 (1991).
12. Chui Y.H. 'Vibration performance of timber floors and the related human discomfort criteria', Journal Institute of Wood Science, 10 (5-issue 59):183-188. (1987).
13. Chui Y.H., Smith I. 'Quantifying damping in structural timber components' Proceedings of the Pacific Timber Engineering Conference University of Auckland, NZ (1989).

10 BEHAVIOUR OF GLULAM BEAMS IN FIRE

R.J. Dayeh and D.R. Syme
CSIRO, Australia

This paper describes two full-scale fire-resistance tests on 150 mm wide x 420 mm deep glulam timber beams 4100 mm long. The tests are carried out in accordance with Australian standard AS 1530.4-1990 which reflects in essence international standard ISO 8334. The effects of loading, timber density and local variation in a full-scale laminar construction. The paper also assesses the extent of critical softening beyond the depth of charring and the factors of safety built into the design recommendations of Australian standard AS 1720.4(1).

INTRODUCTION

The building regulations in Australia, as in many countries around the world, prohibit the use of combustible construction for loadbearing building elements. This exclusion is finally being questioned in Australia by the timber industry. The industry argues that timber, when used independently or in combination with other fire-resisting materials, can achieve levels of fire resistance equivalent to non-combustible construction. In the case of timber beams and columns, fire resistance can be achieved by the timber section alone as the low conductivity of timber and the formation of charcoal contribute to the protection of an effective loadbearing core. If the behaviour of timber in fire is known the magnitude of this core can be estimated and structural timber members designed to carry the applied load for the required period of fire resistance.

BEHAVIOUR OF TIMBER IN FIRE: CURRENT KNOWLEDGE

A large number of factors can influence the burning of timber. It is considered however that the significant ones that have a consequential influence on the behaviour of glued laminated timber can be limited to density, moisture content, size of member, stress level and type of adhesive.

Density
An increase in the density of timber causes a reduction in the rate of charring. This rate can vary from about 0.8 mm/min for softwoods with densities in the range 300-400kg/m^3 to 0.5 mm/min or less for hardwoods with densities greater than 600kg/m^3. In Australia where softwoods have an average density of 550kg/m^3

and hardwoods 800kg/m^3 the formula given in AS 170.2-4 for the charring rate c in mm/minute for timber with 12% moisture content is:

$$c = 0.44 + (280/d)^2 \quad \text{where } d = \text{density in kg/m}^3$$

Moisture content
Schaffer[2] has found that for some softwood species the moisture content of timber can cause a variation in the charring rate of up to 20%.

Member size
As the minimum dimension of a timber member increases it becomes difficult for it to maintain self-sustained burning. In the presence of a heat source the speed of charring is related to the surface area to mass ratio. Many national standards for the design of timber in fire including BS 5268 Part 4[3] and AS 1720.4 prescribe minimum dimension for structural timber members.

Stress level and type
Kordina and Meyer-Ottens[4] have found that the charring rate for the tensile zones of loaded beams was as high as 1.1 mm/min, representing an increase of approximately 100% on unloaded beams. On the other hand Rogowski[5] found a difference of only 20% between the charring rates of loaded and unloaded columns.

Adhesive
It has been found by a number of researchers that resorcinol and phenolic adhesives have little effect on the charring rate whereas urea based adhesives increase the charring rate of laminated section compared with contiguous sections.

Softening effect
Although the formation of charcoal occurs at a temperature between 275°C and 300°C [6,7,8,9], softening of timber resulting in a reduction in tensile and compressive strengths and modulus of elasticity occurs at about 200°C [10,11,12,13,14].

TESTS

General
In order to study both Australian softwoods and hardwoods, two species of timber were chosen, namely radiata pine for softwoods and brush box for hardwoods. These species are commonly used and encompass the range of strengths and densities of commercial timbers. Four glued laminated beams were manufactured so that two beams could be fire tested under load and two identical beams could be structurally tested at ambient temperature and used for comparison. The beams were manufactured by a single manufacturer to AS 1328[15]. Each beam was nominally 150 mm wide x 420 mm deep x 4800 mm long and contained 14

continuous laminates glued together with resorcinol adhesive. The test specimens were only 4100 mm long; the remainder of each beam was used to evaluate the density and moisture content of both the laminates and the block in accordance with AS 1080.1[16]. The results of these measurements are given in Table 1.

Structural

In order to establish the modulus of elasticity and the modulus of rupture of the timber at room temperature, each of the four beams was loaded four times in the elastic range. Subsequently the two structural-test specimens were loaded to rupture. The loading consisted of two concentrated loads applied to the beams at the third points of their effective spans of 4010 mm. The results are given in Tables 2 and 3.

Fire tests

Each glulam beam was simply supported at each end on a 25 mm thick layer of Kaowool ceramic fibre blanket placed on a steel plate. The ends were built into the brickwork which enclosed the volume above the furnace. A gap of 15 mm, filled with Kaowool, was provided between the specimen and the brickwork.

Instrumentation

A series of thermocouples were located at the bottom of 3 mm holes drilled from the top of the specimen to a depth of 105 mm. This depth corresponds to the middle of the fourth laminate from the top. Additionally the temperature at the top of the beam was measured at the centre and quarter points. The thermocouples used were K-type chromel/alumel glass fibre insulated 0.5 mm diameter wire. The location of the thermocouples are shown in Figure 1.

Furnace

The furnace is horizontal and has a nominal opening of 4560 mm x 3660 mm. It is lined with refractory bricks with thermal properties specified in AS 1530.4[17]. It is heated by combustion of a mixture of natural gas and air. The temperature of the furnace is measured by 12 K-type thermocouples and follows the ISO 834 time-temperature curve. The furnace was covered with steel beams partly encased in refractory concrete. The gaps between the beams were sealed with Kaowool.

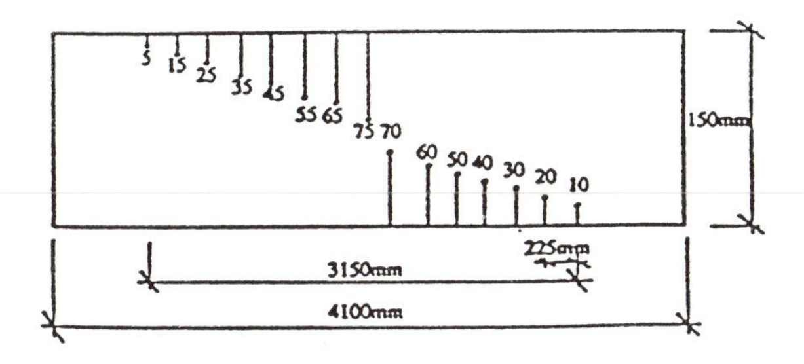

Figure 1. Thermocouple locations

Table 1. Density and moisture content.

ITEM	BRUSH BOX SPECIMENS				RADIATA PINE SPECIMENS			
	NON-FIRE TEST		FIRE TEST		NON-FIRE TEST		FIRE TEST	
	Density (kg/m³)	m/c (%)	Density (kg/m³)	m/c (%)	Density (kg/m³)	m/c (%)	Density (kg/m³)	m/c (%)
Lam. Range	738-1038	10.0-12.5	704-974	10.6-12.3	434-597	10.4-13.6	464-577	9.7-12.7
Mean	859	11.7	823	11.6	503	11.8	524	11.3
S.D.	75	0.7	74	0.5	47	1.0	35	1.0
Whole Block	862	11.3	825	11.3	507	11.4	519	11.1

Table 2. Modulus of elasticity.

TEST	LOAD P (kN)	AV.DEFL δ (mm)	b (mm)	d (mm)	a (mm)	Leff (mm)	MOE (MPa)
Brush box Non-Fire Test	30	6.50	149	415	1336	4010	13,813
Brush box Fire Resistance Test	30	5.87	149	415	1336	4010	15,296
Radiata pine Non-Fire Test	7	1.81	149	419	1336	4010	11,277
Radiata pine Fire Resistance Test	7	1.67	150	419	1336	4010	12,140

Table 3. Modulus of rupture.

TEST	LOAD P (kN)	b (mm)	d (mm)	a (mm)	MOR (MPa)
Brush box	160.0	149	415	1336	49.98
Radiata pine	117.1	149	419	1336	35.88

The pressure inside the furnace is measured by a differential low pressure transducer with a range of ± 50Pa.

Deflection measurements
Deflections of the specimens were measured by linear displacement transducers operating on the potentiometric principle with a range of 300 mm. The transducers were supported independently of the reaction frame and the construction exposed to the heating. Two transducers were used to allow for any rotation of the specimen. The average of the two reading was taken as the deflection.

Test procedure
Each beam was loaded through two hydraulic jacks acting on 15 mm concrete cubes placed at the third points. The jacks were attached to a reaction frame. The beams were loaded for thirty minutes before the fire test and subsequently the loading was maintained constant during the test. The specimens were exposed to the heating condition satisfying AS 1530.4 equation:

$$T - T_o = 345 \text{ Log10 } (8t + 1)$$

where T_o is the initial furnace temperature and T its temperature at time t. The test was stopped when the deflection reached L/20. The specimen was rolled away from the furnace and the beam hosed down. The data logging was continued for 10 minutes after stopping the fire exposure. On the day following each test the specimen was cross cut at its centre and at 655 mm and 1310 mm on each side of it.

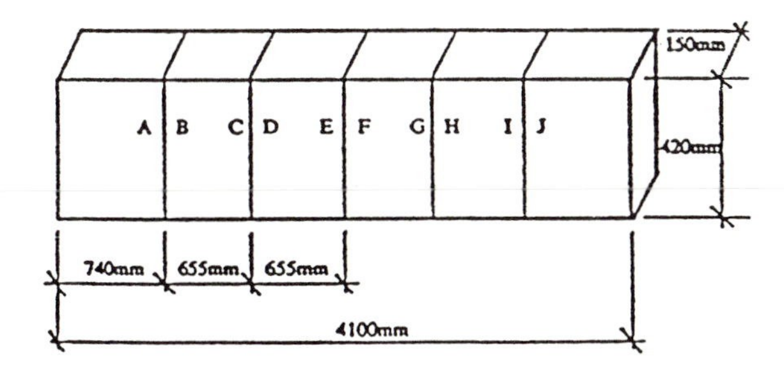

Figure 2. Char measurement sections

RESULTS

Observations during the test

Time minutes	Brush Box Hardwood	Radiata Pine Softwood
2		Smoke
6	Smoke	
8		Flaming, beam visible
9	Charring	Charring
15	Centre charring faster than ends	Char pieces fall from bottom
19	Gluelines charring faster	Charring increased but uniform
24		Gluelines opening at 1/3 depth
27	Gluelines opening up Char pieces fall from bottom	
30	Gluelines opening more pronounced	
34		Large openings, south side
37	Loading frame twisting	
38	Gluelines opened to 10 mm	
40	Large openings, entire length	
45	Large pieces char falling	
46	Deflection greater, south side	
50		More char falling
55		Large openings entire beam
59	Beam fails	
67		Beam fails
69	Hosing stopped	
77		Hosing stopped

Charring

The measured depths of charring at the end of the tests are shown in Table 4 for Brush box and Table 5 for Radiata pine. Figure 2 shows the char measurement sections of the beams. The means, standard deviation and the base charring are given in the tables. Because of the shattering that was associated with the failure of the Brush box beam, some portions could not be reassembled. The associated charring depths at these locations could not be assessed and were not taken into account in the calculation of the mean. The corresponding charring rates were 0.45 mm/minute for Brushbox and 0.75 mm/minute for Radiata pine

Temperature distribution

The temperatures within the timber beams over the periods of the tests are given in Figure 3 for Brush box and Figure 4 for Radiata pine.

Deflection

The mid-span deflections of the beams are given in Figure 5.

Table 4. Charring depths (mm) of Brush base beam.

SECTION	A-B	C-D	E-F	G-H	I-J
Laminate 1	25.5	23.5	30.8	24.2	27.2
Laminate 2	23.4	24.2	29.3	25.0	24.0
Laminate 3	24.7	25.3	28.9	25.5	23.0
Laminate 4	24.2	26.1	-	26.2	25.0
Laminate 5	24.6	27.5	-	27.7	25.3
Laminate 6	23.3	24.8	31.2	25.9	24.0
Laminate 7	22.1	24.9	30.4	27.6	24.0
Laminate 8	26.2	27.2	29.2	27.7	25.2
Laminate 9	28.9	-	30.9	28.7	24.3
Laminate 10	26.9	-	-	28.2	25.6
Laminate 11	26.7	29.5	-	29.0	25.8
Laminate 12	26.9	28.6	-	26.6	26.6
Mean	25.3	26.2	30.1	26.9	25.0
S.D.	1.9	2.0	0.9	1.5	1.2
Base	29.2	29.7	-	33.0	28.8

Table 5. Charring depths (mm) of Radiata pine beam.

SECTION	A-B	C-D	E-F	G-H	I-J
Laminate 1	48.8	50.2	-	-	49.1
Laminate 2	47.0	48.2	48.4	47.6	46.8
Laminate 3	44.9	47.7	48.1	46.8	46.3
Laminate 4	46.8	48.9	49.8	46.1	45.9
Laminate 5	45.8	48.3	49.6	48.4	47.2
Laminate 6	49.2	52.7	51.4	51.8	51.0
Laminate 7	49.1	52.9	54.0	53.7	47.2
Laminate 8	45.8	49.0	53.3	52.5	48.3
Laminate 9	51.1	52.1	53.7	53.8	51.1
Laminate 10	48.8	50.2	52.8	54.3	51.5
Laminate 11	48.0	50.0	54.0	52.8	52.5
Laminate 12	50.2	53.6	55.1	52.8	52.4
Mean	48.0	50.3	51.8	51.0	49.1
S.D.	1.9	2.0	2.5	3.1	2.5
Base	53.9	51.5	56.3	54.7	54.9

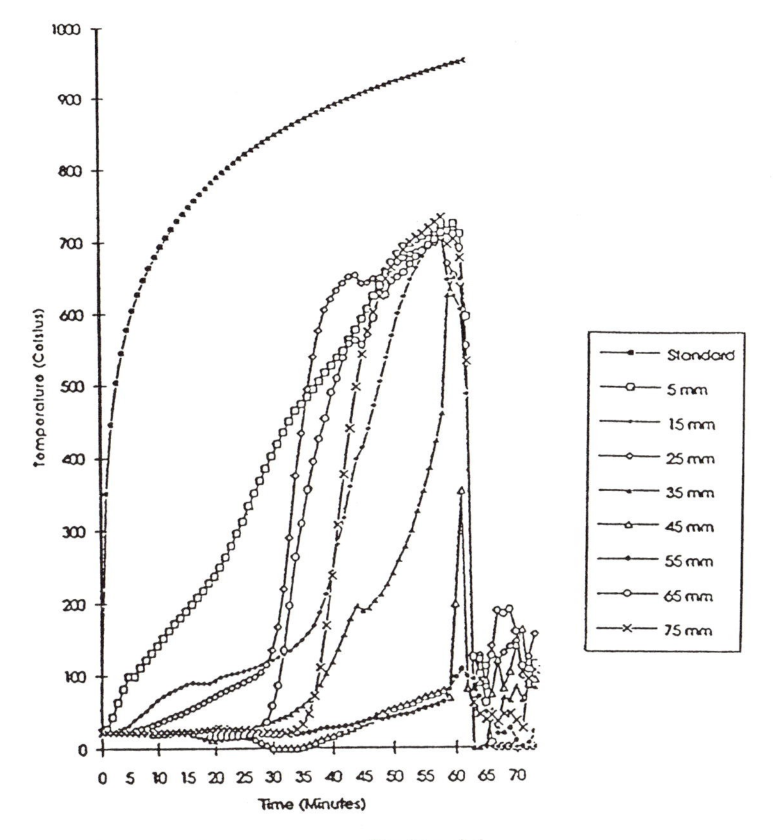

Figure 3. Temperature profile: Brush box

DISCUSSION

Local effects

Although the laminates were in single lengths and all the thermocouples were located at mid depth of laminate 4, it can be seen from the temperature graphs that anomalies exist whereby some thermocouples with greater embedment are showing temperatures sometimes smaller than those with much smaller embedments. The resulting hot spot temperatures are not a good indication of continuous charring.

Effects of Stress

The stress effect is clearly apparent in Tables 4 and 5 where the depth of charring becomes greater as one moves from the top laminate to the bottom one and from the ends of the beam towards its centre. The maximum increase is however only 12%. The anomalies in measured temperature, noted under local effects, could

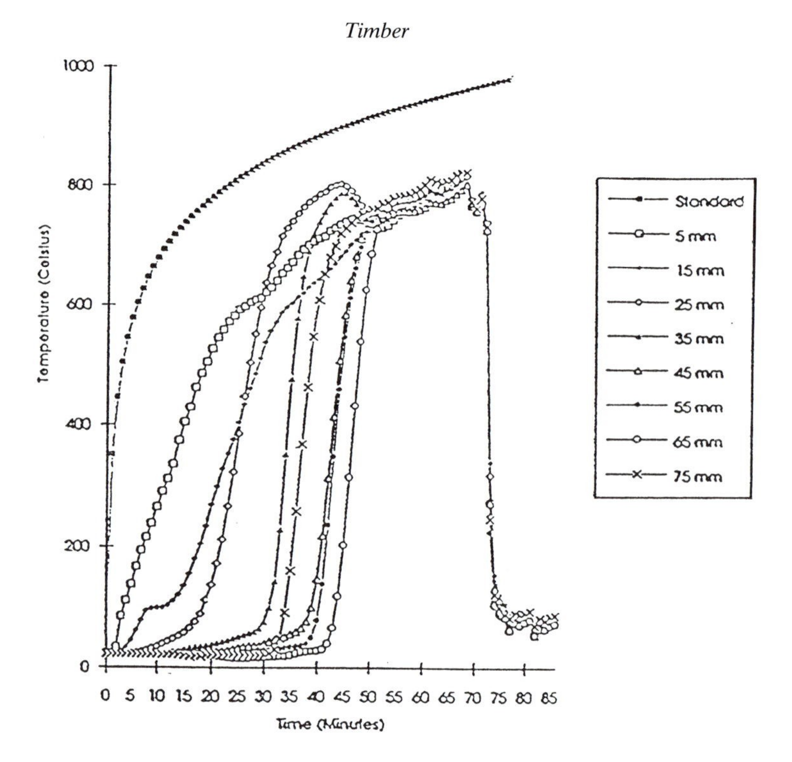

Figure 4. Temperature profile Radiata

be partly due to stresses in the section causing opening of glued joints. It should be noted that the thermocouples with deeper embedments which exhibited unexpectedly high temperatures were those in the highly stressed central portion of the beam. A study by Gardner and Syme(17) on the same species of timber shows regular, nearly parallel time-temperature curves for thermocouples at different depths from the surface. The specimens tested were 270 mm x 150 mm x 1400 mm long and were not loaded. The thermocouples were clustered within a 320 mm length of the beam. The results of these tests do not show the effects of either the spatial variations or the stress effects which are present in the full-scale test.

Softening effect

Based on the measured residual uncharred sections and the tensile strength of the timber obtained from the structural tests at room temperature, the Brush box and Radiata pine beams would have ultimate moment capacities of 112kNm and 36.7kNm respectively whereas their actual failure moments were 101.4kNm and 27.8 kNm. These results suggest a critical softening in the timber beyond charring equivalent to a depth of 5% for Brush box and 12% for Radiata pine.

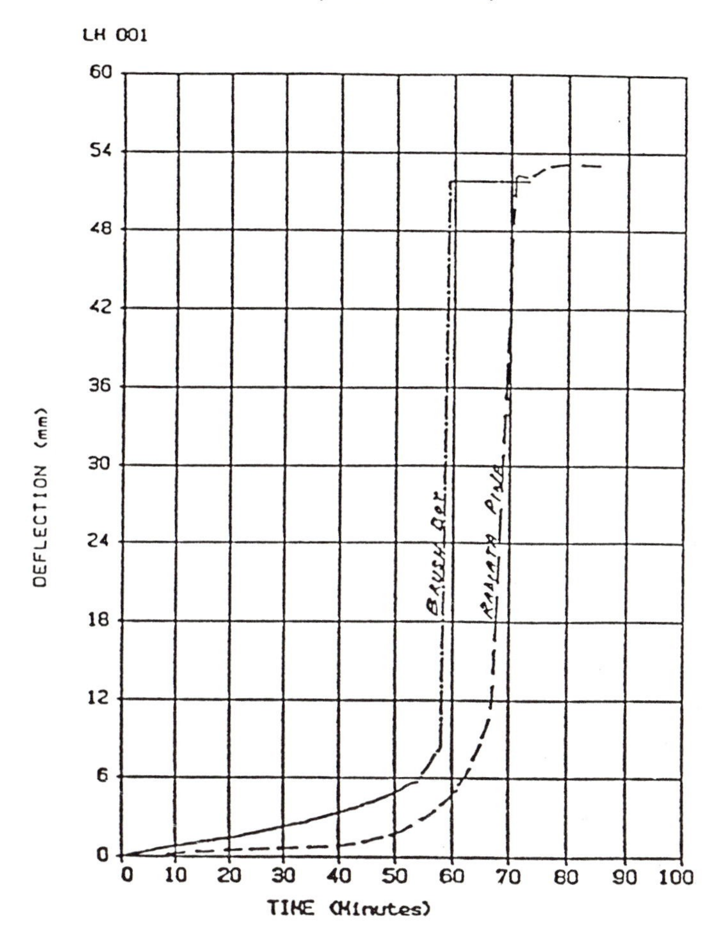

Figure 5. Central deflection

Comparison with AS 1720.4-1990

The charring rate derived from the code formula $c = 0.4 + (280/d)^2$ are 0.51 mm/minute for Brush box and 0.69 mm/minute for Radiata pine. The corresponding experimental values were 0.45 mm/minute and 0.75 mm/minute. The moment capacities calculated in accordance with Australian standard AS 1720.4-1990 are 90.82kNm for Brush box and 34.72kNm for Radiata pine. These results yield safety factors of 1.23 for Brushbox and 1.06 for Radiata pine. This inconsistency in safety factors seems to be due to the fixed allowance for softening which is used in the code formula for the calculation of the residual section irrespective of the timber species, namely $dc = c.t + 7.5$ where dc is the loss of effective timber in mm, c the charring rate in mm/minute and t the time in minutes. A proportional allowance related to the species density would be more adequate.

Soffit charring
Charring at the bottom of the Brush box and Radiata pine beams was respectively 15% and 8% greater than side charring. As depth is highly significant in determining the strength and stiffness of beams, this effect should not be neglected.

CONCLUSIONS

1. Testing of full-scale loaded beams has the potential of reproducing the effects of stress and local variations on charring. These effects are generally absent in small specimen testing. In the case of this study these effects were significant enough to produce hotspots which confused the expected correlation between temperature and average charring.
2. The allowance for softening beyond the depth of charring depends on the timber species and should be reflected in formulas in codes of practice.
3. For the beams tested, the safety of designs to AS 1720.4 for the different species were inconsistent and, for a material as variable as timber, these factors were small compared with those related to other structural materials such as steel and concrete.
4. The effect of stress on charring was obvious in both tests but was limited to a maximum of 12%, unlike other findings which range from 20% to 100%.
5. An allowance should be made in design for greater charring at the bottom of the beam.

REFERENCES

1. Standards Australia, AS 1720 Part 4: Fire-resistance of structural timber members, SAA, Sydney, Australia (1990).
2. Schaffer E.L. Charring Rate of Selected Woods Transverse to the Grain, Research Paper FPL 69, United States Department of Agriculture, Forest Service, Forest Products Laboratory, Madison, United States of America (1967).
3. British Standard Institution, BS 5268 Part 4, Section 4.1, Method of Calculating Fire Resistance of Timber Members, BSI, London, United Kingdom (1978).
4. Kordina K. University of Braunschweig, Institut for Baustoffkunde und Stahlbetonbau, Braunschweig, Federal Republic of Germany (1977).
5. Rogowski B.F. Charring of Timber in Fire Tests, Paper No. 4, Proceedings of Symposium No. 3, Ministry of Technology and Fire Office's Committee Joint Fire Research Organisation, Her Majesty's Stationery Office, London, United Kingdom (1970).

6. Hall G.S., Saunders R.G., Allcorn R.T., Jackman P.E., Hickey M.W. and Fitt R. Fire Performance of Timber - A Literature Survey, Timber Research and Development Association, Hughenden Valley, United Kingdom (1976).
7. Schaffer E.L. Structural Fire Design: Wood, Research Paper FPL 450, United States Department of Agriculture, Forest Service, Forest Products Laboratory, Madison, United States of America (1984).
8. Malhotra H.L. Properties of Materials at High Temperatures, Report on the Work of Technical Committee 44-PHT, Materials and Structures, Volume 15 (1986).
9. White R.H. Analytical Methods for determining Fire Resistance of Timber Members, Section 3, The SFPE Handbook of Fire Protection Engineering, National Fire Protection Association, Quincy, United States of America (1988).
10. Scaffer E.L. State of Structural Timber Fire Endurance, Wood and Fiber, Volume 9, No. 2, Society of Wood Science and Technology, Kansas, United States of America (1977).
11. Knudsen R.M. and Schniewind A.P. Performance of Structural Wood Members Exposed to Fire, Forest Products Journal, Volume 25, No. 2, Forest Products Research Society, Madison, United States of America (1975).
12. Ostman B.A.-L. Wood Tensile Strength at Temperature and Moisture Contents Simulating Fire Conditions, Wood Science and Technology, Volume 19, International Academy of Wood Science, New York, United States of America (1985).
13. Nyman C. The Effect of Temperature and Moisture on the Strength of Wood and Glue Joists, Forest Products Laboratory Report NR 6, Technical Research Centre of Finland, Finland (1980).
14. Springer G.S. and Do M.H. Degradation of Mechanical Properties of Wood during Fire, Report No. NBS GCR 83 433, National Bureau of Standards, Department of Commerce, Washington, United States of America (1983).
15. Standards Australia, AS 1328, Glued-Laminated Structural; Timber, SAA, Sydney, Australia (1987).
16. Standards Australia, AS 1080, Methods of Testing Timber, SAA, Sydney, Australia (1981).
17. Standards Australia, AS 1530.4, Fire-resistance tests of elements of building construction, SAA, Sydney, Australia, (1990).
18. Gardner W.D. and Syme D.R. Charring of Glued-Laminated Beams of Eight Australian Grown Timber Species and the Effect of 13mm Gypsum Plasterboard Protection on their charring, Technical Report No. 5, NSW Timber Advisory Council Ltd, Sydney, Australia, (1991).

11 STRUCTURAL GRADE CHIPBOARD

V. Enjily
Building Research Establishment, UK

This paper comprises a summary of the test programme and an outline of the derivation of the design stresses and modification factors included in BS5268:Part 2:1991[1] for Structural grade chipboard. Also highlighted are, the key areas necessary for consideration by designers and specifiers in the design and construction of structures/components in which structural grade chipboard plays a major structural role. Recommendations are presented for the consideration of concentrated loads and their method of analyses in the design of structural grade chipboard. The strength and serviceability criteria of structural grade chipboard are also discussed and guidance given for the design.

INTRODUCTION

In the past decade or so there has been increasing interest in the potential use of chipboard for structural applications. This highlighted the need to produce specifications and grade stresses for chipboard to enable manufacturers, consumers and designers produce and use structural grade chipboard.

Following the publication of a state-of-art report[2] in January 1984 on the use of chipboard in flooring, a working party was set up (following a suggestion of the Standing Committee on Structural Safety) to provide guidance on the structural use of chipboard for flooring. Their report indicated that while the system for the specifying of chipboard for domestic flooring in BS5669:1979[3] was acceptable, there was very limited provision of data for the design of non-domestic floors and structures. There appeared to be an urgent need for the design of mezzanine floors in industrial premises. The working party comprised practising engineers and representatives from the Health and Safety Executive, Princes Risborough Laboratory(PRL) of the Building Research Establishment(BRE) and the Timber Research and Development Association(TRADA). To meet this urgent need for design stresses, it was agreed that the joint resources of both BRE(PRL) and TRADA should be utilised. Hence, a BRE(PRL)/TRADA committee was set up with representations from both the Chipboard Promotion Association(CPA) and the Wood Panel Products Federation (WPPF), [then called the United Kingdom Particleboard Association(UKPA)]. The committee was given two main tasks:

a) to develop a specification for structural grade chipboard and

b) to determine design grade stresses for structural grade chipboard. BS5669:1979[3] covered only four board types of chipboard which did not permit the expansion necessary to include additional types of particleboard. The revision of BS5669:1979[3] was completed and published by the British Standards Institution in 1989[4] which allowed various particleboards to be manufactured and used according to their specification, environmental condition and use. BS5669:1989[4] was published in five parts. BS5669: Part 2:1989[4] now includes six types of chipboard (C1,C1A,C2,C3,C4 and C5), four of which were already included in BS5669:1979[3]. The additional types are C1A and C5. C1A is for general use and has slightly higher mean quality levels for a number of properties than type C1. C5 is a structural grade chipboard which has the same moisture resistance as C4 chipboard, but with enhanced mechanical properties.

The revised publication of BS5669:1989[4] and the recent BS5268: Part 2:1991[1] now provide enough information for specification, selection, application and design of this material for specific purposes. However, there remains certain areas for which designers and specifiers may need further guidance regarding the design, selection and use of structural grade chipboard. Thus a joint publication by BRE and TRADA[5] was published in May 1992.

TEST PROGRAMME AND THE OUTLINE OF THE DERIVATION OF DESIGN STRESSES

Test Methods

Two phases of testing were carried out: phase one included samples with 12-22 mm thicknesses while phase two consisted of samples with thicknesses in the range of 28-38 mm. Over 4500 test pieces were used in the evaluation. The test methods utilised are those shown in Table 1. All the properties were evaluated for a maximum moisture content of 15%, which is classed in BS5268: Part 2: 1991[1] as 'dry' condition for panel products. The grand mean values were ultimately obtained for the range of tests on each thickness of each board.

Since chipboard is a viscoelastic material (i.e. its performance slowly diminishes with time under load), it was necessary to determine two time dependence factors which could be applied to the derived characteristic values in order to obtain the grade stresses and moduli. The first factor is the duration of load parameter which is required to modify all the characteristic strength data. The second factor is necessary for the reduction of the various moduli of elasticity.

Table 1: Test methods and location of testing

Property	Method of test	Comments	Test lab.
Bending strength and M.O.E. in bending*	BS5669 (1979) A6	Time to failure 90 seconds	TRADA
In plane tension strength and M.O.E. in tension*	BS4152 (1969)	Samples cut to FESYP spec. Time to failure 90 seconds. Gauge length 75 mm for MOE	TRADA
In plane compression strength and M.O.E. in compression*	BS5669 (1979) A15	Time to failure 90 seconds	TRADA
Panel shear strength and M.O.Rigidity*	BS4512 (1969)	Time to failure 90	TRADA
Edgewise bending*	FESYP	Time to failure 90	TRADA
Short span bending	FESYP	Time to failure 90	TRADA
Transverse (rolling) shear strength* - For boards 12-22 mm - For boards 28-38 mm	 BS5669 (1979)A12 BS373	Time to failure 90 seconds Cross head speed 0.5 mm/min.	 TRADA BRE / TRADA
Concentrated load strength*	As in References	Time to failure 90	TRADA
Impact strength*	BS5669 (1979) A8		BRE
Tension strength perp. to the plane*. Repeated after accelerated ageing	BS5669 (1979) A9		BRE
Thickness swelling: - after 24 hours cold soak - after accelerated aging	BS5669 (1979) A17		BRE
Dimensional stability on length and	BS5669 (1979) A18	rh: 30%, 65% &	BRE
Creep behaviour	BS5669 (1989)	- Four point loading - Fixed & variable climate	BRE
Nailed joint strength	Reference 5		TRADA

*These properties were evaluated at two conditions of humidity and temperature: 65% rh and 85% rh at 20°C.

Duration of Load Parameter for Grade Stresses

The relationship between the applied stress level, and log time to failure is illustrated in Figure 1. The regression line was calculated and found to be SL = Stress Level = $96 - 8.89\log_{10}T$ (Figure 1). The mean values, the standard deviation and the 5% characteristic values were calculated from the test results for each property measured, assuming a normal distribution. The grade stresses for structural chipboard used in dry conditions were calculated from the characteristic values using the following factors:

a) Duration of load
The characteristic values obtained related to test duration. It was therefore, necessary to reduce these values before they could be used for long-term design loads. The equation SL = 96 - 8.89$\log_{10}T$ (Figure 1) was used and this produced a SL value of 29.8% for the percentage of the tests load which a board is predicted to sustain for 50 years. The load duration factor was therefore determined as 0.298.

b) Safety factor
A safety factor of 0.855 was used in the derivation of grade stresses for softwood and plywood in BS5268:Part 2:1984[8]. The same factor was adopted for structural grade chipboard.

c) Statistical factor
When CP112[9] was replaced by BS5268:Part 2:1984[8], the stresses derived on a 5th percentile basis were converted to a 1 percentile basis to keep them at a similar level to those formally used in design to CP112[9]. A similar procedure was therefore undertaken for the chipboard stresses, but using a C. of V. of 8% instead of the 16% (which had been appropriate for softwood). The necessary factor was calculated as:
Statistical factor = (1-2.33xC. of V.) / (1-1.645xC. of V.) = 0.937

The characteristic values were therefore converted to long-term dry grade stresses given in the Code[1] by a combined reduction factor of:
0.298x0.855x0.937 = 0.239.

Grade Moduli

The maximum long term applied stress level in practice, expressed as a ratio of the mean test strength, is equal to: The combined factor for derivation of the dry grade stresses x 5% characteristic value / mean test strength = 0.239 x 5% characteristic value / mean test strength = 0.239 x (mean - 1.645 Standard Deviation)/mean.

Using the coefficient of variation of 8%, the stress level is 20.6% of the mean which should be used for driving grade elastic modulus. The variation of M.O.E. with time is presented in Figure 2 using a 5-parameter rheological models fitted to the creep data. The deflection at 50 years was calculated and M.O.E derived to be:

E_{grade} = 560 N/mm2 for 6-19 mm thickness boards
= 555 N/mm2 for 20-25 mm thickness boards
= 550 N/mm2 for 26-32 mm thickness boards
= 540 N/mm2 for 33-40 mm thickness boards

The polynominal from which the time factors for M.O.E. are taken is:
E_{creep} = 3820 - 91.17$\log_{10}T$ - 44.76$(\log_{10}T)$2 ....Equation 1, where T is in minutes. This yields a duration load factor of 5.63 for a life span of 50 years. Therefore, instead of 0.298 (as used for stresses), the factor of 0.178 (ie 1/5.63)

was applied to the mean test data of M.O.E. at 85% rh to produce the grade moduli (a procedure similar to that used for softwood and plywood).

Concentrated Load Tests and Design Values

The test method for determining the concentrated load capacity was developed at TRADA [6][7], and was incorporated into BS5669:Part 1:1989[4]. The aim of the test method was to eliminate the influence of bending so that a pure punching shear failure is obtained. In practice a concentrated load will usually produce bending stresses as well as shear stresses, so a designer must check the resistance of chipboard to both effects, as shown later in this paper.

Five hundred concentrated load tests (Figure 3) were carried out, encompassing various thicknesses of chipboard from 12 mm to 38 mm. Some were conditioned at 20°C and 65% rh and others at 20°C and 85% rh. Two different punch diameters of 25 and 50 mm were used. The procedure which was used for deriving the concentrated design load values was:

a) The 5% characteristic failure loads for each thickness of board were obtained for both punch sizes.
b) The failure area for each specimen was calculated from measurements of the failure perimeter on its under-side.
c) Transverse shear stresses were calculated for each type of board by dividing the characteristic failure load by the failure area.
d) A plot of transverse shear stresses against the board thickness was carried out with a line of best fit by linear regression analysis. Values for each of the thickness classes were then taken from the upper thickness limit of each class. The stresses obtained relate to the area of the defined failure surface.
e) Values of the maximum concentrated load which may safely be applied for any configuration and duration of loading are more helpful to the designers than the grade stresses, so it was decided to present maximum loads for two punch sizes. In order to obtain the values required, the failure areas corresponding to each of the board thickness classes were calculated for each punch size, using a least square regression analysis. The failure areas thus obtained were multiplied by the corresponding transverse shear stresses, to give characteristic failure loads for each board thickness and for the two chosen punch sizes.
f) The characteristic loads were multiplied by the combined duration of load, safety and statistical factors (0.239) as previously explained. The permissible loads obtained were classed as long-term loads.

In the absence of further data, it was not possible to include punch sizes bigger than 50 mm in diameter and smaller than 25 mm in diameter. This is the limitation which BS5268: Part 2: 1991[1] currently imposes.

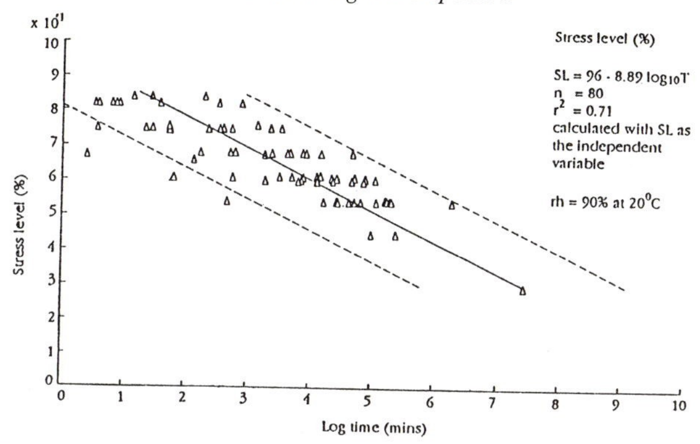

Figure 1. Applied stress and the time to failure relationship.

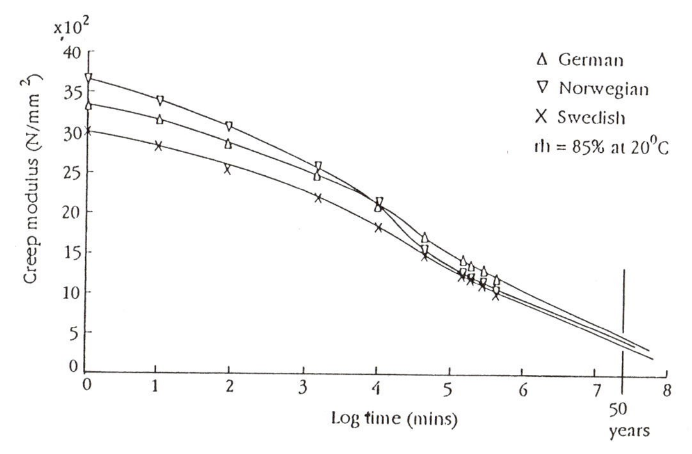

Figure 2. The decrease in M.O.E with log time under load for samples stressed at 20.6% of their short term bending strength.

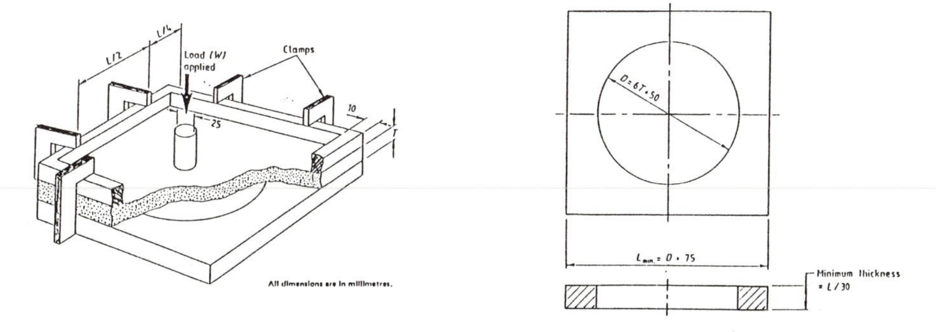

Figure 3. Layout of a concentrated load test rig (ex BS 5669:Part1:1989)

THE OUTLINE OF THE DERIVATION OF MODIFICATION FACTORS FOR STRESSES AND MODULI

Load Duration Modification Factors for Stresses

Stress levels were calculated using SL = 96 - 8.89$\log_{10}T$ (Figure 1) for various load duration factors; their ratios, with the SL of 29.8% (50 years SL) were obtained and are presented in the Code[1] as the modification factor K_{81} given in Table 2.

Table 2. Load duration modification factors for stresses

Load duration	Time (T)	SL	SL/29.8	Code value (K_{81})
Long-term	50 years	29.8	1.00	1.00
Medium-term	30 days	54.8	1.84	1.80
Short-term	1 minute	96.0	3.22	3.20
Very short-term	5 seconds	105.6	3.54	3.50

Load Duration Modification Factors for Moduli

The applied stress level of 20.6% is equal to the permissible long-term bending stress. If the stress ratio of, r, is defined as r = applied stress/permissible stress then an applied stress level of 20.6% corresponds to a stress ratio of 1.0. Thus by inserting appropriate values of T in the Equation 1 for E_{creep}, the load duration modification factors for the M.O.E. in bending can be obtained for a stress ratio of 1.0, as shown in Table 3.

Table 3. Load duration modification factors for moduli (r = 1.0)

Load Duration	Time (T)	E_{creep}	E_{creep}/679.5
Long-term	50 years	679.5	1.00
Medium-term	30 days	2435.5	3.58
Short-term & Very	1 minute	3820.0	5.63
Deflection test	24 hours	3085.6	4.52
Strength test	40	3559.0	5.24

For the moduli of elasticity in tension, compression and the shear modulus, the test values were adjusted to long-term grade values in a similar way, and the same load duration factors of 1.0, 3.58, 5.63 and 4.52 were considered to be applicable.

In the absence of more specific data on the creep behaviour of structural grade chipboard at stress levels below 20.6%, two assumptions were made:

(i) A linear relationship existed between E_{creep} and the stress ratio.

(ii) For all the load durations, the value of E_{creep} for a stress ratio of zero was equal to its short-term test value (This assumption is in accordance with existing design theory for tempered hardboard).

It was therefore possible to calculate a table of factors by which the grade value of M.O.E. might be modified to produce appropriate values of E_{creep} in deflection calculations for various load durations and the applied to permissible ratios as given in Table 4. The values in Table 4 may be calculated from the following equations, where r is the stress ratio: $K_{82} = 5.63 - 4.63r$, $K_{83} = 5.63 - 2.05r$ and $K_{84} = 5.63$. The above equations are presented in Appendix Q of BS5268:Part 2:1991[1].

Table 4: Modification factors for elastic and shear modulus

Ratio of applied stress to permissible stress for each category of load duration*	Long-term loads K_{82}	Nett medium-term loads** K_{83}	Nett short & very short-term loads** K_{84}
1.00	1.00	3.58	5.63
0.90	1.46	3.79	5.63
0.80	1.93	3.99	5.63
0.70	2.39	4.20	5.63
0.60	2.85	4.40	5.63
0.50	3.32	4.61	5.63
0.40	3.78	4.81	5.63
0.30	4.24	5.02	5.63
0.20	4.70	5.22	5.63
0.10	5.17	5.43	5.63
0.00	5.63	5.63	5.63

* Interpolation is allowed

** Nett loads refer to loads of the duration stated, excluding any other loads of longer duration

GRADES IN EUROPEAN STANDARDS

The introduction within the European Community of the Construction Products Directive and its associated set of Essential Requirements has led to the drafting of a series of harmonised standards covering the specification of chipboard; this work is being carried out under the auspices of CEN.

Six grades have been recognised and drafting of the six specifications is almost complete. The requirements for many of these grades are similar in content and level to those in BS 5669, but there is one major difference. Whereas there is only one structural grade in the British Standard, there will be four structural grades in the CEN specifications (Table 5). Furthermore, because the new Eurocode on the structural use of timber and boards is written in terms of limit state design, the values used in design are not the long-term grade stresses as specified for chipboard in BS 5268:Part2:1991, but rather short-term 'characteristic-values' to which are applied appropriate time-modification factors. At the present time the hope is that the new CEN specification for structural chipboard will come into effect in January 1995.

Table 5. Equivalent grades in BS and CEN (approximate) standards

BS 5669	CEN	
	Grade (EN)	Use
C2	312-4	load bearing, dry (65% rh)
C3	-	-
C4	312-5	load bearing, humid (up to 85% rh)
-	312-6	heavy duty load bearing, dry
C5	312-7	heavy duty load bearing, humid

DESIGN CONSIDERATIONS

Strength Criterion

The grade stresses given in BS5268:Part 2:1991[1] apply only to type C5 chipboard in which the effect of creep is included in the mechanical properties. These grade stresses are classed as 'dry' grade stresses, and no modification factors are given for determining 'wet' grade stresses as chipboard should not be used structurally in continuously wet or immersed conditions. It is essential that the recommendations and guidance given in BS5669[4] and BS5268:Part 2:1991[1] adhered to.

In addition to the specification and recommendations given in BS5669[3] and BS5268:Part 2:1991[1], designers should consider the following load cases as appropriate:-

Uniformly distributed load and Concentrated load

- Load acting as punching shear(no bending effect).
- Influence of Concentrated Load on Bending.
- Bearing and shear stresses
- Deflection

Uniformly Distributed Load

The maximum applied bending stress should not be greater than the permissible stress calculated using the grade stress given in the Code[1], taking into account the modification factors for duration of loading.

Concentrated Load

Chipboard's resistance against concentrated load failure is considered more critical than for some other board materials. As a result of this, designers should ensure that wood chipboard adequately resists such a loading.

Chipboard components may be designed to resist concentrated loads perpendicular to the plane of the board in a variety of situations. Principal end uses are flooring and flat roofing, where the concentrated load will be one of the loads to be considered in the design analyses. It is the designers' responsibility as to what method of analyses are used in the design of structural grade chipboard. However, it is important to identify the effect which concentrated loading has on structural grade chipboard; a concentrated load can act as a punching shear alone, and the combination of bending and punching shear. It is this effect of concentrated load on bending which is of a great importance as the structures such as floors undergo such an effect. The BS5268:Part 2:1991[(1)] specifies that the concentrated load capacities given in the Code do not consider the effects of bending and a separate check should be made to ensure that permissible bending stress and deflection are not exceeded due to the combination of concentrated and other loads.

Concentrated load stresses can only be determined by relating a failure shear plane to the ultimate test concentrated load. The ultimate failure surface of chipboard in this type of loading is a frustum of a cone which is generated by the concentrated load punching through the top dense surface layer of chipboard with no signs of distress being observed on the under-side. There is then a short but definite delay, prior to the ultimate failure. The lower dense layer fails due to a combination of bending and tension perpendicular to the plane of the board. Further details of failure are given in references [(6)(7)].

The ultimate failure surface may be idealised by a frustum of a cone, and the overall failure surface area can be calculated. However, the approach adopted for calculating transverse shear stresses was to assume that the design shear failure plane was described by the surface of a cylinder, concentric to the centre of the punch and perpendicular to the plane of the board. For the purpose of calculating the values of punching (transverse) shear, the diameter of the cylinder was determined by assuming that the area of the cylindrical failure plane is equal to the area of the actual conical failure surface. When these punching shear values are ultimately used in design, a theoretical failure plane was defined which is both similar to principles embodied in the design codes of other materials. Maximum concentrated load capacities for certain boards thickness/punch size/span combinations which approximated to actual test values obtained were produced.

The concentrated load may be treated as of medium-term load duration in the deign of domestic flooring, except where the designer decides that the concentrated load has long-term duration; for example partitions on unsupported areas of chipboard flooring (not a good practice).

Load Acting as Punching Shear (no bending effect)
This part of the design would not require extensive analyses as these are already carried out and the results are given in the Code[1]. Designers should simply read the appropriate concentrated load capacity (long-term) given in the Code. For the situations where the concentrated load is assumed to be based on medium, short or very short-term load duration, the concentrated load capacities should be multiplied by the appropriate modification factor given in the Code[1].

The Code[1] gives values for 25 mm up to 50 mm diameter (or non-circular shapes with smallest dimension equal to the specified diameter) punch sizes. However, the Code[1] does not allow extrapolation of the concentrated load capacities because there is no experimental data available at present. For the situations where the concentrated load area of contact is less than 25 mm diameter, designers are recommended to specify and increase the contact area by means of rigid plates.

Influence of Concentrated Point or Patch Load on Bending
For many applications, including flooring, chipboard will be supported on two opposite edges. They are usually used with their long edges across the joists or supports. When a concentrated load is acting on a slab of chipboard, the maximum bending stress varies depending on the position of the load. This situation is common in practice especially in flooring. In order to calculate the applied bending moments and stresses it is necessary to obtain 'effective width' over which the concentrated load is assumed to act. In this situation where the concentrated point/patch load causes bending, designers may take the effective width (measured parallel to the supports) of chipboard as to be equal to span of the board between supports. The applied bending stress can then be calculated using simple bending theory.

Influence of Concentrated Line Load on bending
Concentrated line loads (especially when they are parallel to the supports) should be supported directly by a support such as joist, beam or nogging between the joists. However, it has been known that at some stage during the life of the floor, some people would erect a concentrated line load (such as partitions) without considering the imposed loading, the direction of the concentrated line load to the span of the floor and its position relative to the span of the joists. Therefore, designers should bear this in mind where necessary.

Designers may take the effective width of chipboard (measured parallel to supports) as:

a) Concentrated line load parallel to the supports: 'The effective width' = length of the line load. Note: Where the length of the line load is less than the board span, the effective width should be: The effective width = span of the board between supports + width of the line load.
b) Concentrated line load perpendicular to the supports: The effective width = span of the board between supports + width of the line load. Note : It is not

usually necessary to check chipboard for concentrated line load perpendicular to the supports because the line load (ie partitions) is normally supported by the supports and not the chipboard flooring. However, where partition is not capable of spanning between the joists, the concentrated line load is considered to act continuously over all spans.

The applied bending stress can then be calculated using simple bending theory for all the cases.

Bearing and Shear Stresses
The applied bearing and shear stresses should not exceed the permissible stresses calculated using the appropriate grade stresses and the modification factors given in the Code(1). It is important to note that the permissible bearing stress should be found by multiplying the bearing grade stress given in BS5268:Part 2:1991(1) by the appropriate load duration factor given in Table 17 of BS5268: Part 2:1991(1) and not those given in Table 91e of the Code(1).

Serviceability Criterion
BS5268:Part 2:1991(1) makes allowance for creep in chipboard by establishing a very low long-term modulus of elasticity which may be greatly increased by modification factors where the ratio of applied to permissible stress is small. This makes it possible to allow more precisely for the effect of load duration, but it is important to apply the modification factors correctly to avoid over-design.

BS5268:Part 2:1991(1) recommends that the permissible deflection of flexural members when fully loaded should be 0.003L where L is the span. However, Section 9 of the Code(1) 'Wood Chipboard' does not give any limits for the deflections for the reason that some structural elements including flooring, it is often acceptable for greater permissible deflections to be allowed, provided these are fully taken into account during the design process with due consideration being given to the nature and type of loading. It is the designer's responsibility as to what limit of deflection of the chipboard flooring is used to meet the following criteria:

- Safety is not impaired by the applied deflection.
- The applied deflection would not impair the integrity of the structures and the associated structural and non-structural members.
- The performance of the boards in the joint areas, irrespective of what type of joint is used must be at least as good as for a single board in respect of the loads considered.

The following permissible deflection is given as a guide only. It may be used at the discretion of the designer provided that the above criteria are met:

For loads classed as long-term duration:

D_{adm} = 0.005L for spans up to 800 mm
= L[0.005 - 1.0x10^{-5}(L-800)] for spans > 800 mm to <= 1000 mm
= 0.003L for spans > 1000 mm

For loads classed as medium/short/very short-term duration:

D_{adm} = 0.01L for spans up to 600 mm
= L[0.01 - 1.75x10^{-5}(L-600)] for spans > 600 mm to <= 1000 mm
= 0.003L for spans > 1000 mm

Shear deflection should be considered, especially in the design of 'I' beams. The permissible deflection is then the sum of the shear and bending deflections.

Prototype Testing
Until the publication of BS5268:Part 2:1991(1), any structural use of chipboard was based on manufacturer's data, experience in use and on prototype testing. The use of chipboard, other than C5 grade, in such situations will still rely on this type of information until the publication of Eurocode 5 and its supporting standards. However, if the design analysis for a structure were complicated and tedious, Section 8 of BS5268:Part 2:1991(1) would be useful to the designers in which specifications are laid down for the load-testing of prototype structures or parts of structures. This is a procedure which may be used instead of calculations to verify their structural adequacy. The prototype testing section of the Code(1) includes a strength test and a deflection test, plus criteria for acceptance. In preparing this section of the Code(1) it was necessary to determine what load a structure should be able to support for the 40 minutes of test loading, and how much it may be permitted to deflect during the 24 hour deflection test period.

Composite Structures
In assessing the strength and stiffness of composite structures of chipboard and timber, the proportion of the load which is carried by the chipboard must be determined from the ratio of the stiffness of the two materials. For this purpose the grade modulus of elasticity of the chipboard should be multiplied by a factor of 5.24 and 4.52 for strength and stiffness respectively, as shown in Table 3.

TYPICAL POSSIBLE USES

Structural grade chipboard is yet to be used as one of the following structural components as it is a new grade of chipboard:

- Flooring and shelving (Domestic, Offices, industrial, mezzanine)
- Wall panels (timber frame houses)
- Sarking (bracing)
- 'I' or 'Box' section beams with chipboard webs
- Roof decking and Joints (gussets etc.)

BRE would be interested to receive feedback from any engineer having utilised structural grade chipboard in his/her design as to the performance of the structure.

SPECIFICATION AND WORKMANSHIP

Experience shows that in most cases, the majority of problems with structures are due to bad workmanship as opposed to errors in design. Workmanship has been identified as the single most important factor affecting structural performance. The need for adequate staff training and supervision cannot be overstated.

The apparent simplicity and ease of assembly of chipboard structural components belies the care and attention actually required in construction. Although, the design analysis for structural grade chipboard construction is primarily important for determining adequate sizes and dimensions relating to strength and serviceability, good practice is also very important. Therefore, it is vital that the overall design configuration and the designer's specification for materials are accurately adhered to.

The specification of structural grade chipboard should comply with Type C5 requirements of BS5669:Part 2:1989[4]. Factors such as handling, stacking, transport, storage and installation should be considered carefully for the optimum performance of structural grade chipboard. Readers are recommended to refer to reference[5] which includes some useful information and references for good practice.

ACKNOWLEDGEMENT

The author would like to express his gratitude to Dr J. Dinwoodie of BRE, Mr A. Abbott and Mr A. Page of TRADA for their support.

REFERENCES

1. British Standards Institution. Structural use of timber BS5268:Part 2 Code of practice for permissible stress design, materials and workmanship. BSI London (1991).
2. Institution of Structural Engineers. The structural use of chipboard for flooring: A State-of-art report. The Structural Engineer, 62A(1) January (1984).
3. British Standards Institution. Specification for wood chipboard and methods of test for particleboard. BS5669. BSI, London (1979).
4. British Standards Institution. Particleboard, BS5669. BSI, London (1989); Part 1 : Methods of sampling, conditioning and test; Part 2 : Specification for wood chipboard; Part 3 : Specification for waferboard and oriented strand board (in course of preparation); Part 4 : Specification for cement bonded particleboard; Part 5 : Code of practice for selection and application of particleboard for specific purposes.

5. Timber Research and Development Association. Use of Structural (C5) Grade Chipboard. Design Aid DA9. Joint BRE/TRADA publication. TRADA (1992).
6. Soothill C. Punching shear strength of wood chipboard. Part 1- Initial assessment of influence of board thickness, punch size, joist spacing and tongue and groove joints on the resistance of chipboard to gradually applied concentrated loads. TRADA, Research report 5/83. Hughenden Valley, TRADA (1983).
7. Soothill C. Concentrated load capacity (punching strength), of wood chipboard. Part 2- Assessment of the influence of bending on the resistance of flooring chipboard to gradually applied concentrated loads. TRADA, Research Report 2/84. Hughenden Valley, TRADA (1984).
8. British Standards Institution. Structural use of timber. BS5268:Part 2: Code of practice for permissible stress design, materials and workmanship. BSI London (1984).
9. British Standards Institution. The Structural Use of Timber. CP112:Part 2 1971. BSI London (1971).

12 NEW DEVELOPMENTS IN REINFORCED TIMBER JOINTS

A.J.M. Leijten
Delft University of Technology, The Netherlands

Timber joints are still regarded as the weakest part of a timber structure. In many cases they also govern the size of the connecting timber members due to the spacing requirements. A joint type with a high strength and stiffness capacity as well as a good ductile behaviour is being developed. By reinforcing locally the timber members with densified veneer wood premature failure due to splitting of the timber is being prevented. This leads to new and economic design possibilities.

INTRODUCTION

In many design standards more in particular Eurocode 5 much attention has been given to design rules for joints with dowel type fasteners. General formulas for the ultimate limit state are presented as well as their corresponding failure mode. This approach, the Johanson formulas based on the yield theory, have been accepted in Europe and America and probably will gain acceptation through out the world in the next years. Nevertheless the application of dowel type fasteners has a number of disadvantages. The strength and stiffness is rather disappointing compared to the properties of the connecting timber members.

At present a very promising new type of joint is being developed. Densified veneer wood is used to strengthen the timber. In this paper it will be shown how effective these joints can be in terms of timber sizes reduction and decrease of required number of dowels, etc. How moment transmitting (semi-rigid) joints influence the bending moment distribution and therefore the load carrying capacity and structural safety is assumed to be basic knowledge. A most commonly used portal frame is analysed in its performance.

THE PRINCIPAL OF THE REINFORCED JOINT

The ability of timber to resist high concentrated loads caused by dowel type fasteners is poor. Premature failure by splitting can only be prevented when a minimum spacing between the fasteners is provided. Another possibility is to increase the strength and stiffness capacity of the joint by gluing material on the timber surfaces. In this way the timber is discharged to carry the high loads immediately. This material which in a way reinforces the joints should be able to disperse the high concentrated loads and eventually transmit the stresses to the timber. Now which reinforcement is most suitable?

Steel reinforcement

In the past steel plates have been glued to the timber members to reinforce the joint. The result was a very stiff, strong and ductile joint. However, for the following reasons this application has never left the research laboratory stage.

- the gluing procedure is complicated and requires careful handling because timber and steel are two completely different materials,
- drilling holes is another problem as both materials actually require a different drilling technique,
- tight fitting dowels or bolts can only be obtained applying epoxy injection techniques but manufacturing and quality control problems are foreseen,
- the different response to climatic changes,
- the high difference in the elastic-modulus of steel and timber which introduces stress concentrations and micro-crack propagation,
- the fire resistance.

Altogether a long list of negative factors. Reasons to continue the search for other alternatives. Our final choice has been the densified veneer wood known as Lignostone (trade name) which is being produced by the firm Lignostone in Ter Apel, the Netherlands.

Lignostone: densified veneer wood

This wood product is produced in Germany since ca 1860. The production process can be summarised as follows. The material is being heated to about 120° C and pressure perpendicular to grain is introduced (ca 25 MPa). This weakens the cell wall of the wood fibres. The whole cell structure will finally collapse, Figure 1. The densification is dependent on the pressure, temperature and the moisture content of the veneer. After full collapse the temperature is dropped rapidly

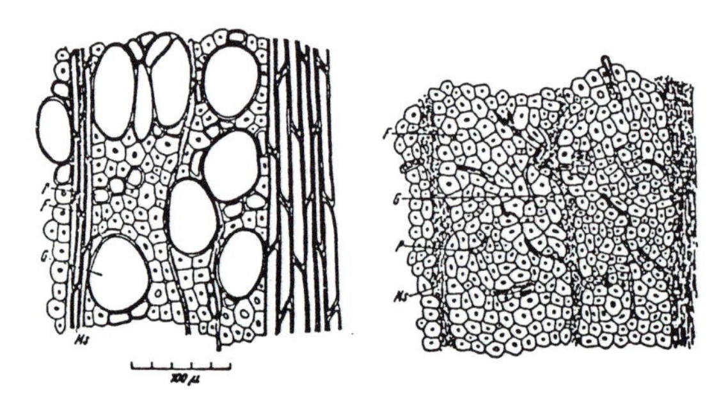

Figure 1. End grain face of poplar before and after densification

freezing the new cell structure. The lignin becomes petrified. This prevents a full elastic recovery. Due to the fact that the substance lignin plays a important part in this process not all species are suitable for this process. The wood species beech is mostly used. Densities up to 1400 kg/m^3 are possible. Prior to densification the material can be impregnated with resin which makes it more brittle but impenetrable for moisture. Because the mechanical properties of the fibres will only be sightly affected by this process, the strength and elastic modulus will increase proportional with the densification ratio. Poplar for instance with a density of 350 kg/m^3 and a bending strength of 40 MPa, will after densification with a factor 3 to 3.5 (density becomes 1225 kg/m^3) possess a strength of 3.5 x 40 = 140 MPa. This is comparable with high strength tropical hardwoods. Despite this process many characteristics of wood are maintained one of which is the ability to glue easy to other wood materials. Not only solid (sawn) timber can be utilised also stacks of veneer sheets can be compressed all together as in plywood production. By changing the fibre direction of the veneer sheets the properties of the final material can be changed from pure orthotropic to almost isotropic. This densified veneer wood is produced up to thicknesses of 6 to 120 mm. Until now civil engineering has almost forgotten this very special material.

DENSIFIED VENEER WOOD

Is this material really an alternative for steel? Let us compare steel with Lignostone for this application. For Lignostone the glue procedure is no longer a problem as generally applied adhesives can be used. Although there is a different reaction on changing climatic conditions both the timber and Lignostone will swell and shrink in corresponding directions although the densified veneer wood react slower. The elastic modulus is of the same order; 10.000 MPa for spruce and 15 000 to 20 000 MPa for Lignostone. Regarding the fire resistance Lignostone will not transfer the heat of fire to the regions were the fasteners are located so quickly as steel. Considering these advantages the steel reinforcement was replaced by the Lignostone.

Mechanical properties of the densified veneer wood (dvw) reinforced joint

In order to give an idea about the effectiveness of the Lignostone reinforced joint in comparison with other timber joints the next table is presented. The basis of the table is that the shear (or contact) surface of the joints considered are approximately equal. The sheared surface (area where the members make contact) is about 105 x 180 mm.

All joints were loaded in tension (mechanical fasteners were laterally loaded). Notice the increase in strength and stiffness of the Lignostone reinforced joint. Illustrative is Figure 2 which show the load displacement curve of a reinforced joint with two types of dowels. Remarkable is the small scatter in strength, stiffness and the enormous ductility for timber joints. These results have a tentative nature because only 2 x 5 tests have been performed. Especially in

Table 1. Comparison of strength and stiffness of three member joints strength per shear plane

Mechanical fasteners	strength kN	ratio	stiffness kN/mm	ratio
6 dowels of 8 mm diameter	23.6	1.0	21	1.0
30 nails of 2.3 mm	23.7	1.0	35	1.7
1 tooth plate (connector)	16.2	0.7	35	1.7
1 split-ring (connector	14.0	0.6	25	1.2
2 dowels 8 mm steelpl. 1.5 mm	40.4	1.7	60	2.9
1 dowel 17 mm Lign. 6 mm	32	1.4	38	1.8
1 dowel 17 mm Lign. 12 mm	38	1.6	62	3.0
1 dowel 23 mm Lign. 12 mm	65	2.7	150	7.2

trusses where the dimensions of the timber members largely depend on the space required by the joints, application of the dvw reinforced joint can lead to economic benefits. The high stiffness can be of much value in transmitting high bending moments as required in portal frame corners as traditional joints fail to do. To outline the capabilities in this respect a comparison will be made between the design of a traditional moment transmitting joint in a portal frame and a Lignostone reinforced joint.

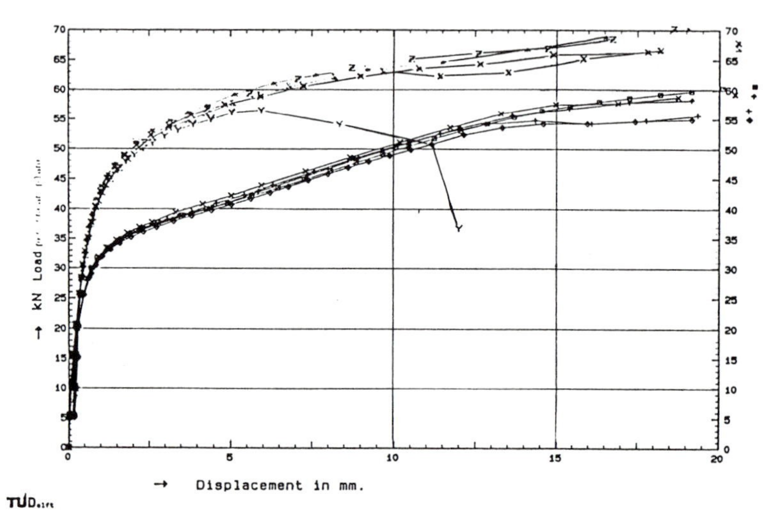

Figure 2. Load slip relation for the Lignostone reinforced joint

It will be shown that less holes have to be drilled when the Lignostone reinforced joint is applied. This also makes manufacturing easier and contribute to the technical and economical acceptance by practise.

Derivation of the properties of the Lignostone reinforced joint
Laboratory tests are necessary ingredients in the developments of the Lignostone reinforced joint. Not all relevant mechanical and physical properties are known yet. Aspects like duration of load factors have never dealt with before. Also the veneer grade might influence the properties of the Lignostone. It will be clear that this all needs a considerable amount of testing. This will be done within the framework of a EC-sponsored research programme named FOREST in which laboratories and industries of five countries participate.

THE DESIGN OF A PORTAL FRAME

To show the benefits of the Lignostone reinforced joint a comparison is made with a traditional design of a portal frame as shown in Figure 3. The basic test data for the design with the reinforced joints is not yet complete however it is felt that extrapolation of certain test results can be justified. In this respect the stiffness modulus found by tension tests of Figure 2 have been modified to be used a rotation stiffness.

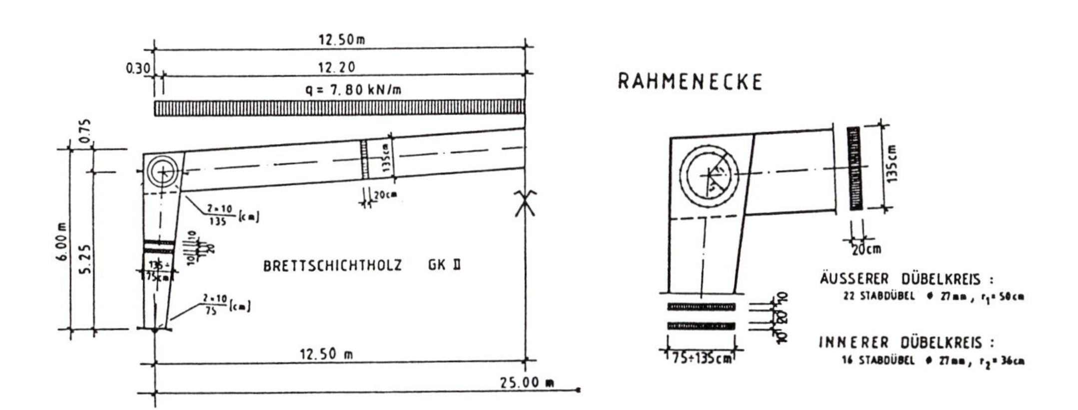

Figure 3. The portal frame

The traditional design with dowel type fasteners
To make an easy start a portal frame completely analysed by Professor Pischl[1] is quoted, see Figure 3. This calculation is made with allowable stresses also called allowable stress design. This method does not allow to take any moment redistribution into consideration and is still being used for timber design throughout Europe. The following data is given:

- portal height H = 5.25 m
- span L = 25.00 m
- vertical load: q = 7.8 kN/m
- column size: 2 x 1350 x 100mm
  EI = 4,92 1014 N/mm^2
- beam size: 1350 x 200mm EIcolumn = EIbeam

In order to derive at a situation where both the bending moment at mid-span and joint are comparable a rotation stiffness is required of 112540 kNm. With the aid of the German standard DIN 1052, in which stiffness values are given for dowels, the size and the number of dowels have been determined. The stiffness requirement is satisfied applying 38 dowels of 27 mm diameter located in two circle patterns, Figure 3.

It will be clear that the traditional joint is not very economical. At mid span the bending stresses are still very moderate. This is due to the spacing requirements of the dowels which dictate the beam sizes.

The design with the Lignostone reinforced joint
Without giving detailed design calculations the result is that the number of fasteners reduces from 38 dowels of 27 mm in the traditional design above to 11 dowels of 33 mm diameter. Therefore the timber sizes may be reduced as the spacing requirements are less sever for the Lignostone reinforced joint. Assuming that the stability conditions (columns have reduced in size) still hold the cross-sectional area is reduced with 33% which is of economical interest.

Until now it has been assumed that the reinforced joint is able to withstand high bending moments. Because research Lignostone moment joints not yet been finished the load-slip characteristic of tension joints have been extrapolated. It could well be that adopting this approach we overlooked something. The capacity of moment joints could be limited by other failure modes which do not occur when joints are tested in tension. One of those limitations have been experienced when steel was used as reinforcement. In the outer most laminations the tension and compression stresses are very high while the end grain face is nearby. Under certain conditions this lead to a pull out (shear) failure the outer laminations. Whether or not the same applies for Lignostone reinforced joint is a question to be answered. Nevertheless this failure mode restrict the bearing capacity Mju of the moment joint. For steel reinforced joints this limitation was found to be about 0.5 Mju. Accepting this limitation will influence the design calculation considerably. This means that applying 11 dowels will result in a brittle failure of the joint, Figure 4a. It follows that the number of dowels should be reduced as to get yielding at about 0.5 Mju. The number of dowels can be reduced when there are only four of them located in the four corners of the joint. This transforms the moment-rotation curve of the joint as given in Figure 4b. Locating the dowels in the corner increases the radius of the pattern and as the stiffness is proportional to the radius square this is pleasant. It will be clear that minimum end and edge distances become important because this affects the maximum bending moment. The yield moment of the joint is reached just before the bending strength of the beam at mid span.

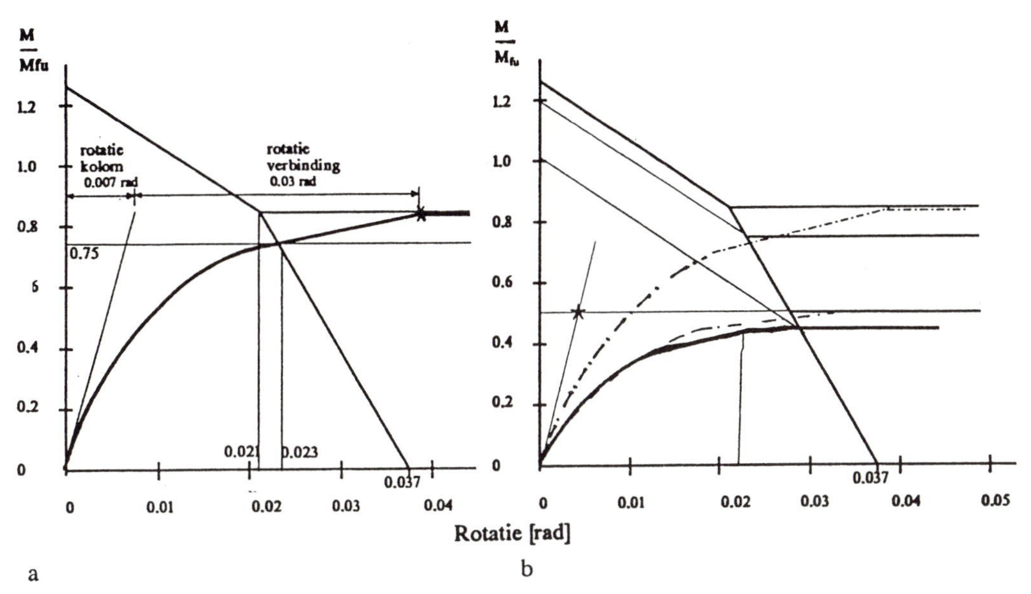

Figure 4. Moment rotation domain of the portal frame joint reinforced with Lignostone
a: Moment rotation line when 11 fasteners are used
b: Moment rotation line for 4 fasteners located in the corners

CONCLUSION

The reinforcement of timber joints can become economical and technically attractive. The application of Lignostone as a reinforcement material is able to satisfy many demands in this respect. Not only for moment joints this type will be effective but everywhere where the spacing requirements of the fasteners force timber members to increase in size, as for instance in trusses. It has been shown that analyses of the capabilities of two types of joints applied in a portal frame can be analysed quickly using the moment-rotation graph. Reinforcing the moment transmitting joint with Lignostone leads to 30% saving in timber while the manufacturing of the joint is cheaper than in traditional design. The role of tests in order to derive information about the mechanical properties is essential.

REFERENCES

1. Pischl R. Zum Einlub der Nachgiebigkeit der Verbindungsmittel, Bundesholzwirt-schaftrat, Wien, SfB 27.1, 1986.

PART THREE

NEW STRUCTURAL MATERIALS

13 CONSTITUTIVE MODELLING OF STRUCTURAL PLASTICS: THE ROLE OF PHYSICAL TESTING

J.C. Boot and Z.W. Guan
University of Bradford, Bradford, UK

The various techniques available for predicting the life of engineering works formed from structural plastics are reviewed, and the importance of constitutive modelling established. The test procedures used to obtained a full-life constitutive model for medium density polyethylene valid over a wide range of stress paths at 20° C are then presented and discussed.

INTRODUCTION

New plastics with improved properties, and a range of applications in the construction industry, are continually being developed. However, the mechanical behaviour of polymeric materials is complex(1); consequently this rapid evolutionary process increases the associated structural design problems in as much as it is implicit that experience of past behaviour cannot reliably be used to estimate future performance, and attention must focus on obtaining long term predictions based on extrapolation of short term behaviour. There are essentially three techniques available to facilitate such information,

(i) elevated temperature testing of representative systems,
(ii) elevated load testing of representative systems,
(iii) mathematical modelling incorporating suitably defined time/temperature/strain dependent constitutive properties.

The effects of time and temperature on structural performance are qualitatively the same for all polymers. Theories providing quantitative relationships for this behaviour have been proposed(1) and these form the basis for the substitution of elevated temperature for extended time testing. However, most actual polymers have complex chain characteristics which result in a number of minor transition points between the major glass transition and melting or flow points. The effect of secondary transition points is often to cause significant sudden variations in certain aspects of constitutive behaviour with temperature(2). Since these transitions are a function of the energy levels within the molecules of the chains they do not affect time dependence in the same way, and any simple quantitative relationship between the two is lost.

Consequently, elevated load testing under temperature regimes closely related to those anticipated under field conditions is often to be preferred, but again straight forward extrapolation to working conditions may be flawed. In particular problems are often encountered if scaling or instability effects are significant, or if any of the prime variables governing system behaviour are related in a highly nonlinear manner.

Accordingly it is preferable not to place total reliance on accelerated testing of representative systems. Mathematical modelling involves appropriate definition of system geometry, external loading, and constitutive behaviour; the system equations can then be solved to yield all the required structural characteristics. Mathematical modelling enables rapid evaluation of performance characteristics over any required time scale. The all important additional capability to study the effects of variations in all system parameters independently is also now available. Particularly in the case of structural plastics, therefore, mathematical modelling provides potentially the most useful source of predictive information. The chosen constitutive model needs only to replicate the relevant relationships between the principal stresses ($\sigma_1 > \sigma_2 > \sigma_3$) and corresponding temperature, strain rate and environment dependent principal strains ($\epsilon_1, \epsilon_2, \epsilon_3$). Consequently, it is now possible to propose generalised testing techniques which establish a range of stress paths in σ_1, σ_2, σ_3 space so as to produce mutually supportive results. In this manner a constitutive model with a high confidence level can be obtained - provided of course all critical conditions have been considered in its derivation.

In this paper we illustrate the role of physical testing in the constitutive modelling of structural plastics by applying the general principles elucidated above to obtain a test regime for a representative problem. The presentation comprises specification of problem, identification of stress paths requiring consideration, design of test apparatus and specimens and discussion of results.

EXAMPLE PROBLEM

A thin-walled tight fitting lining of medium density polyethylene (MDPE)(3) has been proposed as a suitable low cost solution to the renovation of corroding cast iron water mains(4). Figure 1 shows a critical design situation in which the lining has to span a significant corrosion hole in the host pipe subject to long term hydrostatic pressure. A constitutive model is required to enable design guidelines to be formulated for these systems using mathematical modelling(5).

The temperature within a UK water main varies seasonally within the range 0-20° C. Over this temperature range the rigidity of MDPE increases by about 50% with reducing temperature. To achieve a reasonably simple conservative design procedure, therefore, temperature can be eliminated by considering the upper bound situation of a main permanently at 20° C.

We design for adequate factor of safety against failure occurring in 50 years, and thus an ultimate load analysis is required. Failure will be due to accumulating

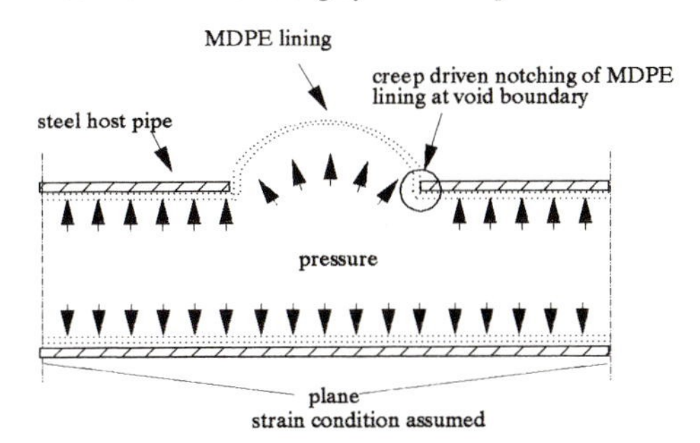

Figure 1. Long section through section of renovated pipe

creep deformations in the mode shown. Strain rates will therefore be generally low, and unloading behaviour not a significant parameter. We define material failure as the instigation of increasing deformation rate under constant applied stress. Thus ductile material failure is characterised by the onset of tertiary creep; structural failure occurs when any part of the structure becomes a mechanism due to the combined effects of plasticity and creep. Prior to failure, however, ductile behaviour will result in a continuous favourable redistribution of stress[4,5,9] due to both plasticity and creep (since the structure is statically indeterminate). Brittle failure of material and structure is precipitated simultaneously by propagation of any unstable crack growth.

Within the specified temperature range MDPE is an extremely ductile material[6]. Any notches accidentally induced during installation will probably have a small effect on behaviour in comparison to the driven notching effect encountered at the void boundaries. Accordingly (in the first instance at least) we consider the possibility of brittle failure due to slow crack growth only at this location. Rapid crack propagation (RCP) cannot occur provided the pipe is full of water[7]. Even if the lining is full or partially full of a gas whilst in service, the lining is predominantly supported by the host pipe and therefore unstressed; again RCP cannot occur. The possibility of RCP due to gas presence during pressurisation is considered extremely remote due to the very low probability of crack initiation under these conditions.

The prime objective is therefore to undertake material tests so as to enable development of a constitutive model valid under the following conditions:

(i) arbitrary biaxial stress in the plane of the shell surface to simulate conditions predominating within the span (see Figure 1),
(ii) biaxial tension in the plane of the shell surface accompanied by transverse shear stresses to simulate conditions at the void boundary.

If the tests are to be able to characterise the material rather than the structure it is essential that the distribution of relevant stress components (or at least their appropriate resultants) and corresponding strains are known; this requires the tests to be statically indeterminate in the stresses or stress resultants of interest. Since long term creep tests are most easily undertaken under dead load it is generally preferable to use small section test pieces rather than full cross-section pipe to reduce the loads required. A further important parameter is that due to the manufacturing process, MDPE pipe is subject to biaxial residual stresses inducing bending in the plane of the shell surface, and which approach permissible working stress levels in the extreme fibres[(8)].

TEST SERIES

Following determination of residual strains[(8)] the series of creep tests described below was undertaken with due reference to the relevant standards [(10,11,12)]. Based on the information available[(6)], except where otherwise stated tests were established at Equivalent Von Mises stress levels[(4)] of 0.0, 1.6, 3.2, 4.8, 6.4, 8.0 N/mm^2 and terminated after 4 months. Each test was undertaken only once at each stress level as subsequent statistical analysis could be used to establish any inconsistencies between the results so obtained. The tests were undertaken at 20° ±1°C. The pipe supplied for testing had an external diameter of 183mm and 6.1 mm thickness.

Recently strain gauges have become available(8) which enable techniques traditionally used to monitor the deformations of more rigid materials to be applied in the present context. The measurable strain in MDPE is in the range 7-13% reducing with age. Additional details of all tests are given in Ref.13.

Tension tests: standard dumb-bell type specimens were cut directly from the pipe wall. The central section (100mm long by 15mm wide) was milled out to a rectangular cross section 15mm x 3mm; the excess material was removed from both the inner and outer walls so as to leave the specimens with no significant residual stress effects. The specimens were suspended through one swivel joint and loaded through two more to minimise eccentric loading effects. Each specimen was mounted with longitudinal strain gauges on both faces (to check for eccentricity effects etc.) and a transverse gauge on one face. Figure 2 shows the results obtained at 8.0 N/mm^2. Both longitudinal gauges have ceased to function properly at a strain of 8%; that the problem lies with these gauges (rather than being symptomatic of material behaviour) is evidenced by the continuing anticipated performance of the transverse gauge. The overall results demonstrate that the secondary creep path was established almost immediately and continued smoothly with Poisson's ratio (υ) remaining in the range 0.43-0.44 throughout all tests. Unloading was undertaken in 6 increments at a rate approximating that observed during the measurement of residual strain.

To establish the onset of tertiary creep at these stress levels, further similar tests were carried out at 10, 11, 12, 13, 14, 15, 16 N/mm^2 nominal stress with sample extension recorded by dial gauge once the strain gauges had failed. Tertiary creep resulting in rupture at 40-50% strain was seen to take place at very

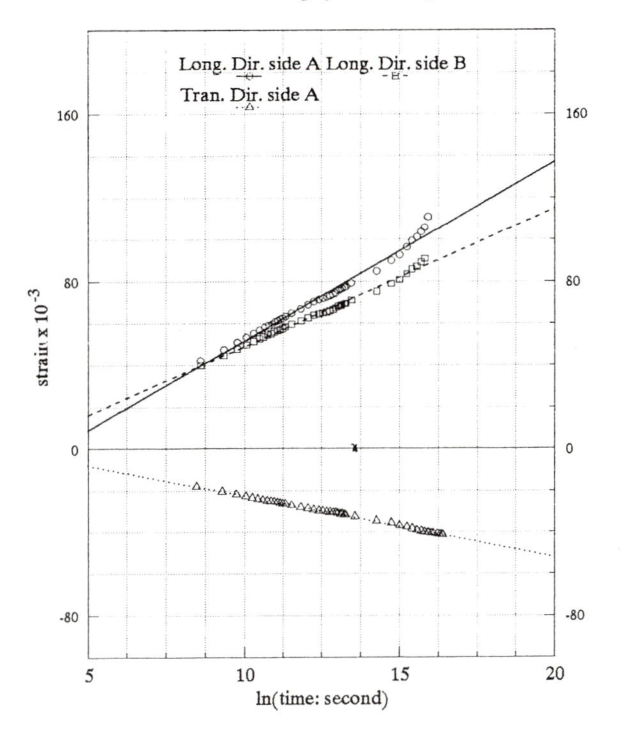

Figure 2. Tensile screep strains for virgin MDPE under stress level of 8.0 N/mm^2

high strain rates in comparison to behaviour on the secondary creep path. Figure 3 shows the projection of the onset of tertiary creep (assumed failure) line in stress-time space.

Compressive tests: Samples 6mm x 6mm x 50mm were cut from the pipe wall and loaded in compression inside a 10mm internal diameter guide tube(13). Within the confines of the guide tubes there was room for longitudinal strain gauges to be located on all four sides of the specimen. Unfortunately release of residual stress left these samples with a significant longitudinal curvature which could not be corrected, and thus an evenly distributed compressive stress could not be achieved. The results obtained(13) clearly demonstrate that in the present case alternative means are required to establish properties in compression.

Longitudinal bending: Specimens were prepared, strain gauged and tested in four point bending as illustrated in Figure 4. Residual stresses were effectively fully released yielding initial deformations of the form induced by the loading. Under these conditions nonlinear changes in geometry are third order effects. In the test

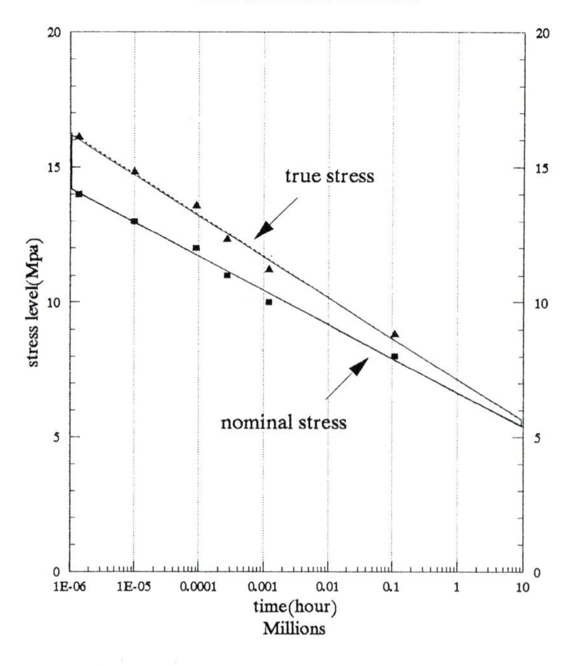

Figure 3. Tertiary creep cut off line

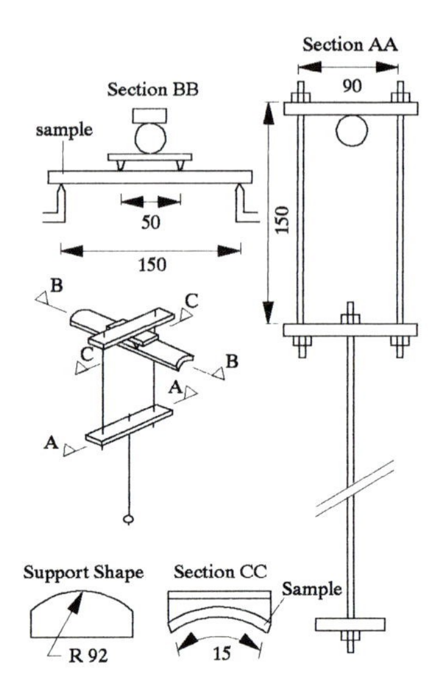

Figure 4. Flexural creep testing apparatus

mode illustrated initial deformations oppose those induced by the loading; two tests were repeated in reverse mode to confirm that the significance of residual stresses on the results obtained was minimal. Nevertheless the large deformations experienced at higher stress levels mean it is important to use knife edge supports, to maintain constant span and minimise support friction, and to use a sufficiently long specimen to allow for recruitment into the span.

In these tests bending moment at the gauged section is statically determinate but the bending stresses are not. The nominal maximum bending stress levels were used to determine test loads assuming elastic behaviour; actual extreme fibre stress levels would be reducing throughout the tests. Since plane sections are constrained to remain plane at the gauged sections, the typical results illustrated in Figure 5 demonstrate (to within the limits of experimental error) identical behaviour in tension and compression with $\upsilon=0.43 \sim 0.44$ throughout the recorded portion of the secondary creep path.

Hoop bending: The previous uniaxial tests were undertaken with samples cut longitudinally from the pipe wall. These tests were therefore designed to establish

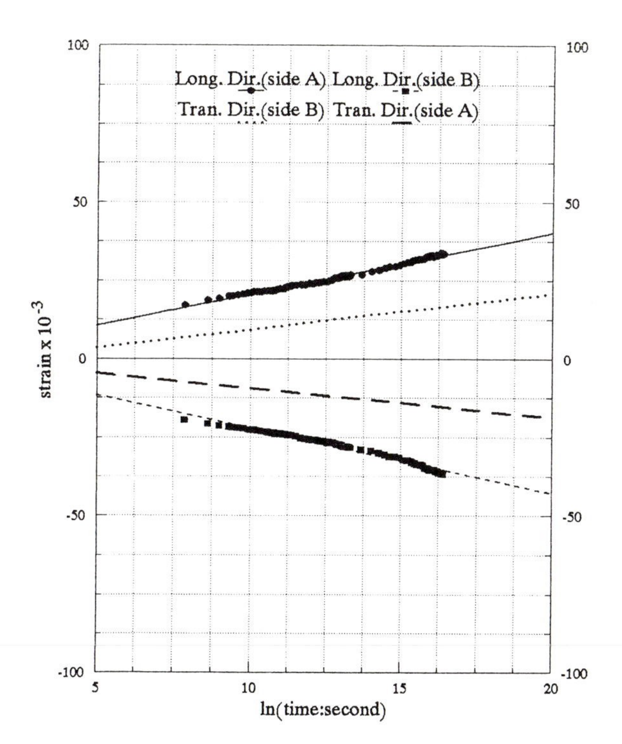

Figure 5. Longitudinal creep bending strains under stress level of 6.4 N/mm^2

whether the combination of pipe manufacture and lining installation[4,8] induce any significant anisotropy in the material. Figure 6 delineates the test apparatus, specimens, and gauge configurations. Nominal stress levels, original pipe dimensions and assumed elastic behaviour were used to calculate the applied loading. Again samples were cut from the pipe wall so as to ensure complete removal of residual stresses[8]. Taking due account of the resulting change in geometry results equivalent to those achieved in longitudinal bending were obtained.

Internal pressure: Here we establish a biaxial stress path which is statically determinate ($\sigma_1=2\sigma_2$; $\sigma_3=0$) assuming thin cylinder theory to apply. We used WASK-RMF (TRANSGRIP) end caps to ensure an even distribution of longitudinal stress, and an aspect ratio of at least 3:1 to ensure the gauged section (longitudinal and hoop gauges at 90° intervals on internal and external surfaces) was remote from end effects. To ensure electrical integrity of the internal gauges compressed air pressure was used; all samples were pressurised from a single source and fitted with a circuit to eliminate them from the system in the event of any significant local pressure drop.

Torsion: To investigate behaviour in shear flow, mechanical engineers commonly test small cylindrical samples subject to combined tension and torsion. Unfortunately in the present case it was not possible to remove suitable samples from the 6mm thick pipe. However, it was felt that a sufficiently representative alternative biaxial stress path ($\sigma_3 = -\sigma_1$; $\sigma_2 = 0$) could be investigated by

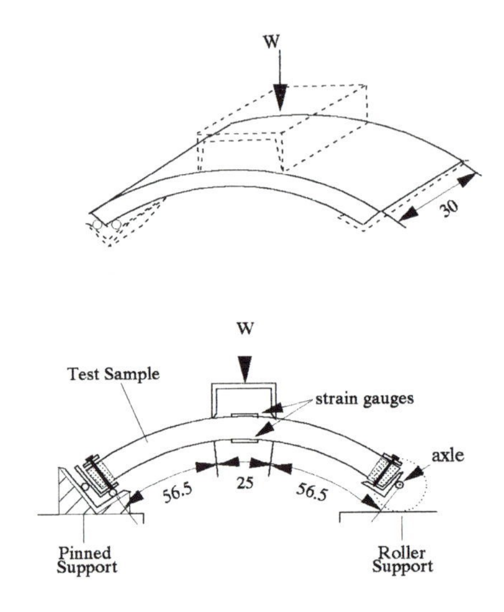

Figure 6. The hoop bending creep apparatus

subjecting a length of pipe to pure torsion. The test apparatus is illustrated in Figure 7.

Driven notching: To evaluate performance under the conditions illustrated in Figure 1 the test shown in Figure 8 was developed. The relationship between tensile force T and notching force R was R=√2T (i.e. ϕ =90° and the lining stresses approximated to a state of uniaxial tension at the void boundary). Penetration of the knife was measured with a displacement transducer, whilst overall and local longitudinal strains were measured using Demec and foil gauges respectively.

DISCUSSION OF RESULTS

Tensile, bending, internal pressure and torsion tests all yielded linear semi-logarithmic plots of the secondary creep path at all strain gauge positions in all tests. Qualitatively these results are consistent[4,5] with classical elastic-viscoplastic theory (isotropic behaviour, equal characteristics in tension and compression, and a high value of υ suggesting largely pressure independent behaviour).

Due to difficulty in obtaining suitable samples for compression testing, it may often be preferable to imply compressive properties from bending and pure torsion testing.

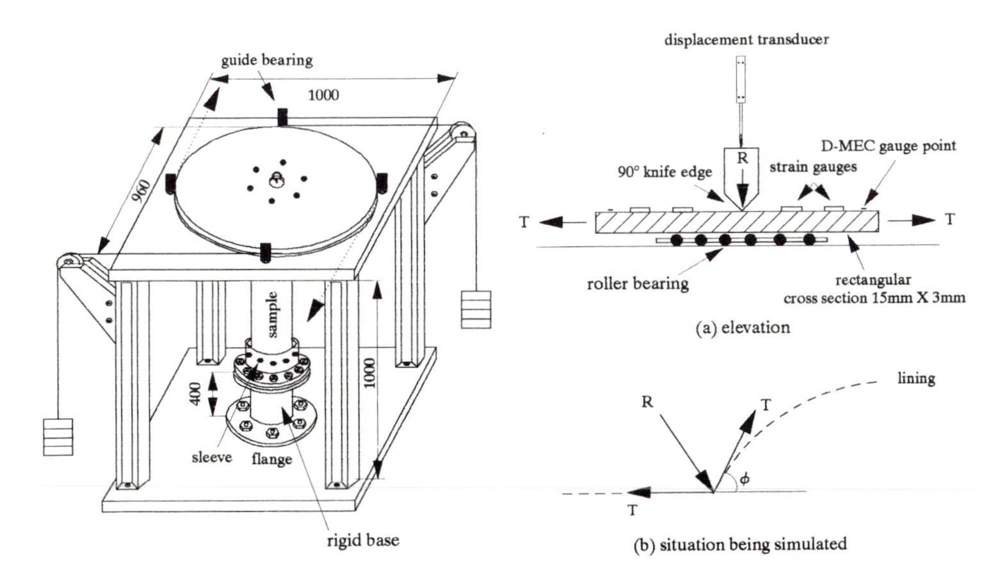

Figure 7. Test rig for torsion creep test

Figure 8. Details of driven notch testing

Results from the driven notch testing were only significant at T≡ 10 N/mm^2 nominal tensile stress. In this case the knife reached a stable equilibrium position half way through the sample, at which point the tensile stress on the remaining ligament was 20 N/mm^2. Since this situation cannot be sustained according to Figure 3, it would appear that MDPE is sufficiently ductile to benefit from notch hardening without undue risk of brittle fracture. The significance of this result is currently under further investigation. The possibility of brittle fracture under pre-notched conditions can be investigated by reference to BS 4618 and ASTM E292-83.

REFERENCES

1. Struik L.C.E. *Physical aging in amorphous polymers and other materials*, Elsevier, Amsterdam, 1978.
2. de Putter W.J. 'Extrapolation of stress rupture data', Plastic Pipes VII, Proceedings of International Conference, Plastics and Rubber Institute, Bath, September 1988.
3. Specification of Blue Polyethylene Pressure Pipe for Clod Potable Water, Information and Guidance Note 4-32-03, Water Authorities Association, May 1987.
4. Boot J.C. and Guan Z.W. 'Structural behaviour of thin-walled plastic pipe linings for water mains', Plastic Pipes VIII, Proceedings of International Conference, Koningshof, September 1992.
5. Guan Z.W. and Boot J.C. 'Elasto-plastic and elasto-viscoplastic large deformations of polyethylene pipe linings', 7th UK ABAQUS User Group Meeting, Cambridge, September 1992.
6. Rigidex PC002-50 R102 for pipes, BP Chemicals Ltd, Polyolefins (UK), Bo'ness Road, Grangemouth, Stirlingshire FK39XH.
7. Marshall G.P., Harry P.G., Ward A.L. and Pearson D. 'The prediction of long term failure properties of plastic pressure pipes', Plastic Pipes VIII, Proceedings of International Conference, Koningshof, September 1992.
8. Boot J.C. and Guan Z.W. 'The measurement of residual strain in plastic pipe'. Plastics, Rubber and Composites Processing and Applications, 16 (1991), pp123-125.
9. Hult J.A.H. *Creep in Engineering Structures*, Blaisdell, Waltham, Massachusetts, 1966.
10. ASTM D2990-77 (1982), Tensile, compressive, and flexural creep and creep rupture of plastics, Vol.08.02, 1987, pp. 692-702.
11. ASTM D1598-86 (1987), Time to failure of plastic pipe under constant internal pressure, Vol.08.02, 1990, pp. 48-50.
12. BS 4618: Recommendations for the presentation of plastics design data, British Standard Institution, 1976.
13. Guan Z.W. and Boot J.C. The Structural Behaviour of Renovated Water Mains, Research Report 41, Dept of Civil Eng, Univ of Bradford, 1992.

14 NEW DEVELOPMENT OF A POROUS WEARING COURSE FOR STEEL ORTHOTROPIC BRIDGE DECKS BASED ON SYNTHETIC MATERIALS

M.H. Kolstein
Delft University of Technology, The Netherlands

The traffic induced stresses in a steel orthotropic bridge deck are significantly reduced by relative thick asphalt surfacings. Because of the wide range of effect on bituminous wearing courses caused by the variation of the parameters in practice, no simple rule for calculating stress reduction could be devised until know. In this paper the stress reduction is studied on a bridge which has been surfaced with a new polyurethane porous wearing course. Field measurements showed a relative constant behaviour with respect to stress reduction in the steel deck plate.

INTRODUCTION

Orthotropic steel bridge decks
Modern steel bridge decks consist of a 10-12 mm thick deck plate stiffened by 6 mm troughs spanning in the direction of the traffic flow between cross girders. Usually the deck plate is surfaced with a 50 mm thick wearing surface of mastic asphalt or asphalt concrete with a weight of about 125 kg/m2. That means, comparing with the ± 350 kg/m2 steel of the box girder of a modern suspension bridge or cable-stayed bridge, 26% of the total weight of the box girder.

The, so called, orthotropic deck is a flexible structure which is highly sensitive to the local bending action produced by the wheel loads of heavy commercial vehicles (see Figure 1). During its lifetime, the bridge steel deck including the wearing course can be expected to suffer many millions of cycles by wheel loading, so that fatigue is an important design criterion.

Composite action between steel deck and wearing course (1,2,6)
The strains that occur in the steel deck plate and wearing course are strongly influenced by the dynamic stiffness moduli of the wearing course and interlayer, the thickness of the deck plate and thickness of the wearing surface (see Figure 2). Besides the dynamic stiffness modulus of bitumen mixtures is strongly influenced by the temperature and loading frequencies. Solutions which increase the dead weight considerable must be prevented.

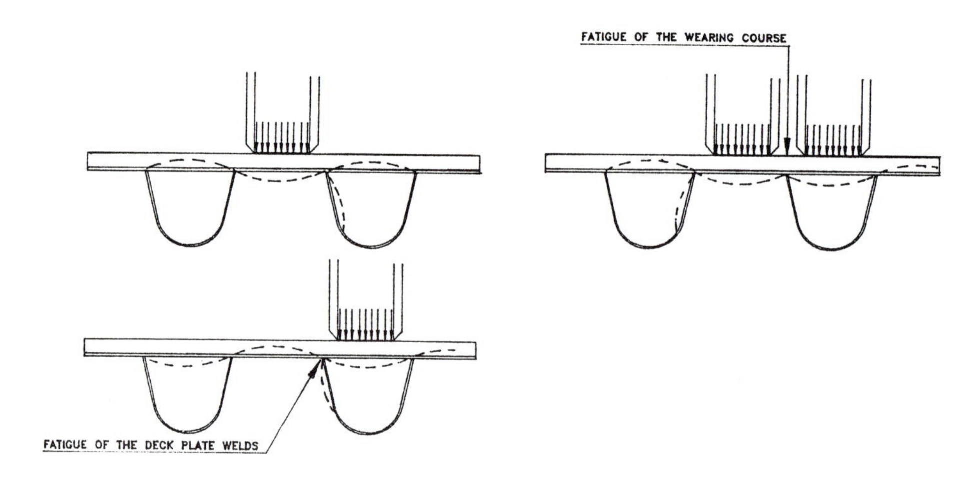

Figure 1. Orthotropic steel bridge deck

Desirable properties and qualities of the wearing course

The pavement on the bridge requires an acceptable flexibility to ensure a good fatigue performance and it is equally important that the pavement possesses enough resistance against rutting of the wheel loads. Extensive deformations of the wearing surface results in high and unacceptable dynamic effects in the steel deck. Next to that the steel deck has to be protected against corrosion by one of the layers in the wearing course.

Nowadays porous wearing courses are very popular in the Netherlands[3]. The reason for this is mainly, that they give comforts to the use of the road. They have a high skid resistance, they reduce the noise nuisance, prevent splash and spray water, while aqua-planing in principle is impossible. The minimum thickness of the asphalt structure ZOAB with one layer of coarse mastic asphalt, one layer of drain asphalt and two membranes on orthotropic steel decks amounts 70 mm which results in a weight of about 175 kg/m2.

New polyurethane porous wearing course ZOK

The recently by Bolidt[4] developed porous wearing course ZOK has a weight of 90 kg/m2 at a thickness of 50 mm. Characteristic of this one layer system is the use of synthetic materials instead of bitumen binders. Bending tests showed a promising flexibility of the wearing course against fatigue. At -20 ^{o}C, yielding of the steel plate occurred without cracking of the ZOK. The dynamic stiffness moduli were measured at temperatures between -20 and +60 ^{o}C at loading frequencies between 1 to 30 Hz. The results (see Figure 4) showed a relative constant value.

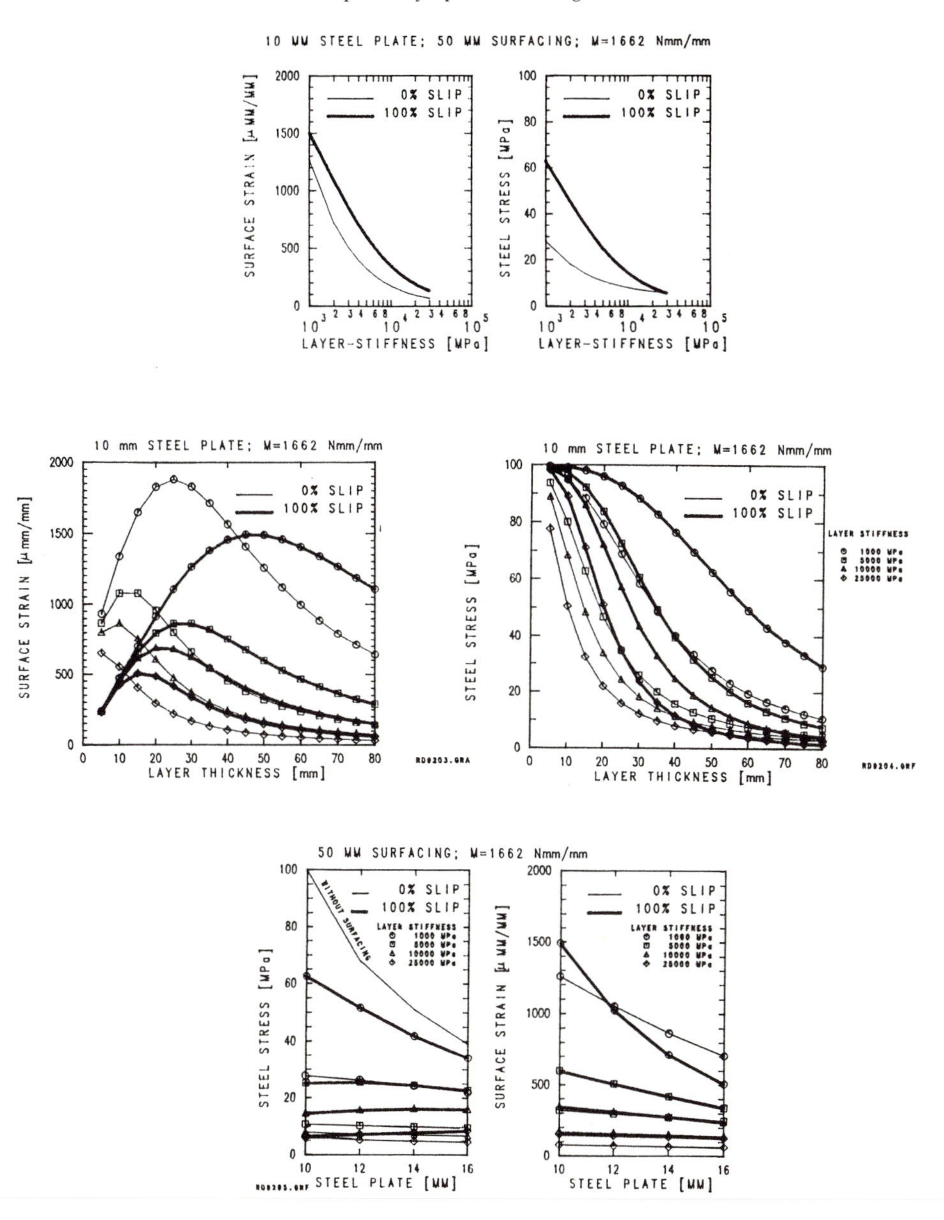

Figure 2. Theoretical ayalysis composite action

Field testing of the ZOK on the Calandbridge
In consequence of e.g. mentioned properties the ZOK wearing course was applied on a part of a heavy loaded Calandbridge in the harbour area of Rotterdam. To study the effects of the wearing course in reducing the stresses in the steel deck plate strain gauges were attached to the orthotropic deck and measurements were carried out with the old mastic asphalt wearing course, without wearing course and after the new synthetic porous wearing course ZOK was applied. Series of static and dynamic load tests were carried out using a 2-axle heavy goods vehicle. Besides, measurements under normal trafficking have been executed. First tests results are available and reported in this paper. More tests are foreseen to study the influence of the temperature.

DESCRIPTION AND INSTRUMENTATION OF THE CALANDBRIDGE

Steelstructure
The Calandbridge built in 1969 is an important link in the harbour area of Rotterdam. It carries four traffic lanes, two railway tracks, a cycle track and foot path. It is a plate girder truss bridge with four spans. One of them is a vertical lift bridge. The road section consists of a steel deck plate with closed longitudinal stiffeners. Typical dimensions are shown in figure 3.

Original wearing course based on mastic asphalt mixtures
First the grit blasted steel deck is protected by applying a thin bituminous primer. Secondly a fine mastic asphalt layer of 8 mm is applied followed by two layers of coarse mastic asphalt with thicknesses of 20 mm respectively 22 mm.

New wearing course based on synthetic materials
After removing the old wearing course and grit blasting of the steel deck, the surface was coated with a 40 mu thick primer, consisting of two components based on polyurethane resins. Then a 2 mm polyurethane membrane is applied

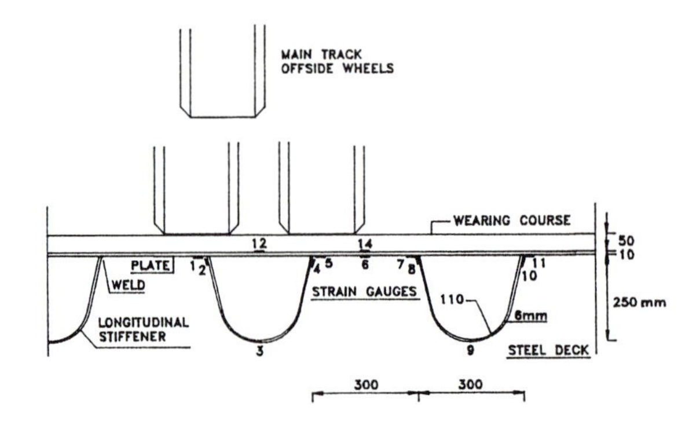

Figure 3. Calandbridge - positions of the strain gauges

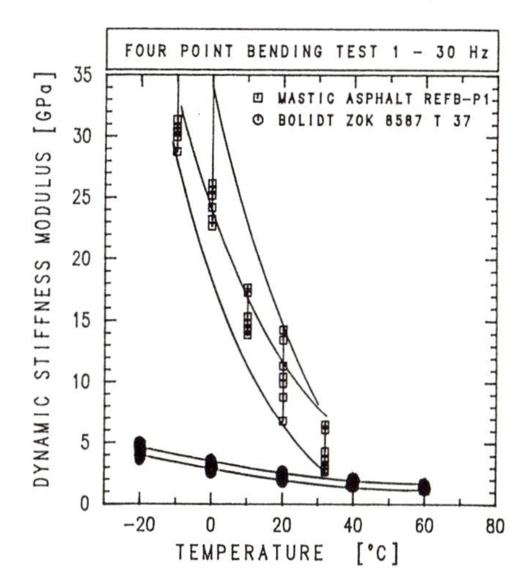

Figure 4. Dynamic stiffness moduli
Mastic Ashphalt and Bolidt ZOK

followed by one layer of a mixture of aggregate and polyurethane binder. Finally a light finishing of fine aluminium bauxite is spread over the surface to ensure an adequate skid resistance.

The total thickness of 50 mm is the same as that of the original wearing course (see Figure 5).

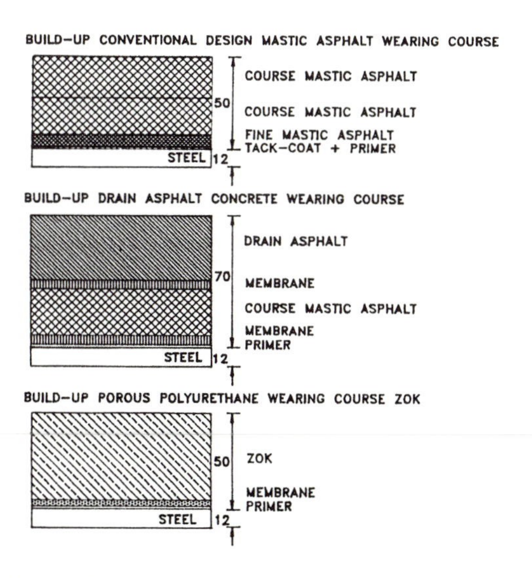

Figure 5. Built-up different wearing courses

Instrumentation

In one cross section of the bridge, various strain gauges were applied in order to get an insight into the effect of the wearing course in reducing the strain in the surfacing as well as the steel structure. Most strain gauges were positioned 25 mm from the welds between the deck plate and the longitudinal stiffeners, and close to the offside wheel track of the heavy loaded traffic lane. Several temperature gauges were attached to the steel deck plate (see Figure 3).

TEST PROGRAM

Next table gives a review of the strain measurements carried out under different conditions of loading and wearing course on the bridge deck.

Table 1. Test program depending on the bridge deck temperature [°C]

Bridge deck condition	Normal Traffic	Test vehicle	
		Static	Dynamic
Mastic asphalt	22.9	18.1	19.4
Without surfacing		13.3	
ZOK Wearing course	23.4 15.7 5.0 -5.0	17.0	21.3

THE PROCESSING OF MEASURED STRESS DATA

Influence lines are a well known method of relating the load on the bridge deck to the stresses in the measuring points. To construct the influence lines from experiments, a dual axle lorry with calibrated wheel loads was placed in a number of fixed positions and at each position the stresses at all the measuring points were taken and recorded. Static as well as dynamic measurements have been carried out.

Furthermore the induced stress variations have been measured under normal trafficking. The range-pair cycle counting method has been used to convert the stress spectra in the form of a complex waveform into a sequence of identifiable cycles to enable Miner's rule to be used to calculate the fatigue damage produced by the spectrum.

TEST RESULTS

Test vehicle loading

Stress reduction with respect to the unsurfaced steel deck

Static loading tests carried on the surfaced deck with *mastic asphalt* at ±18.1°C, showed at various locations at the underside of the deck plate and at the outer

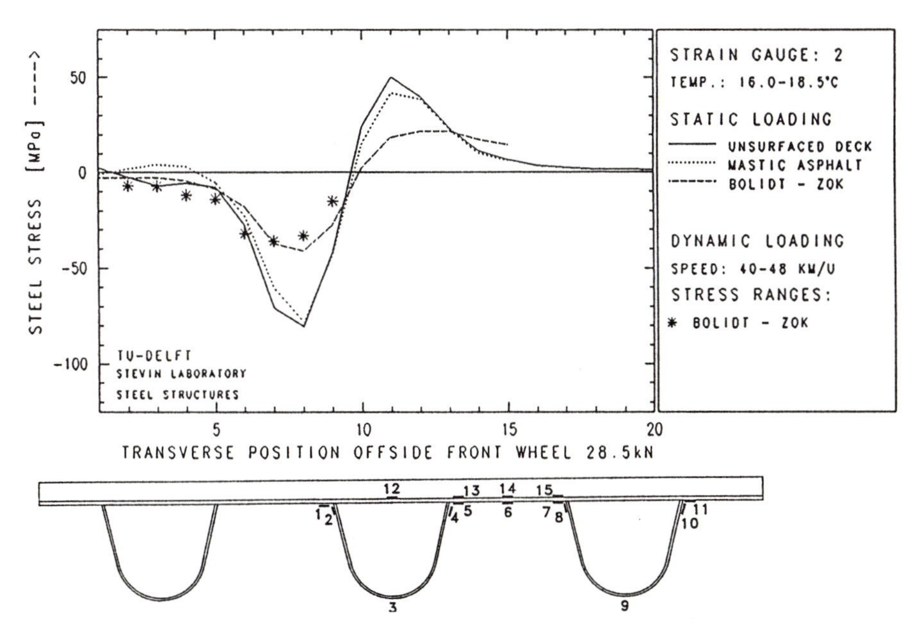

Figure 6. Test results static and dynamic loading test vehicle

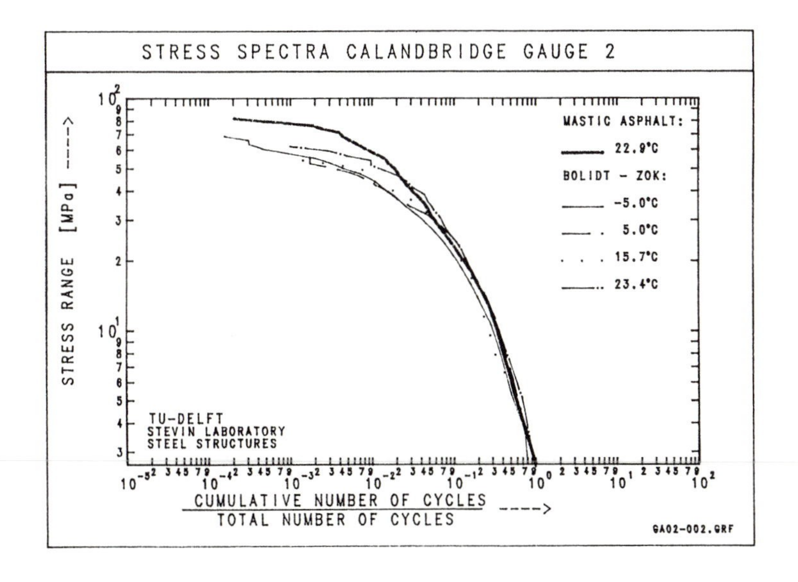

Figure 7. Test results under traffic loading

surface of the longitudinal stiffener a reduction in steel stress varying between 0 and 17%. Same tests on the surfaced deck with the synthetic porous wearing course ZOK at ± 17.0°C, showed at these locations reductions in steel stress varying between 35 and 57%.

Dynamic loading tests carried out on the surfaced deck with the ZOK wearing course at ±21.3°C and a test vehicle speed of ±45km/h showed, that the stress reductions are comparable with those measured under the static loading. Same tests on the deck surfaced with *mastic asphalt* at ±19.4°C results in a reduction of steel stress. However due to the inaccuracy of the position of the vehicle the reduction can not be quantified.

Composite behaviour ZOK wearing course and steel deck
The behaviour of the interlayer between the steel deck and the wearing course can be considered by the measured stresses at the underside of the deck plate and at topside of the deck plate. Results showed that the steel deck and the wearing course acts at ±18°C as two independent beams.

Traffic loading
Comparison stress spectra with different wearing courses
Analysis of the measured stress variations induced under normal trafficking in the surfaced steel deck with the traditional mastic asphalt respectively the synthetic porous wearing course ZOK at a temperature domain of 20 - 23°C showed that the stress spectra at the different locations are about the same.

Influence temperature with the ZOK wearing course
Comparison of the measured stress spectra induced under normal trafficking on the surfaced deck with the synthetic porous wearing course ZOK at a temperature domain of -5.0 - 23.4°C showed that the stress spectra are influenced only slightly by the level of temperature.

Composite behaviour ZOK wearing course and steel deck
Results of the measurements under the situation of traffic loading showed that above ±16°C the composite of steel deck and wearing course acts as two independent beams. Below this temperature they act more in a composite manner. The stress reduction in the steel deck however is always about the same.

CONCLUSIONS

The results of field tests on a real bridge confirm the composite action of the wearing course with the orthotropic steel deck plate in reducing the stresses in the steel from their values in the unsurfaced plate. The effect of the new developed polyurethane wearing course ZOK is hardly influenced by temperature. At the investigated temperature domain -5.0 - 23.4°C the measured stress spectra are almost the same. Based on the measured dynamic stiffness moduli it is to be expected that also outside this temperature domain the stress reduction will be the same. Further field measurements are necessary to confirm this assumption. Anyhow it is to be expected that this polyurethane porous wearing course can result in a simple rule for calculating stress reduction in steel decks.

ACKNOWLEDGEMENTS

The author wishes to thank the Department of Public Works of the City of Rotterdam and Bolidt Synthetic Products & Systems for their permission to publish this paper.

REFERENCES

1. Kolstein, M.H., Dijkink, J.H.,'Behaviour of Modified Bituminous Surfacings on Orthotropic Steel Bridge Decks', 4th Eurobitume Symposium, Madrid, (1989).
2. Smith, J.W.,'Stress Reduction due to Surfacing on a Steel Bridge deck', Int. Conf. Steel and Aluminium Structures, Elsevier 806-814, Cardiff, (1987.
3. Hopman, P.C., Coorengel, R.,'Development of a Drain Asphalt Concrete Wearing Course for an Orthotropic Steel Bridge', 4th Eurobitume Symposium, Madrid, (1989).
4. Bolidt Synthetic Products & Systems,'Product data sheets Bolidt- ZOK, Primer PU 2580, Membrane D60 -', Alblasserdam, (1991).
5. Kolstein, M.H.,'Influence of Different Wearing Courses on the Stresses in the Orthotropic Deck of the Calandbridge', Stevinreport 6-92-4, Delft University of Technology, Faculty of Civil Engineering, Delft, (1992).
6. Kolstein, M.H.,'The Importance of Wearing Courses during the Design of Orthotropic Bridge Decks', Researchdag 1992 Staalbouwkundig Genootschap, Rotterdam, (1992).

15 A NEW TYPE OF HOLLOW BRICK ELEMENT 'ORTHO' FOR GRILLAGE AND RIBBED SLABS

P. Hajek
Czech Technical University of Prague, Czechoslovakia

A new type of hollow brick element for ceramic reinforced concrete floor slabs has been developed. The triangular shape of elements enables realization of flat-slab floors with ribs in two directions (grillage slabs) or with ribs in one direction (ribbed slabs). To prove the viability of the construction method, a test of the loading capacity of a typical grillage slab 'ORTHO' was made. The paper describes the design principle of 'ORTHO' floor slabs and the experimental test. Results from an experiment are presented and discussed.

INTRODUCTION

Ceramic reinforced concrete ribbed slabs with hollow brick elements are widely used in building engineering in view of the flexibility design, good physical properties and reasonable cost. The hollow brick elements are used as a hidden formwork for reinforced concrete ribs and their homogeneous flat ceramic lower face is suitable for plastering. Generally the ceramic construction materials are considered to be ecological.

Commonly used hollow brick elements have a quadrangle shape which enables realization of floors with ribs in one direction (ribbed slabs). A new form of hollow brick element, suitable for realization of ceramic reinforced concrete floor slabs with ribs in two directions (grillage slabs) has been developed at the Czech Technical University of Prague (authors of patent: Petr Hajek, PhD, Vladimir Zdara, PhD). The hollow brick element 'ORTHO' type A has square form with grooves in the diagonal direction. The hollow brick element 'ORTHO' type B represents one half of ORTHO A element i.e. a triangular form. The shape of hollow brick elements type A and B (Figure 1) enables realization of grillage as well as ribbed slabs and their combination in the design of floors (Figure 2). The design principle of grillage and ribbed slab is shown in Figure 3 and Figure 4 respectively(1).

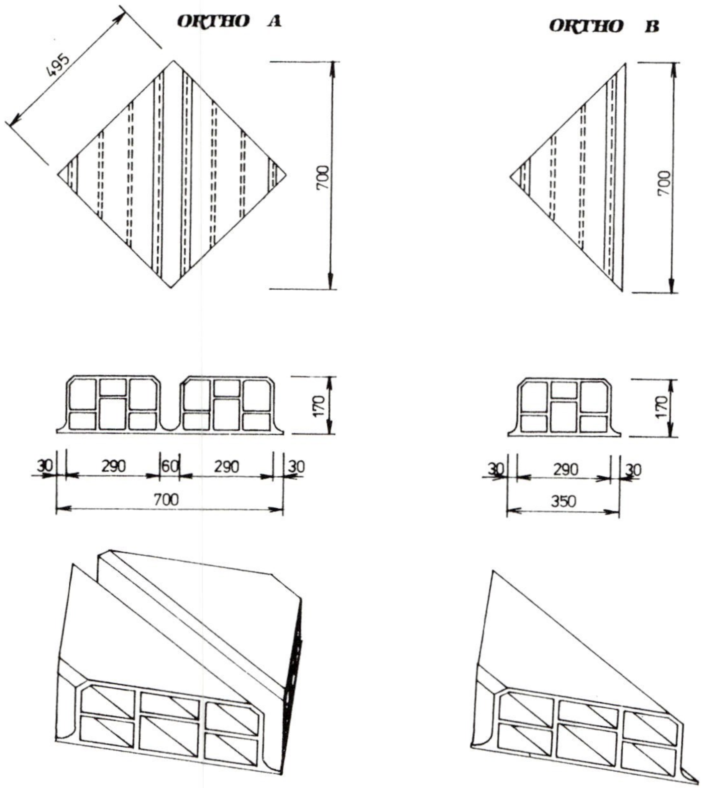

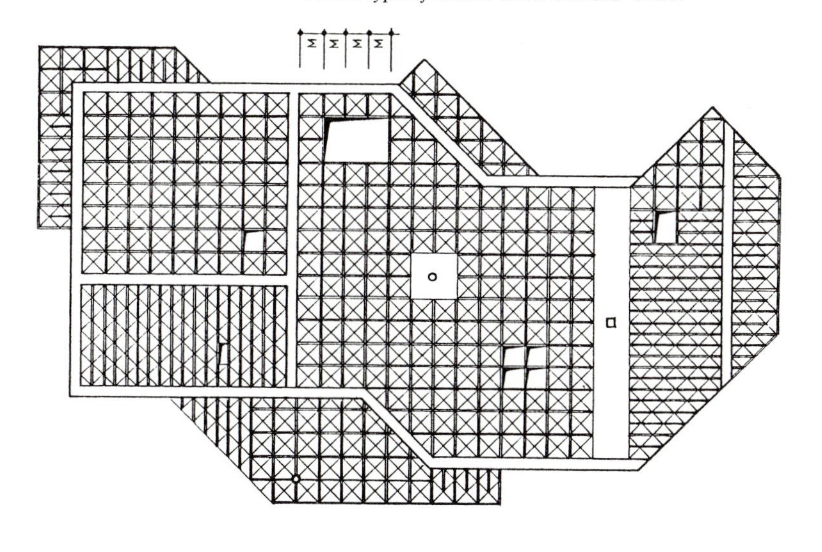

Figure 2. Variability of ORTHO floor slab

Figure 1. Hollow brick elements ORTHO A 700/170/60 and ORTHO B 700/170/60

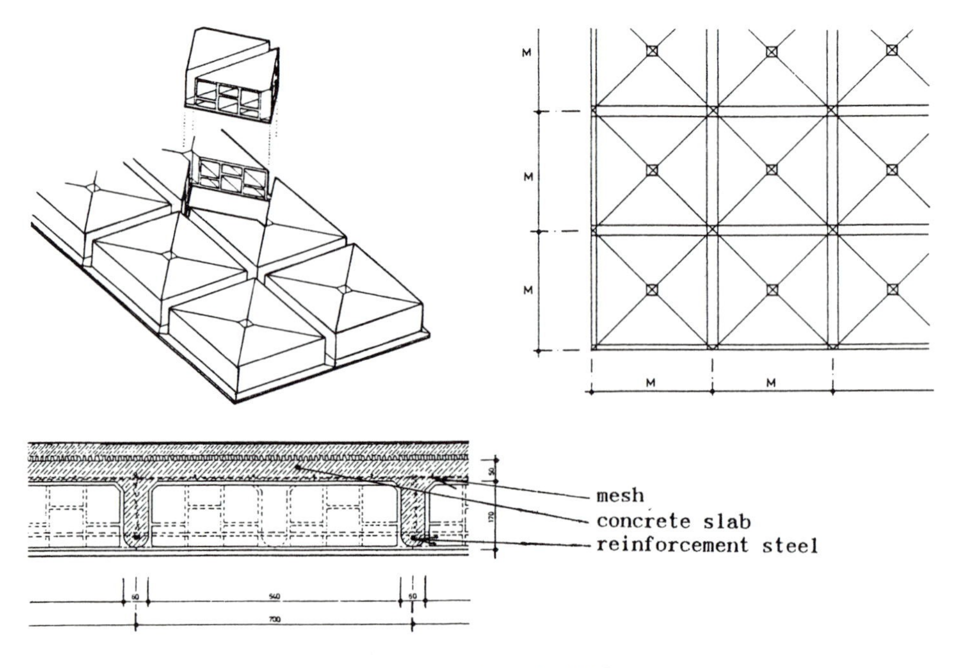

Figure 3. Grillage slab ORTHO

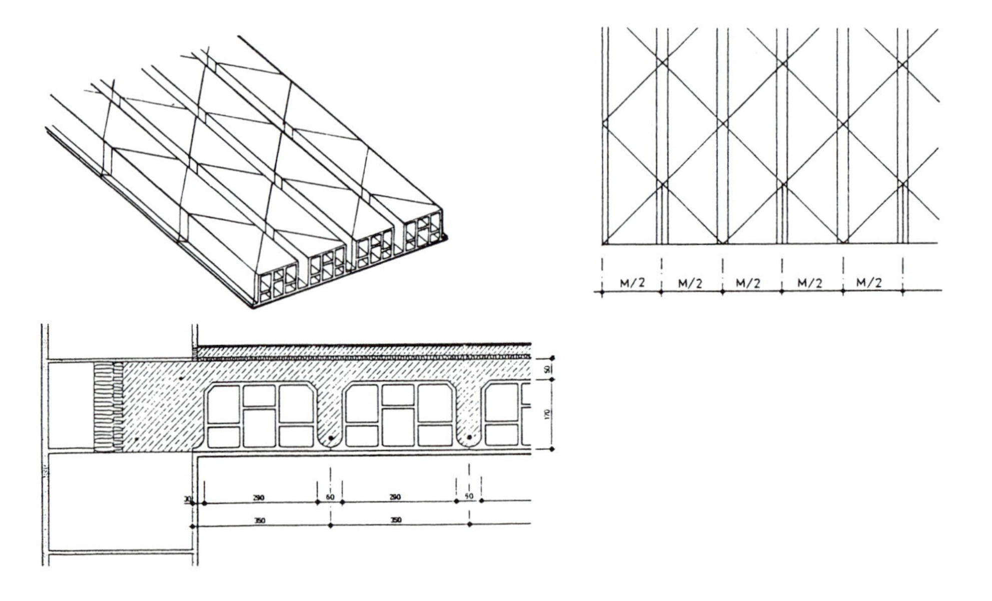

Figure 4. Ribbed slab ORTHO

Grillage slab
The hollow brick elements 'ORTHO' create a hidden formwork for ribs in two perpendicular directions. Using elements type ORTHO A 700/170/60 and/or ORTHO B 700/170/60 the axial distance of ribs in both direction is 700 mm, width of concrete ribs is 60 mm and height of the ribs is 160 mm plus thickness of concrete slab on the top of ceramic elements. The minimum covering concrete layer is recommended for grillage slabs is 50 mm. However using other size of elements, it is possible to create a square axial grid with different distance of ribs and different cross sectional properties. The main bearing steel reinforcement is placed in the ribs on spacer beds. The covering concrete slab is reinforced by a steel mesh 200 x 200 mm from steel bars 3 - 5 mm. Shear reinforcement can be added if necessary.

Ribbed slab
The hollow brick elements 'ORTHO' create a hidden formwork for ribs with axial distance of 350 mm (using ORTHO A 700/170/60 and/or ORTHO B 700/170/60). The thickness of covering concrete slab recommended is a minimum of 30 mm. The other cross sectional parameters of ribs are the same as in grillage slabs.

The grillage/ribbed 'ORTHO' slab structure enables a variety of designs of floors, easy realization of openings of arbitrary sizes, as well as the construction of balcony cantilevers in all direction with effective solution of thermal problems. The design variability of 'ORTHO' floor slab is shown in Figure 2. In comparison with other types of floor slabs, this construction results in very good thermal, acoustic and fireproofing solutions.

The 'ORTHO' floor slab requires relatively low amounts of concrete and reinforcement, and thus is lightweight and the costs are also relatively low.

The test production of hollow bricks 'ORTHO' started in Brickwork Hodonin (CSFR) in the autumn of 1991.

EXPERIMENTAL TEST

In order to prove the viability of the 'ORTHO' floor slab structure, a test of the loading capacity of a typical grillage slab was made in the Testing Laboratory of the Research Institute for Building Structures in Prague in spring 1992 (the test was sponsored by grant T096/42-733, REDES Ltd., Prague).

Parameters of tested structure
Geometry
The design size of tested structure was 4.5 m x 5.2 m, the total thickness of slab was 220 mm (see Figure 5). The ORTHO B 700/170/60 elements were used for creating 5 internal ribs in x-direction and 6 internal ribs in y-direction. The width of the edge ribs was 180 mm in both directions. The thickness of upper concrete slab was 50 mm. The control measurements of geometry were: length 5.23 m, width 4.51 m and thickness of the structure 0.218 m.

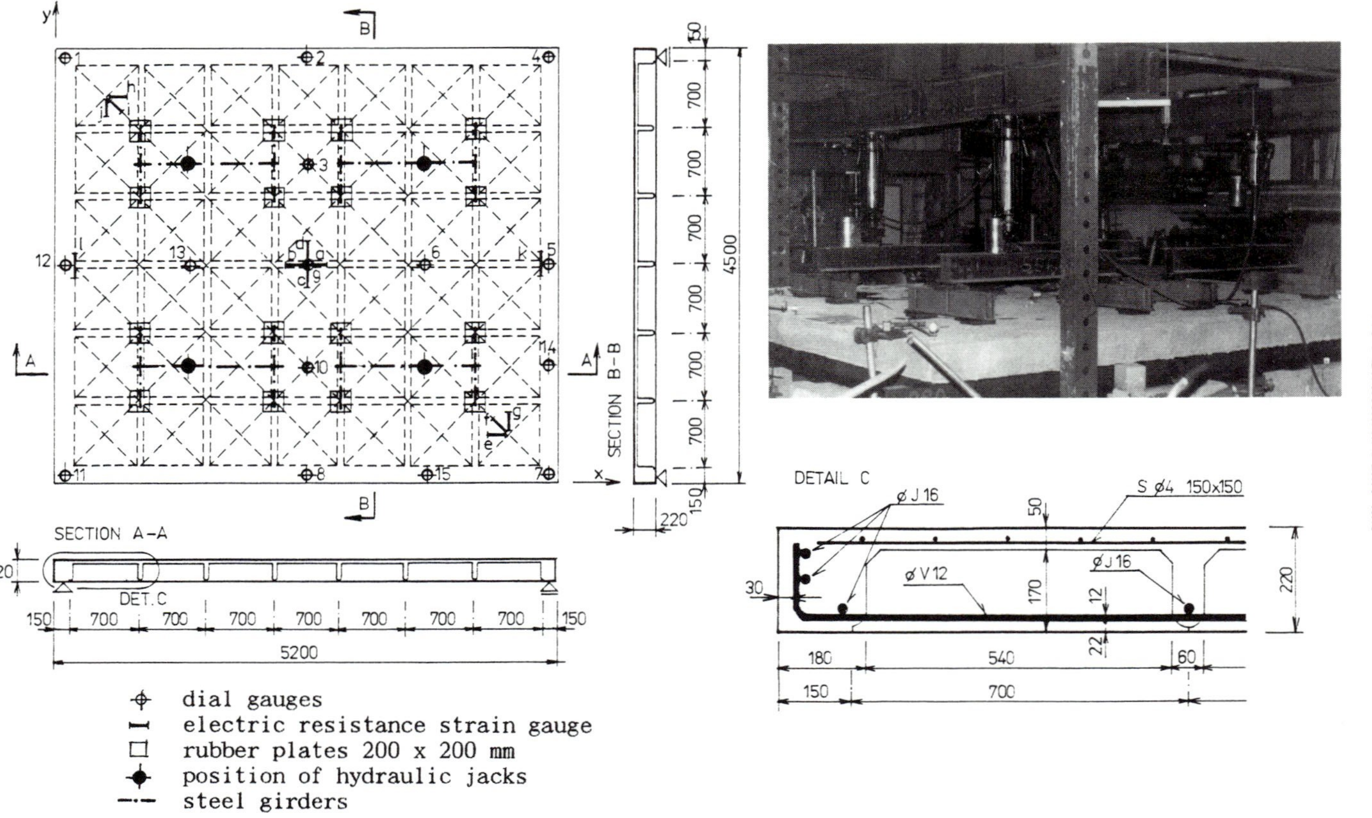

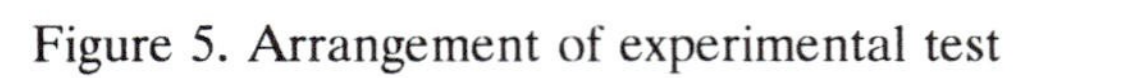

Figure 5. Arrangement of experimental test

Materials

The reinforcement in x-direction was 12 mm type V (10425), in y-direction 16 mm type J (10335). The edge ribs in x-direction were reinforced with one bar V 12 in the bottom and with one bar V 12 in the upper part of the rib. The exact position of steel bars is evident from Figure 5. The edge ribs in y-direction were reinforced with one bar J 16 in the bottom and with 2 bars J 16 in the upper part of rib. The top concrete slab was reinforced by steel mesh 150 mm x 150 mm from bars 4 mm in diameter. There was no shear reinforcement in the tested structure.

The concrete used for the model floor slab was mixed from cement PC 400 (430 kg), aggregate fraction 0/8 (880 kg), aggregate fraction 8/16 (950 kg) and water (175 dm^3).

Material properties obtained from tests of concrete control specimens and steel bars are presented in Table 1. Young's modulus of concrete was obtained from on the prism specimens 100 x 100 x 400 mm.

Table 1. Material properties

Concrete:	specific weight:		2 250 kg/m^3
	cube strength - 14 days:		22.5 MPa
	- 35 days:		37.5 MPa
	Young's modulus:		26 540 MPa
Steel:	type of steel:	V (10425)	J (10335)
	diameter:	12 mm	16 mm
	cross section:	113.1 mm^2	201.1 mm^2
	yield point:	461.3 MPa	400.4 MPa
	breaking strength:	638.1 MPa	594.3 MPa

Test rig

The test rig is shown in Figure 5. The slab was simply supported along all edges by the concrete blocks. The corners were not fixed against lifting. The load was applied on the top of the concrete slab at 16 points (intersection of ribs) using rubber plate of size 200 x 200 mm. The structure was loaded by 4 hydraulic jacks acting through a system of steel girders. This test load approximately represents a uniform load acting on the top of the floor slab structure.

The self weight of the structure, including the test instruments, was 2.571 kN/m^2.

Measuring instruments

The jack loads were applied by means of a controlled hydraulic pump HAPZ. The deformations were measured at 15 points (1 - 15) by means of dial gauges. The strains at the top of the concrete slab were measured at 12 points (a - l) using electric resistance strain gauges. Positions of all measuring instruments are shown in the Figure 5.

Test procedure
The load was gradually applied in 13 steps to the load which represents a uniform load on the top of 15.98 kN/m^2 (load increment 1.23 kN/m^2). After load steps 1 to 5 and 13 the unloading of the structure to 0 kN/m^2 was performed. The loading was then continued up to the 21 kN/m^2. At this stage the deflection was over corresponding serviceability limit. However there was no evidence of structural failure.

DISCUSSION OF EXPERIMENTAL RESULTS

Deflections: Resulting history of deflections in the measured points is presented in the Figures 6 and 7. The symmetry coincidence is evident from the graphs.

Strains: Resulting history of strains in the measured points is presented in an accompanying paper(2). Again very good symmetry coincidence was obtained.

A theoretical analysis of tested structure was made using design method for grillage slab structure (using Czech code CSN 73 1201). The specified design load of this structure was determined to be 6.15 kN/m^2 (excluding the self weight of the structure).

Considering the calculated value of specified design load, it is possible to make a comparison of theoretical and experimental results and to formulate some conclusions.

(a) It is evident from Figure 6 that by the test load 6.15 kN/m^2 the total deflection in the centre of the structure (measure point 9) was 5.28 mm and the plastic deflection was 1.88 mm. Hence the relative deflections of the structure are 1/824 and 1/956 (related to shorter and larger span respectively).

(b) Applying the load 15.98 kN/m^2 the deflection in the centre of slab was 35.7 mm. This load represents 2.6 times higher load than the specified design load. The relative deflection of the structure at this stage was 1/122 and 1/141 (related to shorter and larger span respectively).

(c) The load test was stopped when a load of 21 kN/m^2 was applied, however there were no evidence of structural failure. This load represents 3.4 times the specified design load obtained from the theoretical analysis. At this stage the elastic and plastic deflections were over the corresponding serviceability limits. The structure behaved at this point as a membrane which is fixed to the edge frame represented by the edge ribs.

(d) The proportion of plastic deflections due to elastic deflections was quite high comparing with concrete slab without ceramic hollow brick elements. It is assumed that micro cracks in ceramic elements raised by the deflections of the loaded slab. These micro cracks cannot close entirely during by the

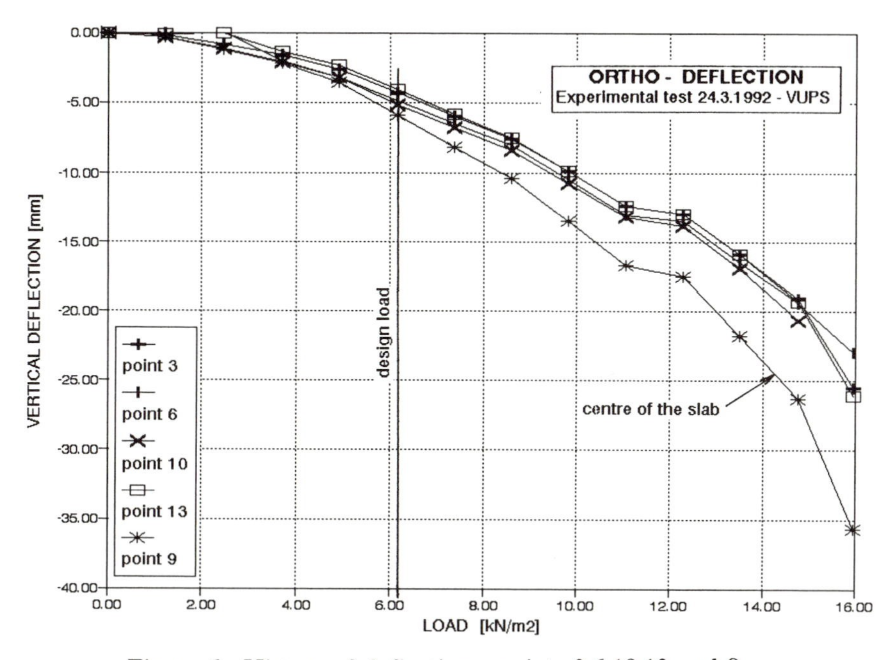

Figure 6. History of deflection - points 3,6,10,13 and 9

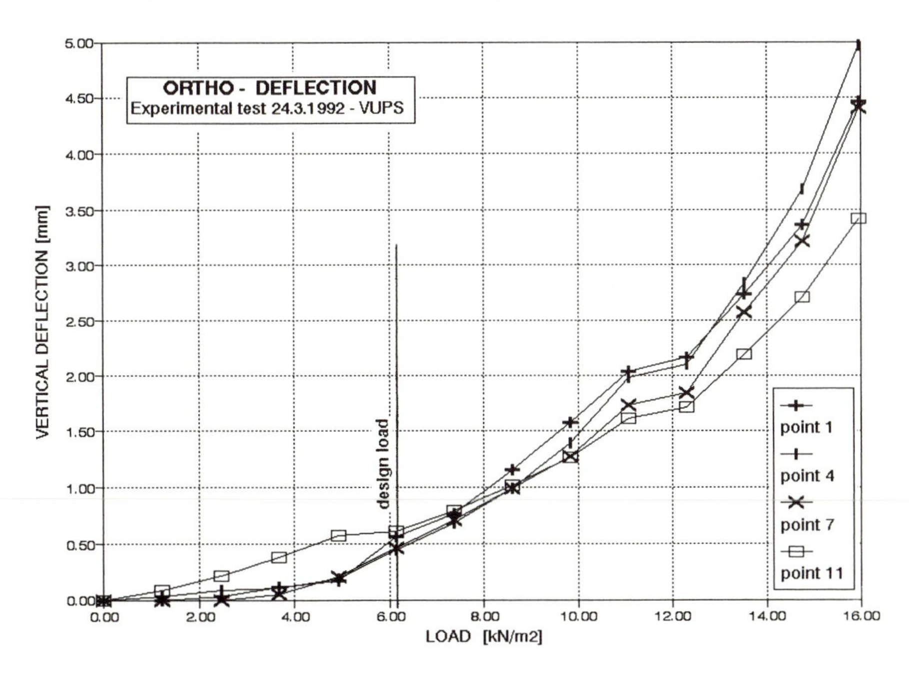

Figure 7. History of deflection - lifting corners, points 1,4,7,11

unloading step, and thus the whole structure displays rather large plastic deflections.

CONCLUSIONS

1. The experimental test demonstrated the very good statical behaviour of grillage 'ORTHO' floor slabs.
2. The statical behaviour of the grillage 'ORTHO' slab is by the extreme loading (over the limit stages) changed to the membrane behaviour. This fact is very important from the point of view of reliability of this structure when used in extreme situations (earthquakes, blasts e.t.c.).
3. The vertical deflections of the structure were relatively very low. This shows quite a significant influence of interaction between ribs and concrete slab which decreases torsional effects in grillage ribs. Also the effect of interaction of ceramic hollow brick elements with concrete is considerable.
4. In order to decrease the cost of the structure, further optimization of the design method is useful and recommended.

Acknowledgements
The research and the test reported in this paper were partially supported by the Czech Ministry of Economy under Grant No. T096/42-733.

REFERENCES

1. Effective floor slab structure for buildings ORTHO, Research report, Redes Ltd, Prague (1991)
2. K S Virdi, P Hajek: Non-linear behaviour of grillage slab 'ORTHO', Building the future., proceedings, Brighton (1993)

16 NEW TECHNIQUES AND DESIGN POSSIBILITIES WITH CAST METALS

C.R. Todhunter
Castings Technology International Ltd (formerly Steel Castings Research and Trade Association) UK

This paper will put into perspective the use of cast metals and introduce new techniques and resources which tilt the balance of cost and performance away from wrought and fabricated steels towards cast iron and steels, especially in structural applications. During the eighties, British foundries took the lead in producing cast steel connections for North Sea oil platforms. They had to prove that their method was more reliable and less expensive than conventional welded connections. In the nineties, a few architects and engineers have been applying similar principles to buildings, both in main and secondary structures.
For some the use of castings has passed beyond the stage of 'innovative and risky' to 'commonplace and safe'. But for most, ignorance and prejudice need to be cleared away before the problems and obstacles of using these materials can even be discussed.
The construction industry has a massive momentum and any potential change of material usage faces a vicious circle: Designers don't know about it so there is no demand. Without demand, there is no market and no marketing. Because there is no information, designers cannot find out about it. Working from the standpoint of a designer, it has been possible to go out and actively find out about the industry and take the information to prominent practices by running seminars and making a fifty minute training video generally available.

INTRODUCTION

All metals when first smelted are in cast form. The notion of rolling the metal out into standard profiles did not really gain ground for a hundred years after Abraham Darby first smelted iron from coke in 1706. Sixty years later his grandson entered the structural field by building the famous Iron Bridge across the Severn. This demonstration of lightness and beauty was so impressive that soon the material was adopted with enthusiasm and brought about a vigorous move to the use of this exciting material in buildings and bridge work.

Right up to the beginning of this century, cast iron was still more cost effective for members in compression such as columns and there are many structures still to be seen with cast columns supporting steel or wrought iron beams.

Why did wrought iron and later rolled steels so completely take over in structural applications? As engineers understood more and more how to calculate the strength of structures, there was a natural move towards high strength and better quality materials. This exaggerated the deficiencies of cast iron, namely its lack of strength in tension and and the risk of brittle fracture as a result of notches or cracks in the material. Malleable Iron was less prone to these defects but in the nineteenth century was expensive. The current method of making malleable or spheroidal graphite (s.g.) iron was invented in 1946; too late to reverse the trend away from using cast iron in building. For many application, s.g.iron costs very much less than steel, and it is now a very under used material.

Twenty years ago, Peter Rice and Renzo Piano built the Pompidou Centre in Paris and they applied a simple principle of using *tubes* for all compression members, *rods* for all tension members and *castings* for all connections. The columns were also cast, perversely, in the form of tubes because standard tubes were not available which would take the loading.

By contrast, the unfortunate separation of engineering and architecture, whereby the building frame is wrapped and deeply hidden, allows architects to get away with poor integration of structure, services and cladding and engineers to design ugly structures. The exceptions to this are noteworthy and include work by Ove Arup and his circle, Peter Rice and the English High Tech school. Integration of construction is also a strong theme among the new traditionalists.

THE INFLUENCE OF DIFFERENT MEANS OF PRODUCTION

Production methods using rollers, gears and cams were proved during the model T Ford era to be continuous, fast and cheap for mass production. These processes completely dominate the building industry today. The formulation of the '*machine aesthetic*' by Walter Gropius and his associates at the Bauhaus, Dessau in 1925 gave people a visual style to go with the rotary processes. Buildings were to look as if they came off the machine in long lengths and were cut to size to fit the site. All joints were to be made by invisibly welding the chopped off ends of the components, which were mostly thin sticks or sheets. Between 1925 and the present day, the form of motor cars have radically changed and the model T Ford is in the vintage class, whereas in the field of building, there are still many buildings being put up which are very close to the Bauhaus style.

This production method, aided and abetted by the style, severely limited the variety of possible structural forms.

But new technologies, such as DNC machining, have been invented to produce *three* dimensional forms, while still allowing efficient production by batch, rather than continuous (rotary) processes. The motor industry and plastics technology have led our tastes firmly away from the thin flat sheets, thin sticks, and the two dimensional materials which we call the *chopped off look*. The construction industry is generally restricted to a palette of two dimensional materials. We are locked into these not only by the common production methods but also by design codes, which are complete when it comes to rolled steel products and the like, but not very useful for cast metals and complex three dimensional shapes. Cast metals flourished in the days of proof testing before calculation methods were well developed.

It is healthy to have a certain number of sceptics in the professions, but as Churchill said referring to the neutrality of the Swedes during the war, if you sit on the fence too long the iron eats into your soul. Finite element analysis combined with proof testing promises to release from their fear those who doubt that the industry can deliver reliable products.

SCOPE FOR CHANGE TO MORE THREE DIMENSIONAL MATERIALS

Underlying demand for more varied structural forms
During the seventies, Prince Charles was not the only person who was expressing dismay at the buildings of the previous decades. There was a common unease about the conventional wisdom of the professionals. This has now grown to a rowdy chorus, with the building professions taking as much part as anyone. The acclaim for Santiago Calatrava's exhibition last year at the RIBA is as much a part of it as Pasternoster Square.

New technologies, old resources
The foundry industry in Europe is a shadow of its former self, in tonnage output terms. The skills are still there but the procurement routes are poorly defined through neglect. The new technologies, which are described below, have many applications.

Misconceptions about castings
Myth 1: Castings are thought to be brittle and unweldable.
Fact 1: Cast steel can have the similar properties to grade 50D steel. S.G.Iron is not brittle but has to be welded carefully in controlled conditions and then heat treated. TIG and MIG welding of stainless steel and aluminium are now commonplace.
Myth 2: Castings are thought to contain unacceptable defects whereas rolled products are not.
Fact 2: Castings are no more susceptible to defects than fabrications and forgings. All three need testing. High integrity steel foundries have proved themselves in the most punishing and critical applications such as the nodes for North Sea oil rigs.
Myth 3: Engineering codes do not cover the materials or complex geometry of cast shapes so the engineer cannot prove them by calculation.
Fact 3: Finite element analysis by computer of cast forms shows up the most highly stressed points. This not only gives the stresses but also allows the form to be optimised.
Myth 4: It is thought that a high degree of repetition is needed.
Fact 4: Foundries will often make a one off prototype just for tender purposes.Patterns are adaptable and their cost is not necessarily high. Many

foundries carry stocks of patterns which can easily be altered. Direct machining of prototypes and/or patterns by DNC link promises to reduce greatly the cost of pattern making for small runs.

CURRENT TECHNOLOGY

Prototyping and pattern making

Castings Technology International have installed software and hardware including three dimensional modelling tools, finite element analysis and a DNC link to several powerful profiling machines and robots. These can produce prototypes, dense polystyrene patterns or models of components. The purpose is to reduce the cost of pattern making for prototypes and to enable patterns to be easily modified. Wooden patterns take up a lot of space in most foundries, whereas to hold them on a compact disk takes virtually none.

Analysis for stresses and cooling shrinkage

Most metals shrink significantly on going from the liquid to the solid state. This means that extra 'feed' metal must be placed at strategic locations on a casting to avoid the formation of unwanted cavities within the casting. Modern simulation technology enables foundries to carry out 'digital prototyping' before any decisions to create patterns or sample castings are made. This allows more freedom to design quickly and gives added assurance of a 'right first time' product.

Structural analysis

The process of casting design allows you to put the material where it is most needed. Current methods of structural analysis deal with block like products, but new methods of finite element analysis can treat a complex form as a minute space frame and thus the load paths can be streamlined. Similarly, if you want a brick to fly, you can either change its shape and streamline it or you can just chuck it harder.

Testing

Many different types of mechanical and physical test can be carried out, ranging from simple tensile tests on sections of castings to full scale structural testing of parts or assemblages of parts. An example of testing on a large scale was the work carried out by the nuclear industry, who use a significant number of castings at the 'hot end' of the power plants. S.G.iron nuclear waste containment flasks have to be exhaustively tested. Many small and medium scale tests are regularly carried out on castings by the motor industry, rail transport industry, the mining sector and the valve and pump business on light alloys and cast irons. The tests can cover a wide variety of parameters including high and low temperatures, corrosive or erosive environments, dynamic or static situations, material transport phenomena (e.g.pumps and valves) and ballistic tests. Details can be obtained from the authors.

Non-destructive test methods include magnetic particle analysis, ultra sonic and X-ray methods.

Nurturing the craft skills in the industry
Much has been said about the high technology, structural aspect of the industry. Foundries are of many different types, and where structural integrity is not the prime consideration, greater reliance is placed on craft skills. This can produce equally high quality products, and provided the necessary testing is carried out they are none the worse for being low on technology, and may have lower overheads to carry.

For example, decorative ironwork is best produced by a skilled foundryman using a loose pattern of wood or resin, or even a previous casting. He will cut his own gates and risers in the mould and the whole process is very quick and simple. In these cases proof testing is often adequate.

The majority of the foundry industry now serves mechanical engineering rather than construction. They are not used to working under the contractual constraints of the building industry. They sometimes do not appreciate that engineering standards of finish will not do for architectural purposes. They are used to quoting costs from a fully dimensioned drawing of a single component; not working from a vague design intent assembly drawing of a complete assembly. All these are barriers to communication. But procurement routes are being opened up by some of the pioneers mentioned above, and by expanding the demands on some of our major subcontractors in the fields of steel fabrication and planar glazing. Some companies who have made a name in a narrow speciality, such as signage, are perfectly capable of expanding their range of products to, say, structural beams and brackets, lights, bollards and street furniture. Leander Architectural, a company near Buxton has done this and has also been making light conservatory structures, gratings and brackets.

Wherever we have an opportunity to nurture these resources, we should do so or they will be lost.

An introductory training video to the subject including technical details of the main processes is available from CERCI Communications Ltd, 19 Store Street, London WC1E 7BT. This will be showing in the lobby at the conference.

APPENDICES

Design and procurement advice

- *Think about it early*
  In the most common cases, foundry work will be procured through a major subcontractor (e.g. a steelwork fabricator). But ample time should be allowed for prototyping and testing.

- *Family planning*
  Repetition of individual elements is *not* important. But early establishment of a standard approach and geometry in the structure *is*. The total value of the

castings in the structure needs to be high to justify the design and verification time needed.

- *Use a prototyping service*
  CTI provide a bureau service which will develop components from design intent drawings through three dimensional modelling, analysis, full prototypes, testing and pattern making, all prior to tender.
  Use this service to establish typical detailing and viability. If the use of castings will save the client money or add value, as it should; the client should pay.

- *Make sure of your sources*
  Contractors' buying departments do not know about foundries. The design consultant should make sure that he can name, or at least vet, the foundries to tender. The wrong choice of foundry will cost all parties dearly. Engineers experienced in the field will be able to help, the authors of this paper will advise. Trade associations will usually only provide lists of members, not recommendations.
  Do not go 'shopping at Harrods' and then let the castings be bought at 'Woolworths'

COSTS

The balance of economies is very sensitive to the particular application and production route. Outside the building industry, very different cost criteria would apply.The table below is an actual comparison of the cost of cast connections between steel tubular trusses and plated ones, with site welding.

Birmingham NEC - Halls 10 to 12
(Source - British Steel, Tubes division)

Item	Cast joint	Plated Joint	Site weld
End connection 2 per joint	£60	£70	n/a
Site welding	n/a	n/a	£ 90
Inspection	£ 5	£ 5	£ 20
Sub total	£65	£75	£110
Estimated remedial	£10	£20	£100
Extra site production costs. (Scaffold etc.)	-	-	£100
Estimated total	£75	£95	£310

Costs in a particular case will vary very markedly, depending on the amount of fabrication and finishing needed. In real life combining castings with fabrication is likely to be the most economical solution for *many* situations.

Types of foundry
Fine Art Foundries, producing sculpture, mostly in bronze, aluminium or resin by sand or investment casting.

Architectural ferrous and non-ferrous foundries with a stock of reproduction Victorian patterns. These foundries are craft based and this gives them a pricing advantage when integrity of the casting is not the first consideration. Most of them do some light engineering work also.

Cast iron foundries specialising in light engineering applications such as gullies and manhole covers, sand cast.

Die casting foundries doing repeats of small items; e.g.taps.

Investment casting foundries using lost wax and ceramic shell processes for light machine items.

Stainless steel specialists in the food, transport and pharmaceuticals industries.

Structural high integrity foundries producing sand cast and machined items, mainly in steel. Weights up to several hundred tonnes are possible.

METALS

Steels
Structural steels can easily be obtained with equivalent strength to grade 50D steel or more. High strength alloys are increasingly being used. Stainless steel can be cast with a fine sand finish and it can be polished or machined afterwards. Gabriel and Company Ltd of Birmingham is one company specialising in stainless. All steels are cast at high temperatures (e.g.1600 degrees C.) and they tend to flow into small cavities with difficulty; so thin sections need skilled treatment.

Spheroidal Graphite Cast Iron
Should be used far more than it is. Invented in 1946. It is non-brittle and is used for such things as heavy duty manholes. It is nearly as strong as steel but may be much cheaper, and it flows well in the mould. Opinions on welding s.g. iron differ. If welded to steel in the factory and then heat treated it is probably satisfactory. Site welding is definitely not on, so it is generally used with bolted connections.

In every case, a comparison should be drawn between fabrication and casting, as any shape which can be fabricated can be cast . The Pie charts below show a *notional* make up of the costs of a fully cast and a fully fabricated element.

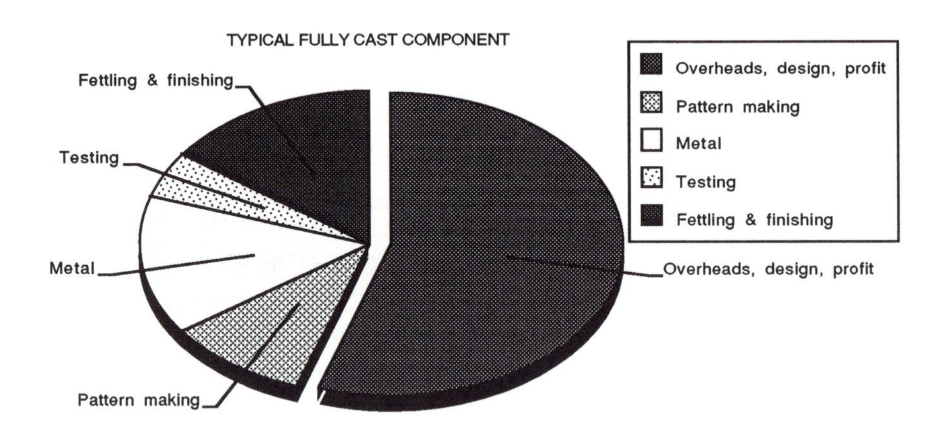

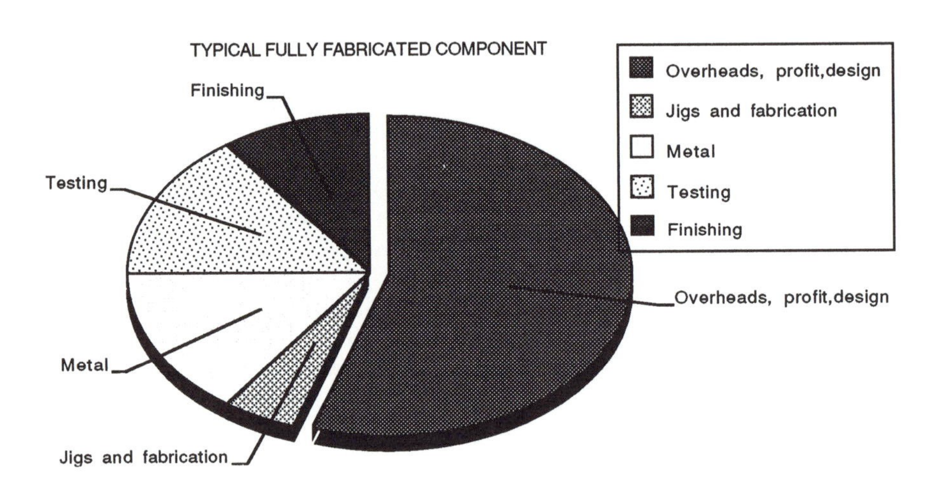

Grey cast iron

The cheapest material for casting at about £1 per kilo, this has the disadvantage of being brittle. Grey cast iron bollards, for example are not satisfactory.

Aluminium

About two and a half times the cost of cast iron - volume for volume - seven times weight for weight. Reduced site handling costs may close the gap still further. There are a number of alloys and most but not all are resistant to corrosive atmospheres. Specify conditions but let the foundry decide which alloy to use.

Aluminium castings take a powder coated finish well but are not suitable for decorative anodising. Polishing tends to expose minor defects.

Bronze

Mainly for sculpture where the casting process itself makes up more of the cost than the metal. Leave the choice of alloy to the foundry unless specifying gunmetal for high stresses or aluminium bronze for extremely corrosive situations. Large sculptures are generally reinforced with stainless steel armatures as building inspectors and engineers will not take the strength of the bronze into account. There is often a good case for proof testing if proving the structure becomes a problem.

A wide range of patination is possible, from polished, through greens, greys and browns to black. Where it will not be touched, resin composites including bronze are often a satisfactory low cost alternative. They take a good patina and will weather like the real thing for your lifetime anyway.

REFERENCES

1. Steel Castings Design Properties and Applications. Ed. W.J.Jackson. SCRATA ISBN 0 900 224 541 (1983)
2. Castings Design Handbook. American Society for Metals. (1962).
3. Copper Alloy Casting Design. Copper Development Association Publication no.76 (1970)
4. Steel Castings Handbook. Steelfounders Society of America. Fifth Edition (1980).
5. Construction Materials Reference Book. Cast Iron by J.Sutherland (Harris and Sutherland) Ed. Doran. Butterworth & Heinemann (1992).

Most current applications for castings are outside the construction industry, but they range from those with high performance, such as Turbines (above) to those with special finishes (below). In between there are plenty of possible applications in building.

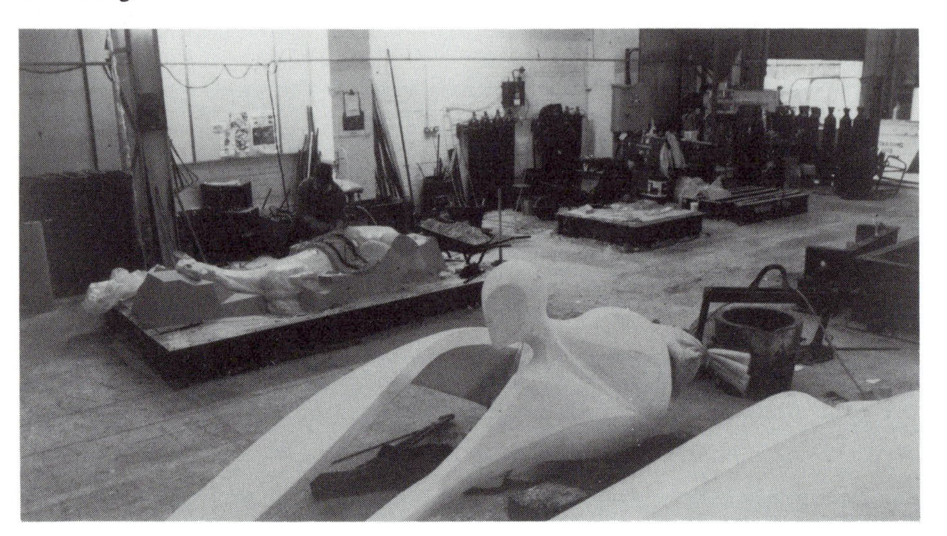

17 NEW MATERIALS AND METHODS OF TESTING DYNAMIC RESISTING ELEMENTS ON THE NEW GENERATION OF SHAPED AIRCRAFT HANGARS

J.A. O'Kon
O'Kon & Company, Inc, Atlanta, USA

Recent advances in structural design and physical testing have enabled engineers to develop a functional and efficient shaped aviation maintenance environment for state-of-the-art aircraft. This new type of shaped hangar structure raises the technology of building construction to that of aviation technology.

SHAPED MAINTENANCE ENVIRONMENTS

Aerospace design and development efforts have produced unique advances in the operational capabilities of aircraft for both military and civilian applications. Advanced aircraft technology has produced changes in the power plants, flight platforms, avionics, and internal guidance systems that have permitted modern aircraft to fly farther and higher, carry more payload economically and enhance their environmental soundness while becoming much safer.

The advances in composite materials for aviation structures, electronics and jet power plants have leaped forward in recent years. However, these advanced aircraft have been maintained in ground based aviation maintenance hangars whose building technology is decades behind the aircraft they maintain. This is like maintaining a BMW 700 series in a log cabin garage.

However, times have changed, recently the design of aviation maintenance environments has taken a great leap forward to bring building advances abreast of the technology of the aircraft which are being maintained.

A new generation of aviation maintenance envelopes have been developed to encapsulate state-of-the-art aviation maintenance activities. These advances include internal environments which balances laminar flow HVAC systems that protect sensitive avionics as well as the health of the workers. AFFF pre-action fire protection systems, specialized 400 Hz and 28 volt electrical generating systems and radio controlled, full coverage, overhead lifting systems. These systems are encapsulated in state-of-the-art maintenance hangars utilizing a geometric configuration that develops a trapezoidal volume to minimize spatial requirements. This truncated trapezoidal volume is made possible through the use of state-of-the-art upward acting hangar doors that create a shaped entrance facade.

The implementation of this new breed of aviation maintenance environments requires a close interface between architectural and structural engineering design, fabrication techniques and physical testing of constructed components.

The new shaped hangar/configuration requires new approaches to structural design in order to satisfy the dynamic loadings on the structural space frame structure, maintain quality assurance and quality control while keeping a tight budget.

Prototypical Design

The hangar used as a prototype for this study was developed as part of a Fuel Cell Maintenance Facility for KC-135 Tanker Aircraft. (Figures 1 & 2) This facility was commissioned by the Mississippi Air National Guard at Meridian Mississippi. This facility is located in the southeastern USA and is part of the U.S. Air Force reserve forces. The KC-135 aircraft has a wing span of 130 ft. 10 in., 136 ft. 3 in. in length and a height of 41 ft. 8 in.

Following are the salient elements which are required to design the unique structure through the use of computer design and drafting, and development of quality control/quality assurance techniques.

Structural Design

The geometry of the 'shaped' hangar dictates a unique configuration of the superstructure in order to accommodate the large dynamic overhanging structure which forms the front facade of the hangar. This overhead structure supports the

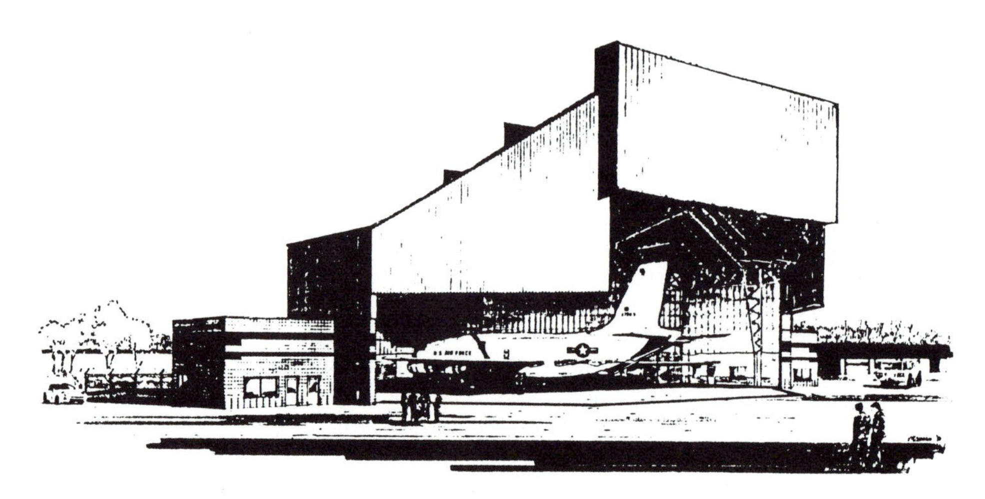

Figure 1. Perspective KC-135 hangar

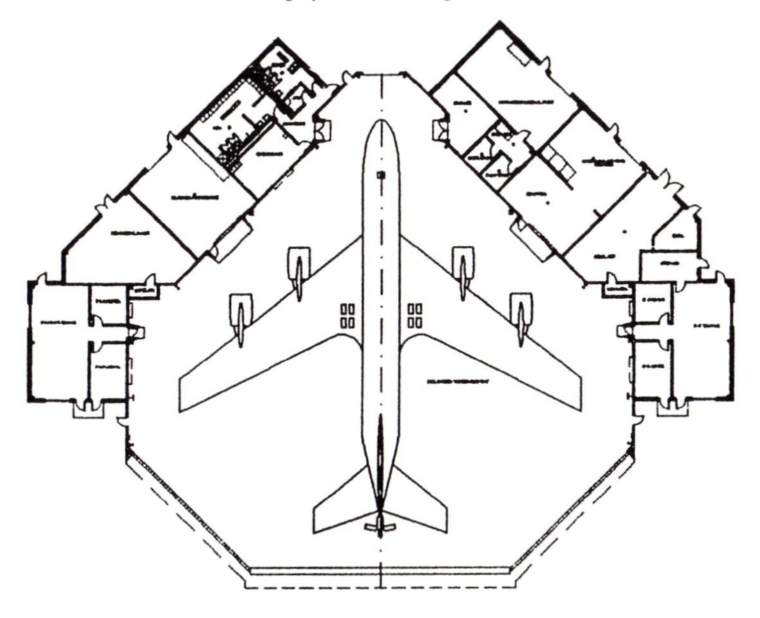

Figure 2. Plan of KC-135 hangar

dynamic door assembly. The structural steel superstructure with its trapezoidal cantilever configuration must resist a large range of loadings including the dynamic loads from the door systems, gravity loadings attendant to dead and live loading, horizontal seismic and wind forces plus the large internal uplift forces generated by hurricane force winds with speeds up to 125 mph.

The geometrical configuration of the hangar envelope is dictated by the size of the aircraft to be maintained. The interior dimensions of the envelope are determined by minimum clearances as required by aviation standards. The standard clearances indicated that the walls and doors are to be located to provide a 15 ft. 0 in. clearance between the aircraft and the enclosure. This formula dictates the shape size of the plan of the hangar and generates the unique truncated shape (Figure 2). The clear height of the hangar is determined by the a 15 ft. clearance above the fuselage and the tail section (empennage). (Figure 3)

When the interior envelope is established and the exterior configuration skin is fine- tuned to include roof slopes, openings and wall finishes. The structural skeleton of the hangar then is configured to follow the vertical and horizontal geometry established by the building envelope. To achieve this goal the optimum structural configuration is developed that will be translated into a low profile three-dimensional space frame.

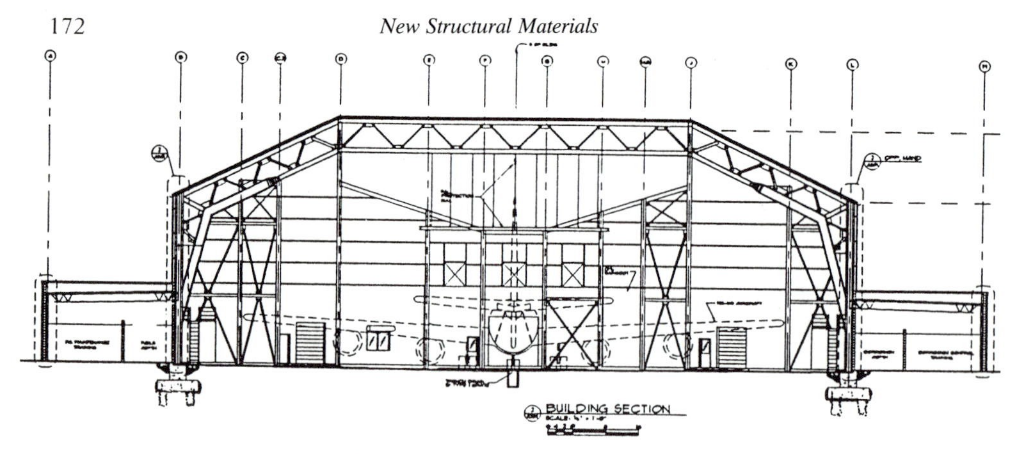

Figure 3. Cross section of hangar

Structural Design Considerations
The plan configuration of the facility was established by the limits of the geometry of the aircraft (Figure 2). The structural system of the hangar uses a three-dimensional space frame whose principal members include:

> *Traverse Bent Truss*: This member spans across the width of the hangar and extends down to the foundation. This member is the principal resisting element for gravity, horizontal and uplift forces. (Figures 3 & 4)
> *Longitudinal Trusses*: These paired trusses occupy the same plane as the transverse truss and cantilever 60 feet across the transverse truss to support the door system (Figure 4).
> *Perimeter Trusses/Tension Trusses*: A series of perimeter trusses were established along door lines, and tension trusses were introduced to resist horizontal forces (Figure 4).
> *Lateral Bracing Trusses*: Trussed bracing were introduced into the side walls to resist lateral forces (Figure 5 & 6).

Computer Applications
The structural system geometry was translated into a three dimensional model using a combination of computer software including STAAD-III and AutoCAD Version 11.0. With this software operating on a 486-50 MHz computer, (Figures 5 & 6) the following criteria were applied.

> *Loading Cases*
> The hangar is located in a region which is subjected to hurricane force winds as well as seismic risk loadings. The structure was analyzed using 18 separate load cases with a combination of dead load, live load, seismic and wind forces. The critical loading case involved uplift forces acting on the superstructure with the hangar doors open to the wind.

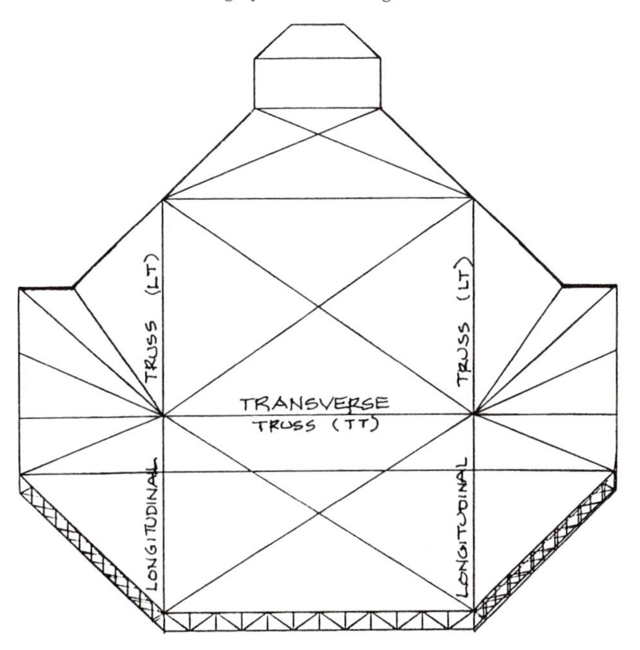

Figure 4. Plan - Space frame

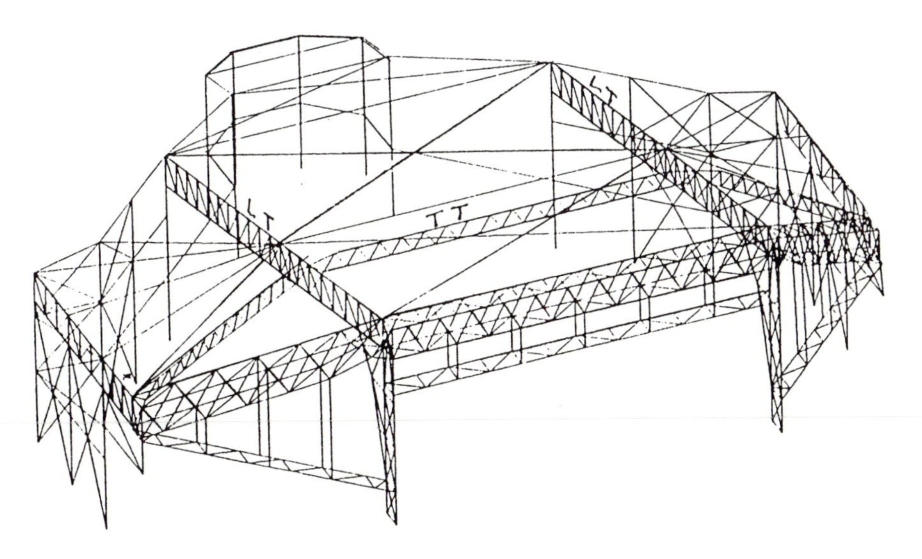

Figure 5. Three dimensional space frame (STAAD-III)

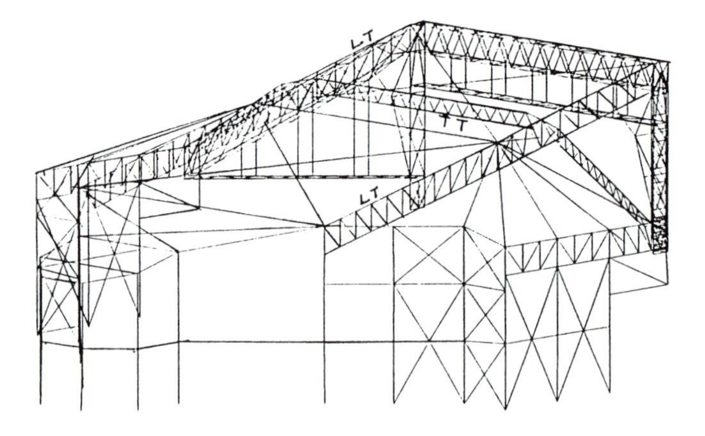

Figure 6. Three dimensional space frame (STAAD-III)

Hurricane force winds produced a more critical loading situation than seismic the force requirements.

Development of Low Profile Structural Members

It is desired to maintain a low profile (minimum height) structural members in order to reduce the height of the hangar. The reduction in height reduces the cost of cladding, reduces wind forces and reduces the interior volume of the envelope for heating and ventilation loadings.

The components of the space frame were analyzed using various structural depths in combination with hybrid structural steel members. The structural components were analyzed using a combination of 36,000, 50,000 and 70,000 psi yield strength steel sections to develop the optimum configuration that produced economy as well as deflection control. A reiteration process was implemented by inserting high strength steel members in areas of high stress in order to reduce the total tonnage while complementing the required geometry. The structural space frame was configured and splice points selected that would produce a self-erecting structure.

Quality Assurance

To assure correctness of design and quality control, two techniques were used to verify the computer analysis.

a) *Approximate Analysis:* A approximate structural analysis was carried out to determine the approximate range of reactions at specific points on the space frame. These reactions were verified by computer results.

b) *Code Checks:* The STAAD-III software includes a code compliance verifications. Each structural member was assessed and evaluated for code compliance including appropriate stress levels for unsupported length.

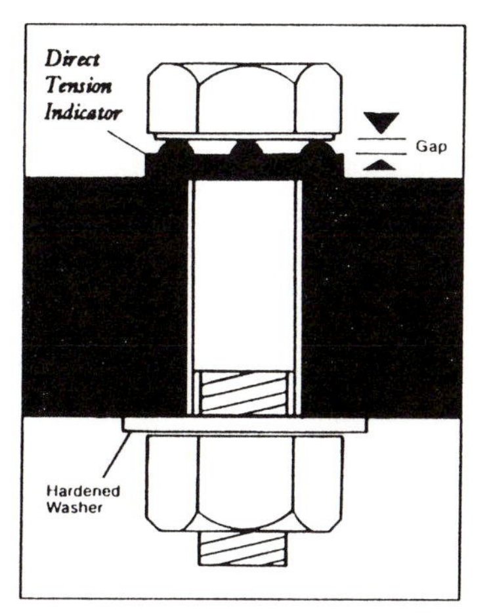

Figure 1. Direct tension indicator before tightening

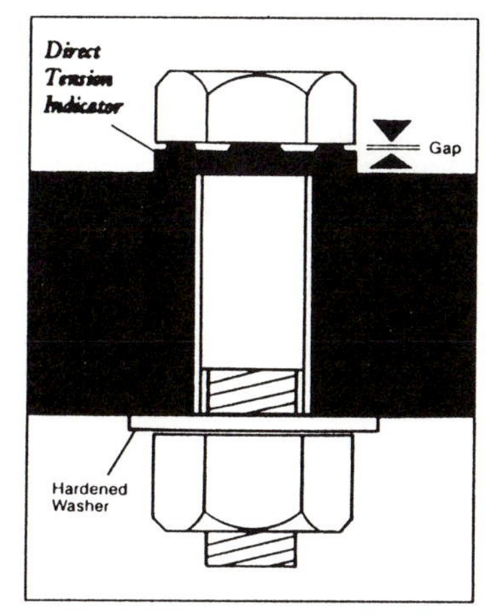

Figure 2. Direct tension indicator after tensioning

Figure 8.

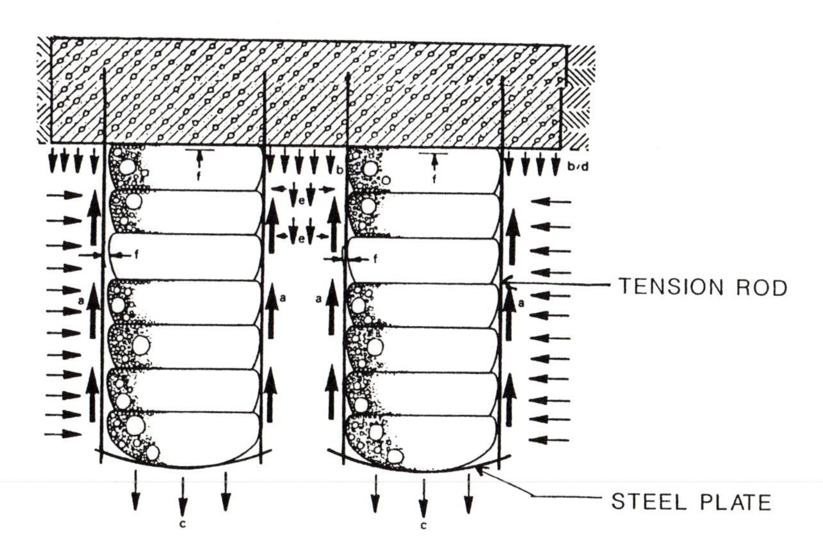

Figure 9. GEOPIER - Free body diagram

Foundation Economics

The optimum steel superstructure system was selected which satisfied geometry, economy and deflection criteria. However, a significant cost factor to be resolved resistance to the large uplift forces generated on the main truss foundation by the transverse bent supports. These forces produced a net uplift of 300,000 lbs. (67,400 newtons) on each side.

Traditional resistance methods such as massive concrete foundations would require 113 cubic yards of concrete on each side. This would require a 10 ft. x 10ft. footing 30 in. in depth. The cost for the total uplift resistance using concrete foundation dead loads was over $200,000.00.

To reduce cost and time, two alternate uplift resisting systems were investigated:

Drilled in helical anchors: These steel anchors are drilled to depths of twenty feet into virgin soil. The shafts are then cast integrally into the foundations. Each anchor would resist 20,000 lbs. of uplift. Fifteen anchors were required per foundation. Cost: $85,000.00

GEOPIER: The GEOPIER foundation soil reinforcement system utilizes a column of compacted crushed stone extending downward approximately 8 ft. into the soil under the foundations (Figure 9). The GEOPIERS are constructed by the densification of crushed stone into a pre-drilled pit. The compaction process densities the stone and interface soil to extremely high density matrix using a high energy tamper. The Geopiers develop 20,000 lbs of uplift resistance. Fifteen were required per foundation. Cost: $32,000.00

Based on economics and time of construction, it was determined to utilize the Geopiers system based on cost and speed of installation (Figure 7).

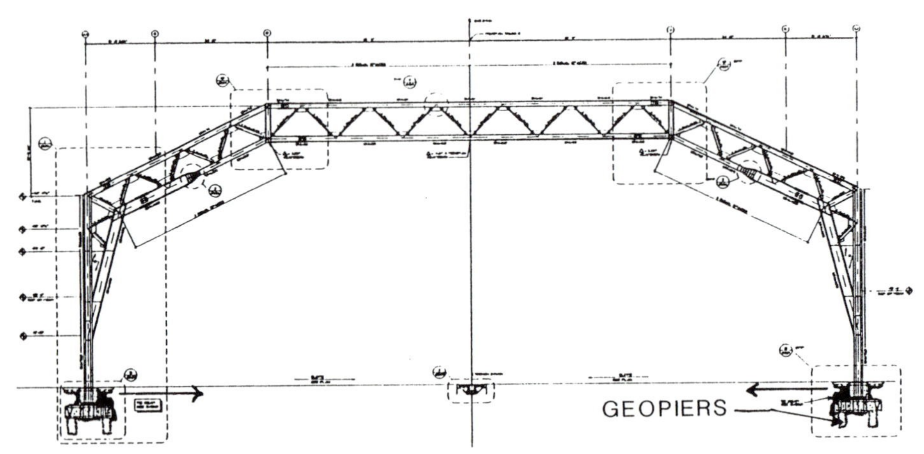

Figure 7. Transverse Truss with Geopiers

INTEGRATION OF DESIGN WITH AUTOCAD DRAFTING

The three dimensional structural design was integrated with AutoCAD software to produce a full set of architectural and engineering drawings. The selected structural design and the foundation uplift system were integrated with the other engineering and architectural disciplines to produce a unique facility (Figures 5,6, & 7).

QUALITY CONTROL/FABRICATION

The use of hybrid materials of construction for the space frame required a high level of quality control during preparation and review of the shop drawings for the space frame. O'Kon & Company developed a color coding system to identify the three grades of structural steel utilized. This color coding system was used on the shop drawings and the system was required to be implemented on the fabricated steel members. The color coding permits identification of the various steel strengths at a glance. The yield strength of each grade of steel was verified by requiring mill test certificates with coupon tests of each grade of steel.

Erection procedures for the steel space frame were carefully reviewed and assessed to assure that a self bracing system would be accomplished. In addition, the Geopier foundation system required extensive shop and placement drawings to assure that the geometrical configuration size and number of the GEOPIERS would satisfy project requirements and could be used for the field verification tests.

It was required that all field connections be bolted, therefore shop welding of connections was maximized. A large number of assembly connections on the steel space frame required full penetration welding procedures. To assure quality control, all welding was carried out under the controlled conditions of the fabrication shop. All field connections were bolted. AWS certified welders were required and a strict program of non-destructive welding was mandated. Ultrasonic and x-ray testing was used to detect flaws or non-conformities in the welding as well as potential de-lamination in plates over 1-1/2 in. in thickness.

CONSTRUCTION METHODS AND TESTING

The project plan was established to minimize the amount of field fabrication and the required physical testing. Therefore, all steel connections were designed to be friction type bolted assemblies. To assure that proper torque was applied to the connection bolts, direct tension indicator washers were specified (ASTM F959). These washers have built-in indicators to verify that proper tension has been applied to each bolt (Figure 8). This system verifies the correct placement of bolts without expensive and dangerous torque wrench testing.

The structural steel space frame was designed to be 'self-erecting' with a minimum of guying supports. Camber was fabricated into the frame in the fabrication shops.

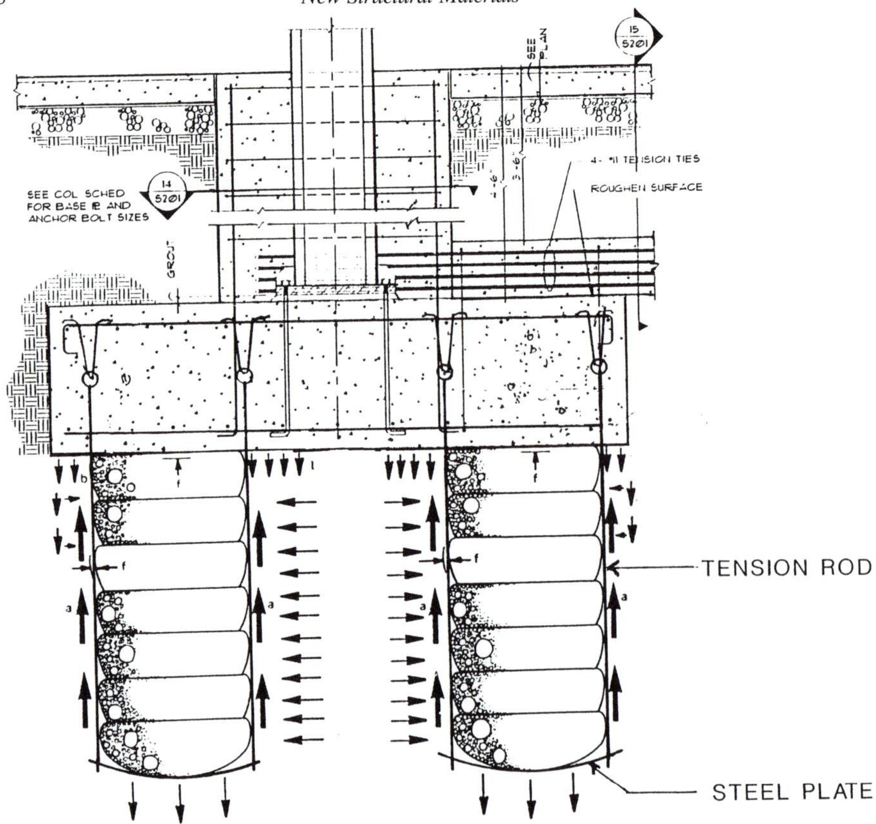

Figure 10. Foundation with Geopiers

An extensive program of physical testing was required for the GEOPIERS system. The subcontractor was required to test a minimum of four GEOPIERS to verify that the system could resist twice the required uplift force.

The uplift tests were carried out as follows:

Installation of Geopiers: Three GEOPIERS were installed in a line. The outside piers would be used as compressive elements and the middle GEOPIERS would be tested for uplift resistance.

Installation of Testing Beams: A pair of structural steel beams were connected to the connecting rods on the compression Geopiers with a bearing plate under the beam. The middle Geopier was used as the tension element

Installation of Testing Jack: A hydraulic jack with a capacity of 100,000 lbs. was placed on a platform on the test beams over the tension pier. The steel rods connected to the bearing plate at the bottom of the Geopiers

were connected to the hydraulic jack
Testing Procedure: The hydraulic jack was activated and the Geopier was loaded in stages. The loading was divided into four loadings over a period of 4 hours. Deflection readings were carried out on the steel rods. The loading was increased until a total of 40,000 lbs. was achieved. The test results of the Geopier were successful.

CONCLUSION

The construction of this unique facility is being completed. The unique design is successfully being implemented with innovative design of the hybrid structural steel space frame which creates a profile/envelope for the aircraft; the use of a self-erecting frame; the use of GEOPIERS and the elimination of field testing have combined to advance the state-of-the-art of aviation maintenance facilities.

PART FOUR

NEW CONCRETE MATERIALS

18 FIBRE COMPOSITES FOR THE REINFORCEMENT OF CONCRETE

J.L. Clarke
British Cement Association, UK

This paper describes the use of resin-fibre composites, in the form of rods or grids, to replace the conventional steel in reinforced and prestressed concrete structures. The materials, which are strong and durable, are particularly appropriate for structures in aggressive environments. Examples are given of applications worldwide. The paper concludes by indicating areas in which study is required before the materials can become more widely used.

INTRODUCTION

When carefully designed, detailed and constructed, reinforced concrete is extremely durable. The reinforcing steel is protected from corrosion by the alkaline concrete environment. However, if the passive layer surrounding the steel breaks down, for example due to penetration of chlorides, corrosion will take place. The resulting rust occupies a volume many times that of the parent metal, cracking and eventually disrupting the concrete protecting the steel. This form of damage can be a significant problem in aggressive environments, particularly in the case of poor workmanship or poor detailing. Examples of structures that may be particularly at risk include marine structures and bridges subjected to deicing salts. The traditional approach adopted by design Codes to overcoming the problem is to use large covers, with relatively high cement contents, and to limit crack widths. However, increasing the cover, while still maintaining a limiting crack width is self-defeating, resulting only in an increasingly limited stress in the reinforcing steel and a lowered load-carrying capacity. In addition, reinforcement is often provided in concrete in situations where it is needed only to control cracking due to shrinkage and early thermal effects. This leads to the paradoxical situation that the reinforcement is required to control cracking but the main reason for limiting the crack width is to protect the reinforcement.

The most significant cause of corrosion is chloride ingress into the concrete. This has been particularly severe for bridges where the use of de-icing salts has led to high levels of contamination. In the U.K., Read(1) reported on the state of the Midlands Link motorway round Birmingham.

'The 11 viaducts stretching over 21 km were originally built in 1972. Within 2 years of construction, decay was observed to have started due to winter salting of the roadway. To date [1989] some £45 million have been spent on repairs, and the total cost of construction was only £28 million when built. It is estimated that over the next 15 years, repairs will top £120 millions'.

Within Europe, the annual cost of corrosion has been estimated at being £1,000 M per year. It should be stressed that while large, these figures represent a small proportion of the concrete construction market.

TRADITIONAL CORROSION CONTROL METHODS

A number of solutions to solving the problems of corrosion are available, each of which has its advantages and disadvantages. It is impossible to conclude that one approach is generally better than another, as one has to consider the type of structure, its method of construction and its intended use along with many other factors. However, they may be divided into three main categories, outlined below.

Improving the concrete

Codes of Practice specify minimum cement contents and maximum water cement ratios for given environments, the intention being to produce a suitably low permeability. Further improvements can be obtained by the use of superplasticisers, that reduce the water demand, or by adding microsilica. The latter is extremely fine and packs into the small voids that usually exist in concrete, resulting in a very low permeability. An alternative approach is to use corrosion inhibitors, such as calcium nitrite which appears to be particularly effective against chloride induced corrosion.

These approaches are all dependent on careful specification and good workmanship will generally ensure a durable structure. However, there may be situations in which, because of extreme environmental conditions or the need for greater durability, more radical approaches need to be adopted.

Protecting the steel

The second approach is to coat the surface of the steel. Galvanizing, dipping standard reinforcing bar into molten zinc, has been available for many years. Applications in the U.K. have been fairly limited though over 1000 tonnes of galvanized reinforcement were used in the National Theatre in London. In the mid-1970s a major study of coating materials was undertaken by the Federal Highway Administration(2) who concluded that fusion bonded epoxy coatings were the most suitable. Production plants have been set up worldwide. The material is used widely for bridges in America and has recently been used for railway bridges in the U.K.

One final option is to apply cathodic protection to the steel. This technique, widely used for steel structures, increases the electrical potential of the reinforcement to a level at which corrosion cannot take place. Two different methods are employed, an impressed current and the use of sacrificial anodes. In the first the structure is connected to the negative terminal of a DC power source, ideally using an anode which does not corrode. In the second the reinforcement is connected to anodes with a more negative corrosion potential than steel, such as zinc or aluminium. The current is reversed and corrosion now takes place at the anode, which is gradually used up. In both cases electrical continuity of the reinforcement is required. Most concrete applications(3) to date have been for the repair and rehabilitation of structures that have already suffered corrosion damage.

Replacing the steel
The simplest approach, though a very expensive one, is to replace the standard reinforcing bar with one made of stainless steel. In the U.K., plain and ribbed bars are available in the same range of sizes as normal reinforcement. Because of the high cost of solid stainless steel reinforcing bars when compared to normal bars, attempts have been made to develop a bar with just a stainless surface but with limited success to date.

The more radical approach is to replace the steel by an artificial fibre, in the form of a continuous strand or as a resin-fibre composite rod. [The use of short, chopped fibres, which chiefly affect the properties of the fresh concrete and have little or no influence on the ultimate behaviour are not considered in this paper]. A number of suitable fibres, with high ultimate strengths and adequate stiffnesses, are available along with a wide range of resins. Some composites have been developed to such a stage that they are produced on a commercial basis. Examples are reviewed below.

RESIN FIBRE COMPOSITES

Figure 1 shows typical stress-strain curves for aramid, glass and carbon fibres. All have ultimate stresses well in excess of that of reinforcing steel. They have elastic moduli that range from about 35% of that of steel to perhaps 50% stiffer. One significant difference between these high grade fibres and steel is that the stress-strain curve does not show any sign of plasticity at high stresses.

Although the fibres tend to be resistant to corrosion, most applications have been with bundles of fibres encased in a resin. This improves the protection against mechanical damage and may simplify the problem of bonding it into concrete. However, because the fibres occupy only a proportion of the cross-section, the effective ultimate stress in the element and the effective elastic modulus are both reduced. Table 1 gives the ultimate strengths of various materials.

Because the elastic modulus of the resin-fibre composites is generally significantly less than that of steel, most of the major applications to-date have been for prestressing. Examples include the Ulenbergstrasse bridge in Dusseldorf, a two span highway bridge built in 1986, and the Marienfelde footbridge in Berlin(4). Both were post-tensioned using Polystal glass fibre tendons, developed by SICOM of Cologne. In Japan a small highway bridge has been built at

Table 1. Comparison of materials in terms of ultimate stress

Material	Ultimate Stress N/mm^2
Parafil F and G [Aramid rope]	1930
Tokyo Rope [carbon fibre strand]	1770
Prestressing steel	1600-1860
Arapree [Circular cross-section]	1700
Polystal [glass-fibre composite]	1670
JITEK [glass or carbon fibre composite]	1000-1600
Arapree [Rectangular cross-section]	990-1190
Prestressing bar	1000
Kodiak [glass-fibre reinforcing bar]	760-1030
Nefmac [glass & carbon fibre composite]	625-700
High yields reinforcing steel	460

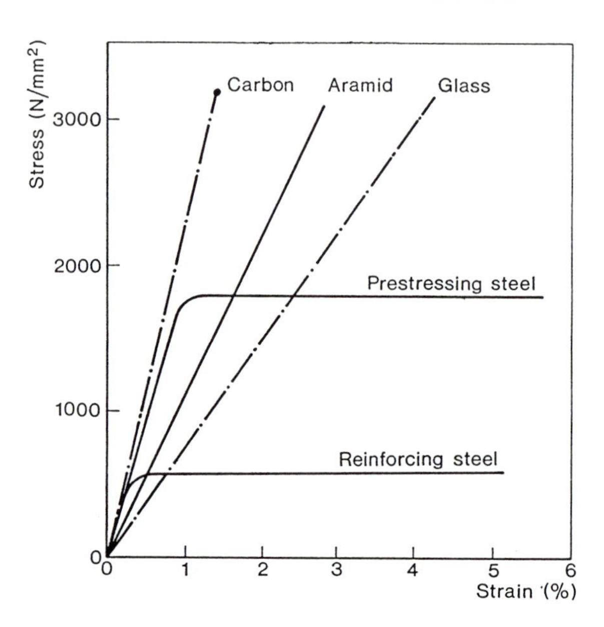

Figure 1. Typical stress-strain curves

Shinmiybashi using precast beams pretensioned with carbon fibre composite strand, supplied by Tokyo Rope[5]. In Holland precast units, pretensioned with aramid fibre composite strips, have been used in a number of applications by the Hollandsche Beton Groep[6]. In the U.K., Linear Composites supply Parafil aramid ropes which were used for the stressed repairs to cooling towers at Thorpe Marsh power station. Experimental work has shown that the material is suitable for prestressing concrete and the company has recently formed a partnership with VSL International to market suitable tendons.

However, prestressing is only used for a very small proportion of concrete structures. There is therefore a growing interest in the use of fibre composites as unstressed reinforcement. Glass fibre reinforced plastic reinforcing bars have been used in the USA for about 15 years. Often they have been used in situations where non-corroding reinforcement is required, such as below sensitive electronic equipment[7]. More general applications would be structures in aggressive environments with low levels of reinforcement. A typical example might be an abutment or a retaining wall beside a road treated with de-icing salts. Another situation is where reinforcement is needed only to control shrinkage or early thermal movements. An example of this is sprayed linings for tunnels; in Japan several kilometres of tunnel have been reinforced with Nefmac, a glass fibre composite grid designed to be a direct replacement for welded steel fabric[8].

DESIGN USING ALTERNATIVE MATERIALS

One limiting factor when considering the use of alternative reinforcement is the lack of guidance in design codes. They are written assuming that steel will be used and many of the rules are based on experience rather than on scientific principles.

Design for reinforced concrete is carried out at the ultimate limit state, to ensure that the structure has adequate strength, and at the serviceability limit state to ensure that its behaviour is satisfactory during the intended design life.

In some areas, it is relatively easy to predict the behaviour when a composite rod is used as reinforcement. In others, the present rules are purely empirical and cannot be adapted easily when using a new material.

Bending

Failure of a reinforced concrete member in bending is due to yield/rupture of the reinforcement or crushing of the concrete. Knowing the stress-strain behaviour of the alternative material, the bending response could be easily determined from first principles. However, there will be two significant differences that must be allowed for. The first is that the alternative materials have a straight elastic stress-strain response throughout and show no significant yield before failure. This may limit the rotation capacity of the reinforced section and hence the amount of 'plastic' behaviour that can be assumed in the design process.

The second difference is that, under sustained load the alternative materials will creep to failure [known as stress rupture]. However, for a given design life, the reinforcement stress to cause failure can be determined and this value used as the maximum permitted stress under service load conditions. The significance of this limitation could be determined readily by parametric studies.

Shear and torsion
The shear failure of beams, or the punching of slabs, occurs with little or no warning. Current design methods are empirical. For reinforced concrete these are based on both the strength of the concrete and the amount of tensile reinforcement present at the section under consideration. Extensive testing would be required to determine similar empirical design approaches for alternative reinforcement.

Deflections and cracking
For an uncracked section the deflection will be a function of the gross concrete cross-section only and hence not influenced by the type of reinforcement. For the cracked situation, the deflection will depend on the distribution of cracking which will be influenced by the bond between the reinforcement and the concrete. It is likely that deflection, rather than strength, would be the governing criterion for many structures reinforced with non-ferrous material.

Crack widths have traditionally been limited because it was thought that they influenced the corrosion of the reinforcement. It is now generally accepted that crack widths have little or no significant influence on the long term durability of reinforced concrete structures. Hence the major reason for controlling crack widths would appear to be from the point of view of aesthetics. For many structures, that will be viewed from some distance away, realistic limiting widths with alternative reinforcement could be greatly in excess of those currently specified. Assuming that some crack width calculation would still be required, experimental work would be required to check the validity of the current design methods or to propose others.

Cover to reinforcement
The cover concrete is primarily to provide protection to the reinforcement, either from the point of view of durability or from considerations of fire. In addition a minimum cover, depending on the bar size and the maximum size of aggregate, is required to ensure adequate bond between the bar and the concrete. This latter will still be required for the alternative reinforcements but additional cover for durability will no longer be necessary. Limited test data available suggests that, with the covers currently specified for fire resistance, some alternative materials will behave satisfactorily. More information is required on the behaviour of both the fibres themselves and the fibre/resin composites at elevated temperatures. However, fire resistant resins are being developed which should lead to reduced cover requirements.

Priorities for research
In the light of the above, it may be seen that significant research and development work is required before non-ferrous reinforcement can be fully incorporated into

Codes and Standards. To date most of the work has been carried out by individual manufacturers on their specific products. Before fibre composites can be widely accepted, work is required to compare the behaviour of different materials to formulate general design rules. The author is aware of only one limited series of such tests, which looked at the bending behaviour of slabs[9]. Two types of glass-fibre rod, made by International Grating in the USA and by Cousin Frères in France respectively, were tested along with the Japanese Nefmac resin fibre grids and Tensar polymer grids. The results were compared with those from similar slabs reinforced with conventional welded steel fabrics, see Figure 2. It may be seen that the ultimate bending capacity could be predicted from the strength of the reinforcing material in all cases.

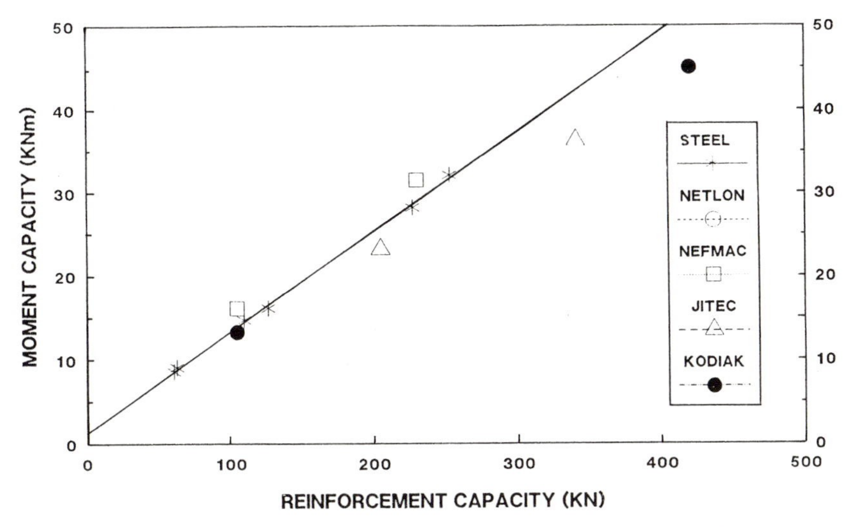

Figure 2. Strength of slabs reinforced with alternative materials

NEW DESIGN APPROACHES

Resin-fibre composite materials have been developed for reinforcement as direct replacements for conventional steel rod. This is a very restricted approach. There are various manufacturing methods used, which will produce an infinite range of forms and geometries. The pultrusion process, in which resin-impregnated fibres are drawn continuously through a heated die, can form not only rods but also plates and open and closed sections. Some would be suitable for externally reinforcing concrete members, either for new structures or for strengthening existing structures. The latter, typically done at present by bonding steel plates to the soffits of beams, is being studied at a number of research institutes. For example, Meier[10] reports on the use of carbon fibre ribbon for strengthening bridges. More complex shapes would act as permanent formwork for decking systems or composite columns.

One limitation on the use of the present generation of resin fibre rods is that they cannot be bent as readily as steel ones. However, complete two or three dimensional reinforcement assemblies could be formed by a variety of techniques: the Japanese Nefmac is an example of a two dimensional grid. Additionally, developments in resins are leading towards materials that can be heated and bent to form more complex shapes.

The inability to bend rods would necessitate a new approach to reinforcement detailing. Currently straight steel bars are bent to a variety of shapes and sizes to suit the geometry of the structure. However, the reinforced plastics industry is geared to repetitive production. The solution may be to use a restricted range of standard reinforcing shapes, for example closed rectangular units, in a restricted range of sizes. The required reinforcement cage would be built up from the standard units, probably bonded together rather than being tied as at present.

ECONOMICS

At present, fibre-composite material for reinforcement or prestressing is made in small quantities and hence is considerably more expensive than the steel it is replacing. It is unclear how much the price would fall if the material was widely used but it is likely to remain an expensive option. However, it is essential that the cost of the total structure be considered and not just of the reinforcement itself. By way of an example, epoxy coating roughly doubles the cost of steel reinforcement. However, the additional cost for a complete bridge superstructure is only about 2%, a small price to pay for improved durability.

SUMMARY AND CONCLUSIONS

Design codes and standards currently attempt to ensure durability by requiring an adequate thickness of good quality concrete, generally combined with some limitation of the surface crack width. However, this approach has often proved inadequate, particularly in highly aggressive environments and when only first costs have been considered. The designer has at this disposal a range of special products that enhance the behaviour of the concrete or protect the steel from corrosion. These may be described as 20th century solutions. Now reinforced and prestressed concrete structures are set to move into the 21st century with a range of new non-ferrous materials to replace the steel. The alternative materials offer huge benefits, but these will only be realised fully if designers break away from traditional approaches, take full advantage of the properties and consider the whole-life cost of structures.

REFERENCES

1. Read J.A. FBECR, 'The need for correct specification and quality control', Concrete, Vol 23, No. 8, [September 1989].

2. Clifton J.R., Beeghly H.F. and Mathey R.G. Final report, 'Non metallic coatings for concrete reinforcing bars', Federal Highways Administration, Report RD-19, 1974.
3. Weyers R.E. and Cady P.D. 'Cathodic protection of concrete bridge decks', Journal American Concrete Institute, [Nov-Dec 1984], pp. 618-622.
4. Wolff R. and Miesseler H.J. 'Application and experience with intelligent prestressing systems based on fibre composite materials', FIP Congress, Hamburg, [June 1990].
5. Zoch P. et al. 'Carbon fiber composite cables: a new class of prestressing members', Paper at Annual Convention of the Transportation Research Board, Washington D.C. [January 1991].
6. Reinhardt H.W. et al, 'Arapree, a new prestressing material going into practice', FIP Notes, 1991/4.
7. Randall F.A. 'Plastic based rebar meets special demands, Concrete Construction, Vol. 32, No. 9, September 1987, pp. 783-789.
8. Nefcom Corporation, Technical Leaflet 1, 'New fibre composite material for reinforcing concrete'.
9. Clarke J.L. 'Tests on slabs with non-ferrous reinforcement', FIP Notes, 1992/1 and 1992/2.
10. Meier U. 'Carbon fibre-reinforced polymers: modern materials in bridge engineering, Structural Engineering, 1/92, pp 7-12.

19 FLEXURAL BEHAVIOUR OF CARBON FIBER REINFORCED CEMENT COMPOSITES WITH CONVENTIONAL STEEL REINFORCEMENT

Y. Mitsui
K. Murakami
Kumamoto University,
T. Kage
Ministry of Construction,
H. Sakai
Mitsubishi Kasei Corporation, Japan

An experimental study was conducted to clarify the effect of steel reinforcement on flexural behaviours of Carbon Fiber Reinforced Cement Composite (CFRC) beam. The ordinary reinforced concrete (RC) beams and the reinforced CFRC ones with three different effective depths were tested under pure bending loads. The RC beams with CFRC-overlay both including and not including the steel reinforcement were also tested. The CFRC in cooperation with the steel reinforcement had a beneficial effect on increasing the cracking load and the yielding one of the beam.

INTRODUCTION

Carbon Fiber Reinforced Concrete (CFRC) is a new cement composite material with high flexural and tensile strength. However, in practical uses such as a beam member or wall panel of the building, steel reinforcement is needed to maintain the security against failure.

This paper presents an experimental study on flexural behaviours of the CFRC beam with conventional steel reinforcement (reinforcing steel bar). Firstly, flexural tests of the ordinary reinforced concrete (RC) and reinforced CRFC beam with three different effective depths were conducted. The difference in flexural behaviours between both types of the beam was clarified. Secondarily, flexural tests of the RC beam overlaid with CFRC were conducted. The CFRC-overlay both including and not including the steel reinforcement was performed on the tension side of the beam. The effect of the CFRC-overlay and the steel reinforcement on flexural behaviours of the beam was investigated. Finally, scale effect of test specimen on the flexural strength at first cracking and also the flexural strength at yielding of the tensile reinforcement of the beam were discussed.

MATERIALS

Materials used for concrete and CFRC, and their proportion are shown in Tables 1 and 2 respectively. The properties of carbon fiber (CF) are given in Table 1. An omunimixer and forced mixing type of mixer were used for mixing of CFRC and concrete respectively. Mixing procedure for CFRC was a follows: first, cement, aggregate, CF and dispersing agent were charged in the mixer and fry mixed for 3 minutes and second, with water added, wet mixed for about 6 minutes. Flow of CFRC and slump of concrete was 15.1cm and 20.5cm respectively. Flow test was conducted in accordance with JIS R 5201 1). Results of material test of CFRC and concrete at the age of 28 days are shown in Table 3. The standard cylinder of 10cm diameter by 20cm long and the rectangle of 10cm wide by 10cm high by 40cm long was used for the compression and bending test respectively. Loading condition was 3 points bending with span of 30cm long. Deformed bar of 10mm diameter of SD30 grade steel (equivalent to BS 4449) was used as a main reinforcement. Its mechanical properties are shown in Table 4.

Table 1. Materials used for concrete and CFRC

	CFRC	Concrete
C	Early-strength portland cement	Normal portland cement
S	Silica sand	River sand and river gravel
Mc	Methyl cellulose	-
CF	High performance pitch based carbon fiber. Diameter=18μm, Length=18mm Tensile strength=190kgf/mm^2, Modulas of elasticity=21 tonf/mm^2	

Table 2. Proportioning of CFRC and concrete (weight percentage)

	Fiber volume ratio Vf(%)	Water-cement ratio W/C(%)	Aggregate-cement ratio S/C(%)	Dispersing agent-cement ratio Mc/C(%)
CFRC	2.0	55	60	0.25
Concrete	-	60	40	-

Table 3. Results of material tests (CFRC and concrete

	Compressive strength fc (kgf/cm^2	Modulas of elasticity E (kgf/cm^2	Bending strength fb (kgf/cm^2
CFRC	320	150	93
Concrete	384	300	75

Table 4. Results of material test (main reinforcement:deformed steel bar)

Grade	Diameter (mm)	Yield point σy ($tonf/cm^2$)	Tensile strength σu ($tonf/cm^2$)	Elongation (%)
SD30	10	4.02	6.00	23.5

SPECIMENS AND TESTING PROCEDURES

Two types of the beam specimens were presented for bending tests. The one was the ordinary reinforced concrete (RC) and reinforced CFRC beam with the effective depth, d of 12.5, 17.0 and 22.0cm as shown in Figure 1. The other was the RC beam with the CFRC-overlay. The overlay was performed on the tension side of the beam both including and not including tensile reinforcements as shown in Figure 2. Thickness of the overlay, were (a) 6.0, 12.0 and 18.0cm for the former

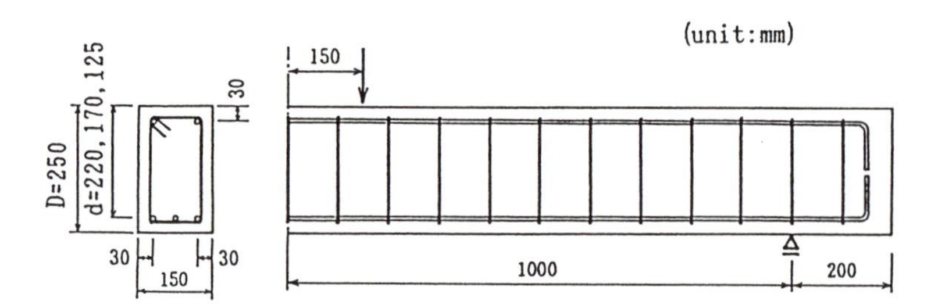

Figure 1. Configuration of test specimen

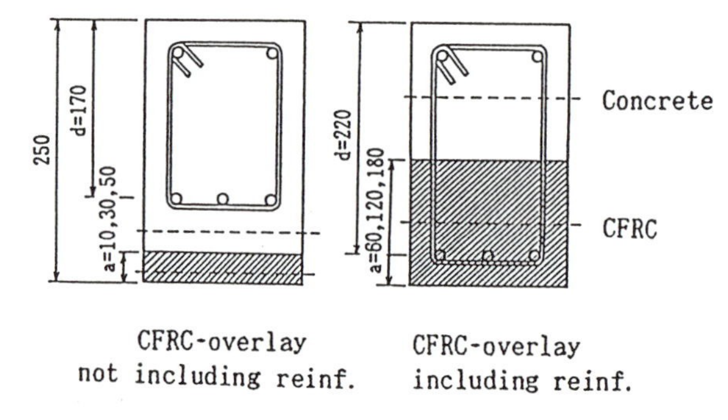

Figure 2. Section view of the specimen with CFRC-overlay

and 1.0, 3.0 and 5.0cm for the latter. All of the specimens had same configuration and overall dimension of 15cm wide by 25cm high by 240cm long. The deformed steel bar of 10mm diameter was used for the main reinforcement. The round bar of 5mm diameter was used for the stirrups with spacing of 10cm long. Summary of the beam specimens is shown in Table 5 with the test results.

Procedures for forming the specimens with the CFRC-overlay were as follows. Firstly, CFRC was cast at the specified height. After CFRC hardened, laitance on the surface was taken away by wire brush and then fresh concrete was filled up to the brim of the formwork. The specimens removed from the form work were seal-cured under wet conditions for three weeks and afterward air cured.

Test set up is shown in Figure 3. Loading condition was four points bending with span of 200cm long. Deformations of the specimen and strains of the main reinforcement were measured by deformation transducers and strain gauges respectively.

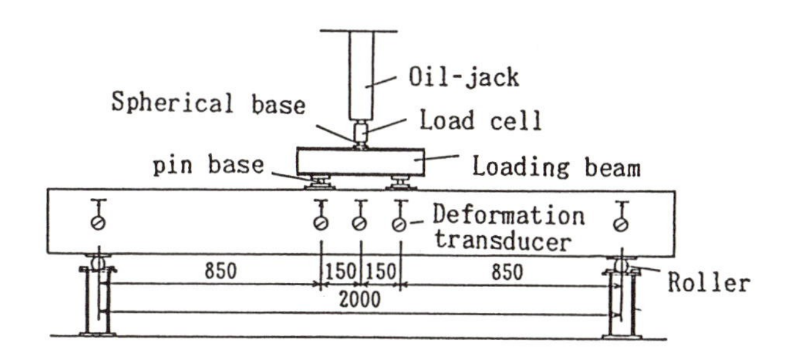

Figure 3. Set up of bending test

TEST RESULTS

Figure 4 shows typical cracking patterns of the beam specimens at test end. Multiple flexural cracks occurred and some of them developed in the RC beams. On the other hand, a few cracks occurred and one of them developed in the CFRC beams. The beams with CFRC-overlay showed almost the same crack development behaviours as the RC ones.

Figure 5 shows load-deformation (at midspan) curves of the specimens. The CFRC beams had the smaller initial flexural rigidity and the larger cracking strength Pcr and yielding strength Py than the RC ones. The beams with CFRC-overlay not including reinforcement had the same rigidity and strength Py and the large strength Pcr as compared with the RC ones. The beams with CFRC-overlay including reinforcement showed the different deformation behaviours as the overlay thickness, (a) increased. Namely, the beams with (a) =6 and 12cm and the beams with (a) =18cm showed the behaviours similar to the RC beams and the CFRC ones respectively.

Table 5. Summary of test results

Specimen No.	Effective depth	Thickness of CFRC	First crack		Yielding of tension bar	
			Bending moment	Nominal bending stress	Exp.bending moment	Cal.bending moment
	d(mm)	a(cm)	Mcr(t cm)	σb(kg/cm^2	Mye(t cm)	Myc(t cm)
RC-1	12.5	-	34.0	21.2	72.3	94.1
RC-2	17.0	-	38.3	23.2	114.8	127.9
RC-3	17.0	1.0	51.0	31.0	119.0	127.9
RC-4	17.0	3.0	76.5	46.5	119.0	127.9
RC-5	17.0	5.0	68.0	42.3	114.8	127.9
RC-6	22.0	-	42.5	24.3	157.3	165.6
CFRC-1	12.5	25.0	72.3	43.8	93.5	94.1
CFRC-2	17.0	25.0	89.3	51.3	136.6	127.9
CFRC-3	22.0	6.0	106.3	54.4	163.6	165.6
CFRC-4	22.0	12.0	97.8	50.1	157.3	165.6
CFRC-5	22.0	18.0	112.6	57.7	198.9	165.6
CFRC-6	22.0	25.0	114.8	58.8	201.9	165.6

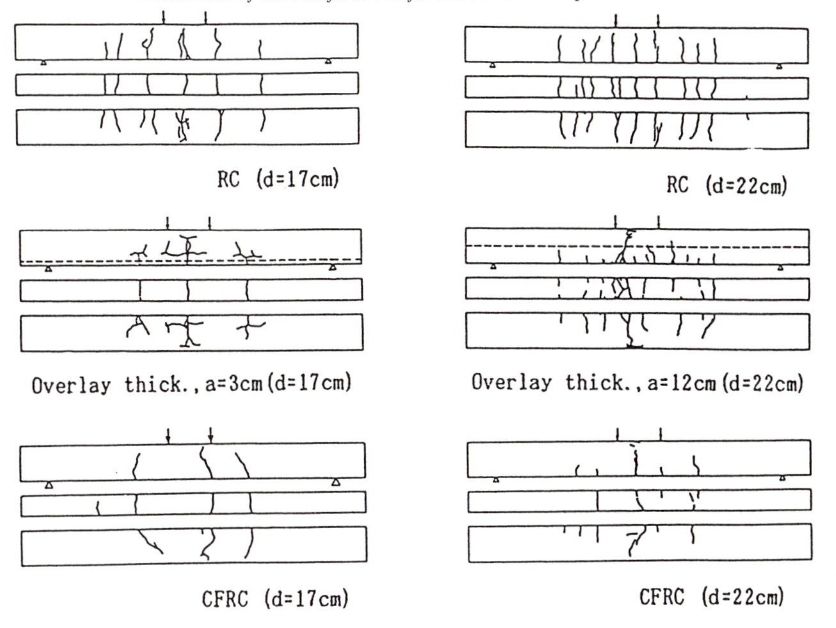

Figure 4. Cracking patterns at test end

Summary of the test results is shown in Table 5. Figures 6 and 7 show the ratio of cracking load Pcr and yielding load Py of the CFRC beams and the beams with CFRC-overlay to those of the RC beams respectively. As the effective depth, (d) increased, the cracking load Pcr of the CFRC beams increased from 2.1 to 2..7 times as high as that of the RC beams. The yielding load Py of the CFRC beams was 1.3 times as high as that of the RC beams. The cracking load Pcr of the beams with CFRC-overlay both including and not including reinforcement was from 2.3 to 2.7 and from 1.3 to 2.0 times as high as that of the RC beams respectively. From these results it can be found that the CFRC in cooperation with the steel reinforcement has a beneficial effect on increasing the cracking strength of the beam.

DISCUSSIONS

In Table 5, bending unit stress in extreme tension fibre of cross section of the beam at midspan subjected to cracking bending moment of each specimen is shown under the name of nominal bending stress, σ_b at first crack. The calculations were based on theory of elasticity in consideration of tensile stress of concrete and CFRC. The ratio of modulas of elasticity of steel to that of concrete and CFRC was selected to be 14 and 7 respectively (see Table 3). The beam with CFRC-overlay both not including and including reinforcement was supposed to

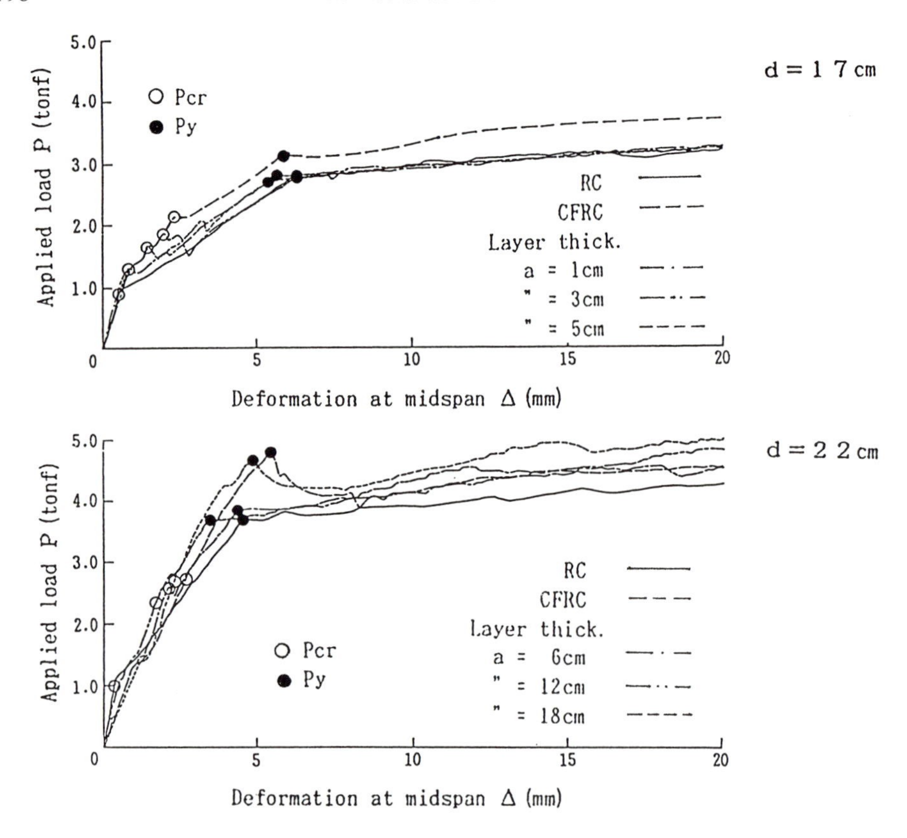

Figure 5. Load-deformation curve (Pcr:cracking load, Py:yielding load)

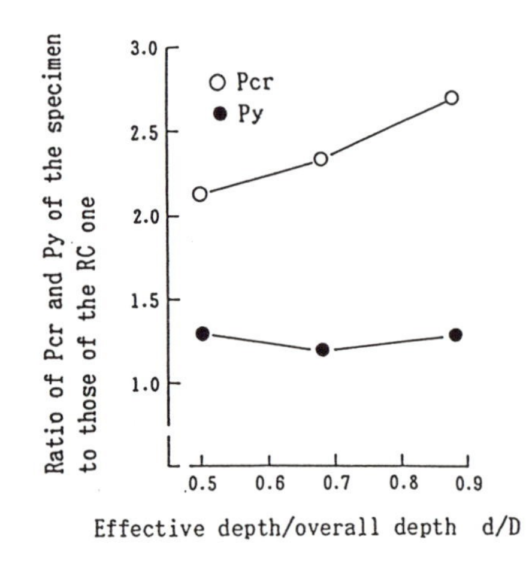

Figure 6. Effect of effective depth, d on cracking load Pcr and yielding load Py

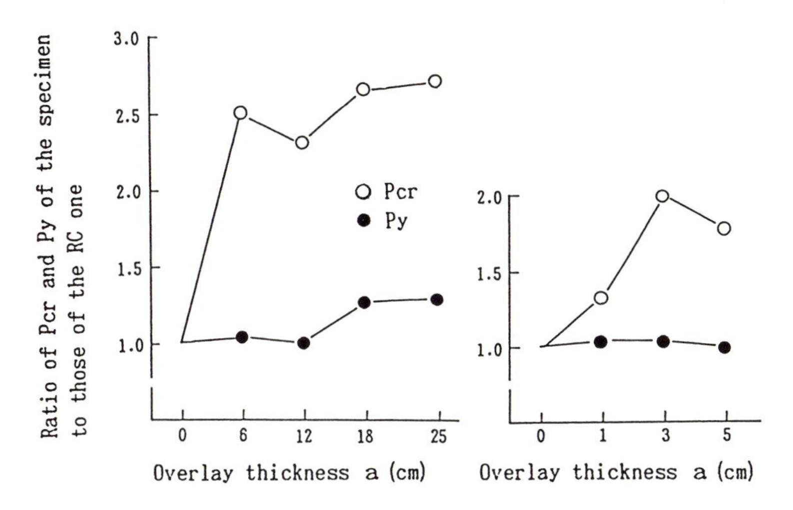

Figure 7. Effect of CFRC overlay-thick, a on cracking load Pcr and yielding load Py

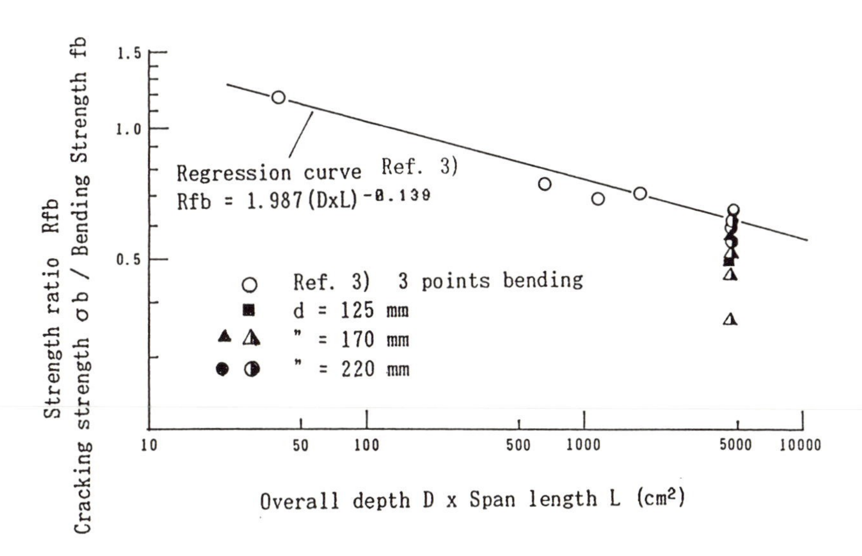

Figure 8. Relationship between specimen geometry and cracking strength

be the RC and CFRC beam respectively for the reason that flexural rigidity of both types of the beams was similar to that of the RC and CFRC beam respectively (see Figure 6).

According to the past test results 2) and 3), bending strength of CFRC is greatly affected by specimen size and decreases with increase in width, depth and length of a beam. Figure 7 shows relationship between ratio of cracking bending strength σ_b of each beam specimen to bending strength f_b of material test specimen and depth of the beam D by span length L. The past test results of the unreinforced CFRC beams composed of the same materials and proportioning in this test 3) are also shown in Figure 7. From this figure, it can be found that results of the CFRC beams and the beams with CFRC-overlay including reinforcement correspond with the past test results, that is bending strength of the reinforced beams as well as the unreinforced beams is affected by specimen size.

In Table 5, measured bending moment Mye and calculated one Myc at yielding of tensile reinforcement for each specimen are presented. The equation to be used in the calculation is:

$$Myc = 0.9 \times a_t \times \sigma_y \times d,$$

where

a_t = total section area of tensile reinforcement,
σ_y = yield strength of reinforcement and
d = effective depth.

Good agreement is observed between Mye and Myc for the CFRC beam with d-12.5 and 17cm. For the CFRC beam with d=22cm, the value of Mye is about 1.22 times as large as that of Myc. From these results it can be found that contribution of tensile resistance of CFRC to the yield bending moment of the beam increases with increase in effective depth.

CONCLUSION

An experimental study was conducted to clarify the effect of conventional steel reinforcement on flexural behaviours of Carbon Fiber Reinforced Cement Composite (CFRC) beam. The ordinary reinforced concrete (RC) beams and the reinforced CFRC ones were tested under pure bending loads. The RC beams with the CFRC-overlay both including and not including the reinforcement were also tested. Main results obtained are as follows.

The CFRC in cooperation with the steel reinforcement has a beneficial effect on increasing the cracking strength of the CFRC beam.

The CFRC-overlay has a beneficial effect on improving the cracking resistance of the RC beam.

Contribution of tensile resistance of CRFC to the yielding strength of the beam increases with increase in effective depth.

REFERENCES

1. 'Physical testing of cement'. Japanese Standard Association (in Japanese).
2. Akihama S., Kobayashi M., Suenaga T., Nakagawa H. and Suzuki K. 'Experimental study of carbon fiber reinforced cements composites (Part 4) Scale effect of CFRC test specimen on flexural properties'. Annual reports of Kajima Institute of Construction Technology, No 29, Kajima Corporation (in Japanese) (1991)
3. Urano T., Murakami K., Mitsui Y. and Sakai H. 'Study on scale effect on tensile characteristics of fiber reinforced concrete'. Summaries of technical papers of annual meeting. Architectural Institute of Japan (in Japanese) (1991)

20 DEVELOPING FRC THROUGH INTEGRATED MATERIAL–STRUCTURE OPTIMIZATION

C. Pedersen and H. Stang
Technical University of Denmark

The present paper describes a model for the prediction of crack width in concrete structures where the traditional concrete is replaced with Fiber Reinforced Concrete (FRC). The paper furthermore describes the verification of the model through experiments and the experimental procedure used. Finally, the paper discusses some industrial applications of the model and how a model like the proposed can be used both in the development of new FRC-materials and new structures which fully exploit the enhanced properties of FRC-materials as compared to conventional concrete.

INTRODUCTION

Research on Fibre Reinforced Concrete (FRC) materials has been going on since the mid-sixties and literally hundreds of papers has been published on material composition, production techniques and material properties of FRC-materials. Judging from these papers FRC-materials are clearly superior to conventional concrete in a number of aspects. Yet, in praxis the use of FRC-materials in engineering structures is very limited. The reason for this discrepancy - as the authors see it - is that most of the research on FRC materials has been conducted strictly within the materials research community. Thus, the research results are often reported in terms of toughness index, modulus of rupture, etc. These properties, however, are not very useful in world of the structural engineer.

The present paper represents an attempt to show that materials and structures can be developed and optimized provided the materials are characterized in such a way that the characterization can be applied in structural calculations.

In a number of conventionally reinforced structures a large part of the reinforcement is placed only to reduce and control the crack widths in the structure in the serviceability limit state. This is especially true for structures in aggressive environments and for structures where the water-tightness is critical such as reservoirs and containers.This crack controlling reinforcement is often difficult and expensive to place and tie. Furthermore, it is difficult to cast and vibrate the concrete around the often very compact crack controlling reinforcement. Thus, in such structures it is desirable to be able to reduce the crack controlling reinforcement.

It has been shown experimentally that fibre reinforcement reduces the crack widths in structures which also contain conventional reinforcement, see e.g.[1] and [2]. Thus, fibre reinforcement could be an attractive alternative to conventional crack controlling reinforcement. The price for FRC is high compared to conventional concrete, however, casting premixed FRC is a much less labor intensive operation than placing and tying conventional crack controlling reinforcement. Conclusively, one can say that fibre reinforcement could be a new cheap crack controlling reinforcement.

Stang[3] and Stang and Aarre[4] have presented an analytical FRC design tool which makes it possible to predict crack widths in conventionally reinforced FRC structures. The experimental verification of this design tool, which is presently taking place at the Department of Structural Engineering, Technical University of Denmark (TUD), is not yet finished, however, preliminary results will be presented here. Furthermore, it is shown how the analytical tool can identify new applications of fibre reinforcement. Finally, the paper points out some material and structural optimization possibilities by combination of fibre reinforcement and conventional reinforcement.

CRACK WIDTH MODEL

Material characterization of FRC

It is characteristic for most of the fibre reinforced concretes that have structural potential with respect to price, workability and mechanical properties that the mechanical *strength* parameters are only slightly changed when compared to a corresponding plain concrete. However, measuring toughness as the area under the load-displacement curve of a given test specimen loaded to complete collapse (zero load), it is found that the toughness of fibre reinforced concrete is very high compared to that of plain concrete. Thus, the first crack strength (and strain) is more or less unchanged by the fibre reinforcement, but, during crack opening, the mechanical behaviour is changed significantly. These observations lead to the following modelling of the mechanical behaviour of FRC-materials.

Analytical characterization

The fibre reinforced concrete is assumed to be linear elastic in tension up to maximum stress $\sigma^u{}_{frc}$ with Young's modulus E_{frc} and Possion's ration ν_{frc}.

After the maximum stress has been reached, a discrete crack is formed in the material. This crack is characterized mechanically by the stress-crack width relationship which is assumed to take the following simple form:

$$\sigma_{frc}^{crk} = f(w) = \frac{\sigma_{frc}^{u}}{1 + (\frac{w}{w_0})^p} \tag{1}$$

where σ^{crk}_{frc} denotes the stress transferred across the crack surfaces, w_0 represent a characteristic crack opening and p is a curve shape factor. Thus, the FRC-material is characterized by five material constants, three related to the pre-peak behaviour: the elastic constants and the ultimate tensile strength, and two related to the post-peak behaviour: the characteristic crack opening w_0 and the shape factor p.

Experimental characterization

The stress-crack width relationship can be determined experimentally. The experimental setup used at the Department of Structural Engineering, TUD is described in(4) and (5). Examples of experimental results (0.5 and 1.5 vol.% steel fibre reinforced high strength concrete, V/C=0.37 at age > 28 days) for the stress-crack width relationship is shown in Fig.1 along with the fitting of the analytical expression(1). The figure represents a small part of a large experimental investigation, see(5). It has been found that eq.(1) is well suited to describe the behaviour of both brittle un-reinforced concrete and tough concrete with a high fiber content.

The structural crack width model

To calculate crack widths in FRC structures with conventional reinforcement under different external loadings a plane model of a prism consisting of a conventional reinforcing bar with thickness 2t surrounded by a layer of FRC-material with thickness b is modelled as shown in fig. 2. The geometrical reinforcing degree is $\varphi = t/(b+t)$. The total length of the element is 2(l+d) corresponding to the spacing between discrete macro-cracks in the FRC-material. It is assumed that there is displacement continuity on the FRC/reinforcing bar interface for $-l < x < l$ while the interface is stress free on the interface for $x < -l$ and $l < x$, allowing for the modelling of FRC/reinforcing bar debonding.

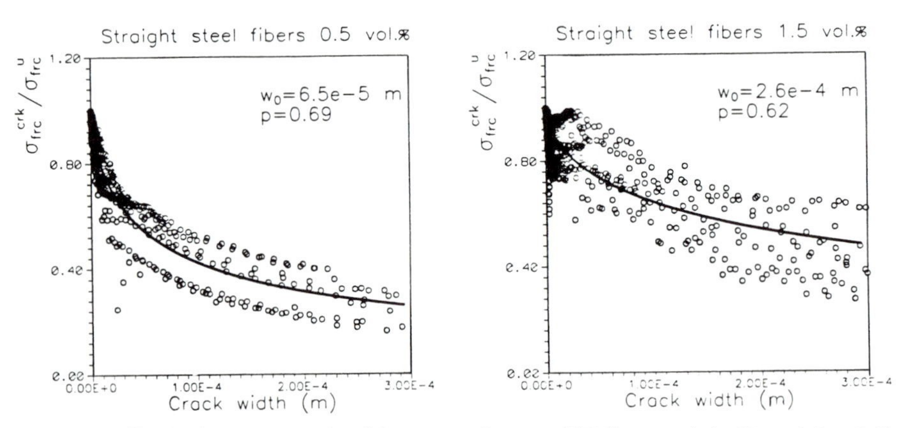

Figure 1. Typical stress-crack with curves for two FRC-materials (1 and 2 vol.% steel fibers in normal strength concrete) and the corresponding analytical fit.

By solution of geometrical, constitutive and equilibrium equations using a shear lag model,and taking into account the boundary conditions prescribed by the stress-crack width relationship put forth in eq.[1] maximum FRC-stress due to a given overall strain is determined. In this solution procedure the crack *spacing* is considered unknown. The crack spacing is then determined in such a way that the maximum stress in the FRC-material σ_{frc}^{max} is equal to the ultimate tensile strength of the FRC-material $\sigma_{frc}^{u.}$ The result of the calculation is corresponding values of overall strain, maximum stress in the reinforcing bar, stress in the FRC-material at the cracks and average crack width and spacing. A complete description of the model and the solution can be found in[4].

Structural application

The model presented above makes it possible to calculate crack widths in conventionally reinforced structures using fibre-reinforced concrete. The geometrical reinforcing degree φ to be used in the model is in accordance with the Danish Code of Practice for the Structural Use of Concrete, DS 411[6] given by:

$$\varphi = \frac{A_s}{A_{cef}} \tag{2}$$

where A_s is the total steel area and A_{cef} is the effective area of concrete. A_{cef} is the part of the concrete area which has the same centre of gravity as the total steel area. Given a sample beam as illustrated in Fig.3, the reinforcing degree is:

$$\varphi = \frac{A_s}{2b(h-h_e)} \tag{3}$$

By use of the model it is also possible to predict crack widths due to shrinkage.

Experimental verification

The presented model is currently being verified experimentally at the Department of Structural Engineering, TUD. Both prism shaped specimens tested in uniaxial

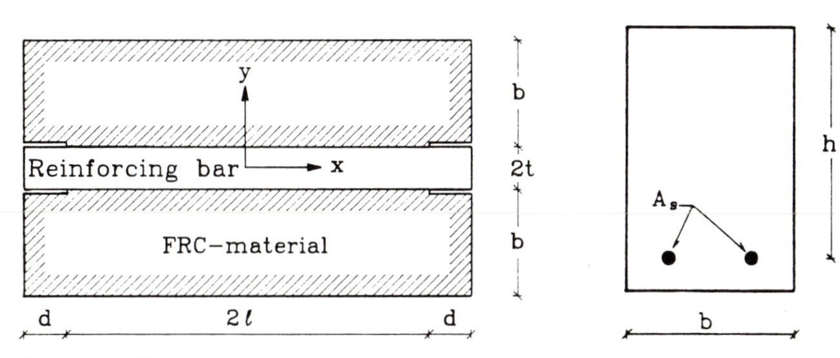

Figure 2. Outline of structural model

Figure 3. Effective concrete area

tension and beam shaped elements tested in bending are being investigated. Here, preliminary results of the prism elements each containing one central steel reinforcing bar will be mentioned.

Three normal strength concrete recipes have been tested: a plain concrete called '7000' tested for reference, and two FRCs: one called '9000' with 2 vol.% of polypropylene (pp) fibre and one called '8500' with a combination of 1 vol.% of hooked steel fibers and 1 vol.% of pp fibres . See[7] for full details. All prisms were quadratic (60 mm x 60 mm) in cross section and 1200 mm long. The specimens reported here were all reinforced with one 16 mm deformed steel bar, resulting in a reinforcing degree of 5.9 %. Observation of cracking patterns was done using a combination of manual registration and recording with a high resolution video camera equipped with a macro lens and connected to a high definition VCR. Using this equipment images containing 2 mm long crack segments were stored on VHS video tapes for later processing. This processing was done on an image analysis and processing system grabbing each digital image from the VCR. The results from the image analysis contain information on average crack width, maximum crack width, average crack width distribution, and maximum crack width distribution. Furthermore, average crack widths were determined from overall deformation measurements and crack counts. Here, only results showing the average crack width will be shown. In fig. 4 the experimental results for crack widths in 15 different prisms as function of stress in the reinforcing bar is shown along with the theoretical predictions based on the model described above. It should be noted that all the material parameters used in the model are mean values obtained directly from material testing of the three different concretes with the following modification. It was found that the following relationship could be established between the tensile strength σ^u_{frc} used in the model and the tensile strength measured from uniaxial tensile testing of the plain concrete or the FRC-material, σ^u_{exp}:

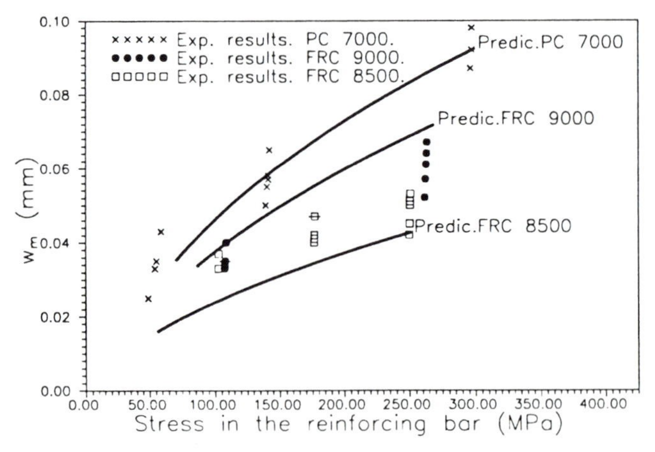

Figure 4. Experimental\theoretical results for crack width in prism specimens

$$\sigma^{u}_{frc} = \alpha \sigma^{u}_{exp} \tag{4}$$

where $\alpha = 1$ for smooth re-bars and $\alpha = 0.5$ for deformed re-bars.

It is clear that the model is well suited to describe the influence of fibre reinforcement on crack widths. This conclusion is further emphasized in the following table where the observed and predicted crack width reductions introduced by the fibre reinforcement is shown:

Table I Average crack width reduction in the two FRC materials relative to the plain concrete for 2 different load levels.

Re-bar stress (MPa)	176	249
FRC 9000/PC 7000		
Exp. results:	71 %	70 %
Prediction:	81 %	82 %
FRC 8500/PC 7000		
Exp. results:	66 %	58 %
Prediction:	51 %	54 %

OPTIMIZATION

The structural crack width model now constitutes - together with the experimental material characterization presented above - a complete design tool which can be used when designing FRC structures where the fibre reinforcement is used as a crack controlling reinforcement while the conventional reinforcement has load carrying function.

At the Department of Structural Engineering, TUD, a large number of different FRC-materials has been tested and stiffness parameters, $\sigma_{frc}{}^{u}$, and the post-peak parameters w_0 and p reported[5]. Among these parameters the three last-mentioned can be considered functions of fiber volume concentration: $\sigma_{frc}{}^{u}$ showing a weak dependency of fiber volume concentration (V_f) and w_0 and p showing a strong dependency of V_f. Typical relationships between V_f and w_0 and p for different fibre-matrix systems are shown in fig. 5. By introducing the parameters w_0, p and $\sigma_{frc}{}^{u}$ as functions of V_f for a given matrix type the experimental stress-crack opening relationship[1] yields:

$$\sigma^{crk}_{frc} = f(w,V_f) = \frac{\sigma^{u}_{frc}(V_f)}{1 + \left(\frac{w}{w_0(V_f)}\right)^{p(V_f)}} \tag{5}$$

By use of equations[3] and [5] and the crack width model, prediction of crack width is possible under external loads for given values of φ and V_f assuming that the stress in the re-bars can be calculated.

Furthermore, it is possible given the crack width requirement and the load in the serviceability limit state to determine the optimal amounts of fibre and conventional reinforcement.

Fig. 6 shows corresponding values of fibre volume concentration and degree of reinforcement for a beam loaded to simple bending for various values of the ratio M/b, where M is the bending moment and b is the width of the beam. Hence the cost of the total amount of reinforcement can be calculated and finally the optimal amounts of fibre and conventional reinforcement can be determined minimizing the total material costs.

The total material costs for the fibre and the conventional reinforcement for a beam is shown in Fig. 7 as function of the fiber volume fraction.

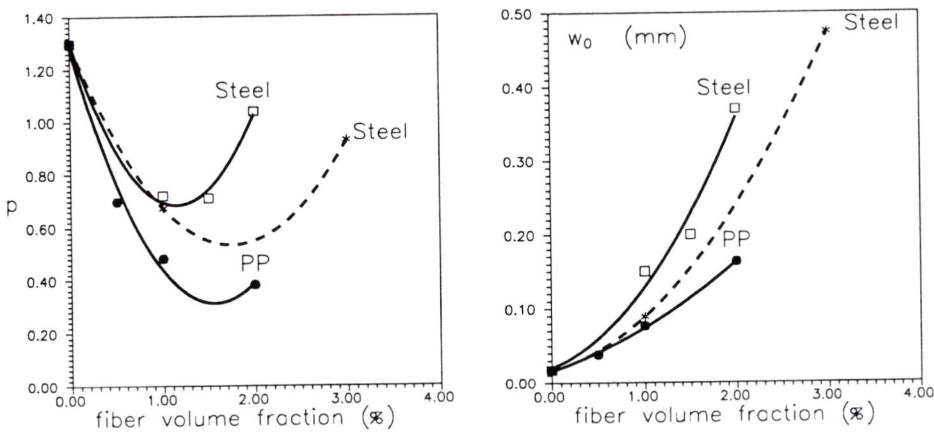

Figure 5. Typical example of the post-beak parameters as function of fibre volume concentration for different fibre/matrix systems

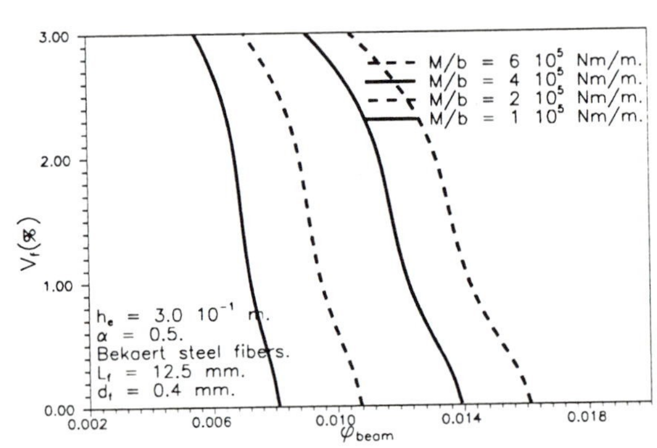

Figure 6. Corresponding values for fibre volume concentration and amounts of conventional reinforcement for beam given loading and crack width requirement

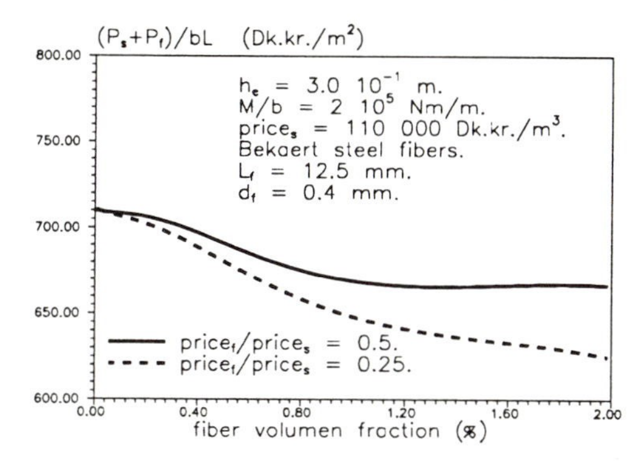

Figure 7. Example of prices for total reinforcement (fibres and conventional reinforcement) for beam with given load and crack width requirement

DISCUSSION

An analysis carried out at Department of Structural Engineering, TUD indicates that significant savings can be obtained in structures where severe restrictions to crack widths are imposed. These savings are obtained by substituting parts of the conventional reinforcement with fibre reinforcement. This is possible even though the following is not taken into account:

- the possible reduction in the cross-sectional area due to the reduction of conventional reinforcement degree (e.g. two layers of reinforcement in a concrete wall reduced to one).

- the decrease of stress in the steel due to the tensile stress in the FRC.
- that casting premixed FRC is a much less labor intensive operation than placing and tying conventional crack controlling reinforcement.
- that the shrinkage of some fiber reinforced concrete is probably reduced compared to conventional concrete.

Unfortunately, no detailed experimental results are yet available to confirm the predictions presented here, however, a research program is currently carried out at TUD. This program is intended to investigate:

- how α in eq.(4) is influenced by the steel bar type (deformed, smooth)
- the applicability of relation(3) and (4) for a variety of flexural members.

The development of this structural design tool has made design and optimization of FRC-structures possible in a traditional sense. This opens up new possibilities for developing economical structures which fulfil strict requirements to crack widths, durability, and load carrying capacity.

However, it has to be realized that FRC-materials cannot be expected to be used on a large scale before the structural engineers and the contractors have confidence in all aspects of practical handling of the material. More work and experience is needed to ensure that FRC construction work can be carried out on the building site and in the factory.

ACKNOWLEDGEMENTS

The authors acknowledge the support from the Research Program on Fiber Reinforced Cementitious Composites sponsored by The Danish Council for Scientific and Industrial Research and The Danish Ministry for Industry.

REFERENCES

1. Kanazu T., Aoyagi T. and Nakano T. Mechanical Behaviours of Concrete Tension Members Reinforced with Deformed Bars and Steel Fiber. 'Trans. Japan Concr. Inst.' **4**, 395-402, (1982).
2. Naaman A.E., Reinhardt H.W. and Fritz C. Reinforced Concrete Beams with a SIFCON Matrix. 'ACI Struc. J.' **89**, 1, 79-88, (January-February 1992)
3. Stang H. Prediction of Crack Width in Conventionally Reinforced FRC. In: *Brittle Matrix Composites 3* (Eds. A.M. Brandt and J.H. Marshall). Chapman and Hall, 388-406, (1991).
4. Stang H. and Aarre T. Evaluation of Crack Width in FRC with Conventional Reinforcement. 'Cem. Concr. Comp.', **14**, 143-154, (1992).
5. Hansen S. and Stang H. *Experimentally Determined Mechanical Properties for Fiber Concrete*. Report, Department of Structural Engineering, Technical University of Denmark, (1992). (Under preparation, in Danish).
6. *Dansk Ingeniørforening's Code of Practice for the Structural Use of Concrete*, DS 411. 3rd ed. Teknisk Forlag. (August 1991).
7. Rasmussen T.V. and Stang H. *An Experimental Method for Determining Crack Widths in Concrete and FRC*. Report, Department of Structural Engineering, Technical University of Denmark, (1992). (Under preparation, in Danish).

21 FIBRE REINFORCED COMPOSITE MATERIALS FOR REINFORCED CONCRETE CONSTRUCTION

J. Anderson
International Grating Inc

The information in this paper has been provided by Professor H V S GangRao and Dr S S Faza based on the research and development work which they have carried out at West Virginia University, Morgantown, West Viginia, USA.

The load-deflection behaviour of concrete beams reinforced with fibre-reinforced plastic (FRP) rebars is investigated extending the current methods used for steel reinforced beams. In addition, certain recommendations to compute post-cracking deflections in beams reinforced with FRP rebars are presented.

INTRODUCTION

Concrete reinforced with mild steel rebars is commonly used in the construction of bridge decks, parking garages, and numerous other constructed facilities. The extensive use of de-icing chemicals is reducing the service life of these facilities. In addition, concrete structures exposed to highly corrosive environments, such as coastal and marine structures, chemical plants, water and wastewater treatment facilities, have been experiencing drastically reduced service life and causing user inconvenience.

The gradual intrusion of chloride ions into concrete leads to corrosion of steel reinforced concrete structures and concrete cracking; which may be due to shrinkage, creep, thermal variations or some unexpected design inadequacies due to external loads (Nawy, 1990). A corrosion protection system that would extend the performance of our constructed facilities would have a payoff in billions of dollars, since the replacement runs twice the original construction cost (America's Highways, 1984).

Several recommendations have been adopted in the design of concrete structures to prevent the corrosion of steel reinforcement such as use of waterproofing admixtures in concrete, impermeable membranes, epoxy-coated steel rebars, and others without a complete success. Therefore, use of noncorrosive fibre reinforced plastic (FRP) rebars in place of mild steel has been researched as an alternative to improve the longevity of structures.

RESEARCH SIGNIFICANCE

The performance of FRP rebars embedded in concrete is not fully understood, even though FRP rebars have been used in structural applications for over ten years (Nawy, 1977). The current mathematical models and design equations of concrete beams reinforced with mild steel cannot be applied directly to beams reinforced with FRP rebars without fully understanding the following issues:

1. Lower modulus of elasticity of FRP than steel
2. Bond behaviour
3. Long term degradation
4. Post-cracking behaviour of the concrete beams with FRP rebars

The primary focus of the research conducted at the Constructed Facilities Centre, West Virginia University is to study the behaviour of reinforcing beams and bridge decks with FRP rebars. The main objectives of this paper include two aspects of the behaviour of FRP rebars used as reinforcement in concrete:

(1) Investigation of the pre- and post cracking deflection behaviour under bending of concrete beams reinforced with FRP rebars;
(2) Development of post-cracking deflection design equations for FRP reinforced concrete, which are practical and simple to use for structural design applications.

TEST PROGRAM

Fibre-Reinforced Plastic Rebar Characteristics
Since the 1950s, glass has been considered a good substitute to steel for reinforcing or prestressing concrete structures. For example, a typical E-glass fibre reinforced plastic rebar has minimum 55% glass volume fraction embedded in a matrix of vinylester and an appropriate coating to prevent alkaline reaction (Rubinski 1954).

In order to develop good bond strength between FRP rebars and concrete, different surface conditions for rebar were developed. Among them, 45 degree angular wrapping or helical ribs produce a deformed surface on the rebar. Coating FRP rebars with epoxy and rolling them in a bed of sand creates a roughened surface and is one of the alternatives that will improve bond strength.

Experimental results indicate that the average tensile stiffness depends on the fibre type and volume fraction and is virtually independent of manufacture, bar size, bar type (with or without ribs), test procedure, and type of resin. A mean tensile stiffness of 48.26 GPa for 55% fibre volume fraction was reported by Wu (1991). However, ultimate tensile strength is sensitive to bar diameter, quality control in manufacturing, matrix system, fibre type, and gripping mechanism. The ultimate tensile strength of continuous glass fibre reinforced rebars with vinylester resin decreases rapidly with an increase in bar diameter (Table 1). Considering the fact that FRP rebars do not exhibit a yield plateau as steel, it is necessary to assume a 20% reduction in effective yield strength of the rebar for reasons of safety; hence minimum of six rebars have to be tested to obtain an average ultimate rebar strength, f_{ult}.

The reduction factor was arrived at after testing samples from a variety of manufacturers (Wu, 1991, Faza, 1992). If testing of rebar samples is not possible, the following results outline in Table 1 can be used as the effective yield strength, f_{yf} of FRP rebars.

Table 1. Effective yield strength of FRP rebars (MPa)

REBAR SIZE	#3	#4	#5	#6	#7	#8
f_{ult}*	897.4	738.6	655.8	621.3	586.8	552.2
f_{yf}	717.9	590.9	524.6	497.0	469.4	441.8

* Based on test data on Kodiak Rebars

Concrete Properties

In order to take advantage of the high tensile strength of the FRP rebars, concrete strength of 34.5 - 69.0 MPa was used in the testing program. Class K concrete from a local mixing plant was used in all the specimens. For each batch of concrete delivered, eight 101.6 x 203 mm (4 x 8 in.) cylinders were cast, cured and tested with the specimens.

Beam Specimens

Twenty five rectangular beams, 152.4 x 304.8 mm by 3048 (6 x 12 in. by 10 feet) were tested under pure bending (as simply supported under four point bending), using different configurations of FRP reinforcement and concrete strength. The variables are:

a) Rebar size (#3, #4, #7, #8)
b) Type of rebar (smooth, ribbed, sand coated)
c) Type of stirrups (steel, smooth FRP, ribbed FRP)
d) Reinforcement distribution (3#4 versus 5#3)
e) Concrete compressive strength, 29-69 MPa (f_c' = 4.2, 5.0, 6.5, 7.5 and 10ksi)

Experimental Results

The major emphasis of the test program was to investigate the FRP reinforced concrete beam behaviour and compare with the steel reinforced beam, in terms of

- Pre- and Post-Cracking Behaviour
- Load-Deflection and Stress-Strain Variations
- Elastic and Ultimate Load Carrying Capacities
- Crack Patterns (spacing, width, propagations)
- Modes of failure

The simply supported rectangular beams were tested under pure bending using different configurations of FRP reinforcements. In order to take advantage of the high tensile strengths of FRP rebars, beams with higher strength concrete ($f_c^{/}$ = 34.5 - 69 MPa) were tested for the purpose of maximising the bending resistance of the beams. In this paper, the pre- and post-cracking deflection behaviour are emphasized.

Pre- and Post-Cracking Behaviour

The precracking segments of load deflection curves in all specimens are essentially straight lines indicating the full elastic behaviour. The maximum tensile stress in concrete beams in this region is less than the tensile strength of concrete. The flexural stiffness E I of the beams can be estimated using Young's modulus E_c of concrete and the moment of inertia of the uncracked reinforced concrete cross section. The load-deflection behaviour before cracking is dependent on the stress-strain relationship of concrete.

When flexural cracking develops, contribution of concrete in the tension zone is neglected. Thus the slope of a load-deflection (or stiffness) curve is less steep than in the precracking stage. The stiffness continues to decrease with increasing load, reaching a lower limit that corresponds to the moment of inertia of the cracked section, I_{cr}.

The post-cracking experimental deflection of FRP reinforced beams (Beam #B & #H1) were about four times larger than the beam reinforced with steel rebars (Beam #11) as shown in Fig. 1 The larger post-cracking deflections were expected due to the low modulus of elasticity of FRP rebars, which is about 4.83GPa. The deflection behaviour was vastly improved when sand coated rebars were used as shown in Fig. 2 This behaviour is attributed to the reduction in crack widths, and the improvements in the distribution and propagation of the cracks when sand coated rebars are used (Faza, 1992).

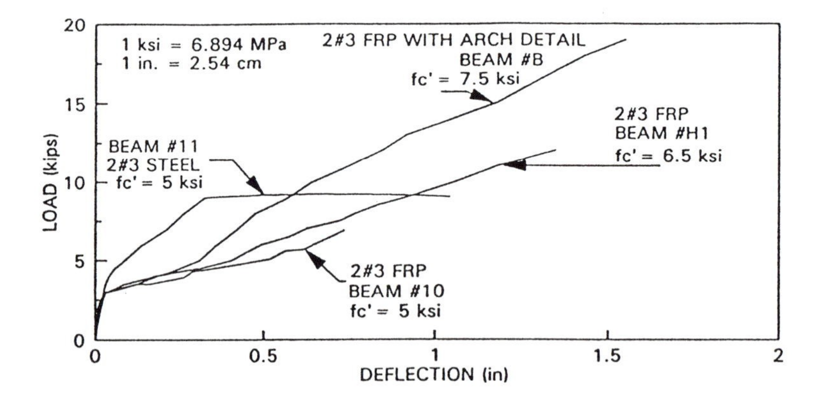

Figure 1 Experimental Load vs Deflection

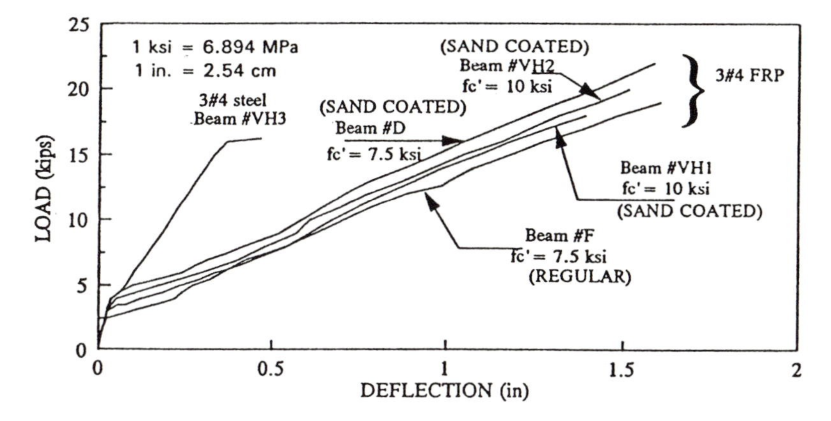

Figure 2 Load vs Deflection (using sand coated FRP bars)

Theoretical Correlation of Experimental Deflections
The theoretical correlation of the experimental deflections utilizes, as a first step, the current mathematical models and design equations for concrete reinforced with mild steel rods. The results from these equations are checked with the experimental deflections and modified as necessary to accommodate FRP reinforcement.

Various methods have been considered by researchers in an attempt to calculate post cracking deflections of concrete beams reinforced with steel (Nawy, 1990). The differences among the various methods consists mainly of the ways to compute the modulus of elasticity, E, and the moment of inertia, I. Both quantities are difficult to define in a steel reinforced concrete member. Considering that cracking behaviour of concrete beams reinforced with FRP bars is different from that of steel reinforced concrete beams, the effective cracked moment of inertia I_e, would be different from that of conventional steel reinforced beams. Such difference can be attributed mainly to the extent of cracking, which is a function of the bond between concrete and rebar.

Precracking Stage

The precracking segments of load-deflection curves in all specimens were essentially straight lines indicating the full elastic behaviour. The load deflection behaviour before cracking is dependent on EI of the beams and the stress-strain relationship of concrete from which the value of E_c can be calculated either using the ACI 318 code expression

$$E_c = 57{,}000 \sqrt{f_c'} \quad \text{where } E_c \ \& \ f_c' \ f \ (\text{psi, 1psi} = 6.903 \text{ KPa}) \tag{1}$$

or ACI 363R committee recommendation

$$E_c = 40{,}000 \sqrt{f_c'} + 1 \times 10^6, \text{ where } E_c \ \& \ f_c' \ (\text{psi, 1psi} = 6.903 \text{ KPa}) \tag{2}$$

An accurate estimate of the moment of inertia I necessitates the consideration of FRP reinforcement area A_f in the computations. This can be done by replacing FRP rebar area by an equivalent concrete area $(E_f / E_c)A_f$. However, the use of gross moment of inertia resulted in acceptable results in the precracking stage with the uncracked section and neglecting additional stiffness contribution from the FRP reinforcement.

The precracking stage stops at the initiation of the first flexural crack when concrete reaches its modulus of rupture, f_r, which is typically $7.5\sqrt{f_c'}$ (f_c', psi). The ACI 363R recommends a value of $11.7\sqrt{f_c'}$ for normal weight concretes with strengths in the range of 3000 to 12,000 psi (1 psi = 6.903 KPa). For curing conditions such as seven day moist curing followed by air drying, a value of $7.5\sqrt{f_c'}$ is closer to the full strength range.

Postcracking Stage

When flexural cracking develops, tensile strength of concrete is neglected. The slope of a load-deflection (or stiffness) curve is less steep than in the precracking stage as shown in load-deflection curves in Fig 3 and 4. The stiffness continues to decrease with increasing load, reaching a lower limit and corresponds to the moment of inertia of the cracked section, I_{cr}. The moment of inertia of a cracked section can be obtained by computing the moment of inertia of the cracked section about the neutral axis resulting in the following relationship after neglecting the concrete section below the neutral axis:

$$I_{cr} = \frac{bc^3}{3} + nA_f(d-c)^2 \tag{3}$$

where n = Modular ratio, (E_f/ E_c)
c = Distance from top fibre to the neutral axis

In actual cases, only a portion of a beam along its length is cracked. The uncracked segments below the neutral axis possess some degree of stiffness which contributes to the overall beam rigidity. The actual stiffness of the beam lies between $E_c I_g$ and E_cI_{cr}. As the load approaches the ultimate value, beam stiffness approaches E_cI_{cr}. The major factors that influence the beam stiffness are:

1) Extent of cracking
2) Contribution of concrete below the neutral axis

The ACI 318 code specifies that deflection shall be computed with an effective moment of inertia, I_e as follows, but not greater than I_g.

$$I_e = (M_{cr}/M_a)^3 I_g + [1 - (M_{cr}/M_a)^3] I_{cr} \tag{4}$$

where, $M_{cr} = \frac{f_r I_g}{Y_t}$ M_a = Applied Moment, and $Y_t = h/2$

The effective moment of inertia adopted by the ACI 318-89 is considered sufficiently accurate for use in control of deflection of beams reinforced with steel. I_e was developed to provide transitional moments of inertia between I_g and I_{cr} and it is a function of $(M_{cr}/M_a)^3$

By investigating the experimental versus theoretical load-deflection curves using I as prescribed by eq. (4), a large discrepancy is found in deflections after the first crack as shown in Fig. 3 and 4.

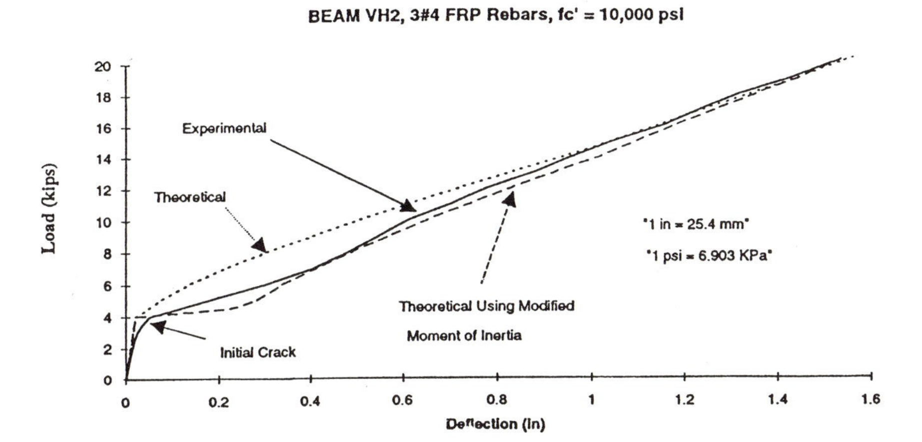

Figure 3 Load vs Deflection (Theoretical vs Experimental)

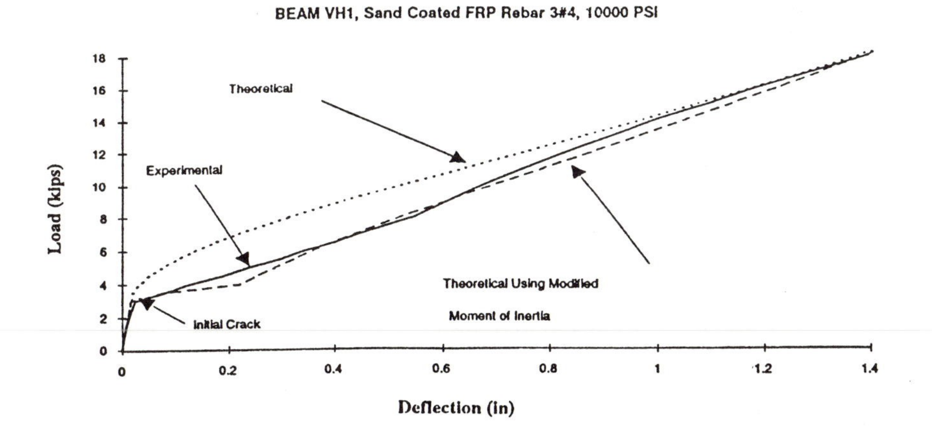

Figure 4 Load vs Deflection (Theoretical vs Experimental)

The equation for deflection of a simply supported beam of span L, loaded with two concentrated loads P (Fig. 5), in kips, at a distance a from each end is written as:

$$\Delta_{max} = \frac{Pa}{24E_cI_e}(3L^2 - 4a^2, \text{(in, 1 in = 25.4 mm)} \quad (5)$$

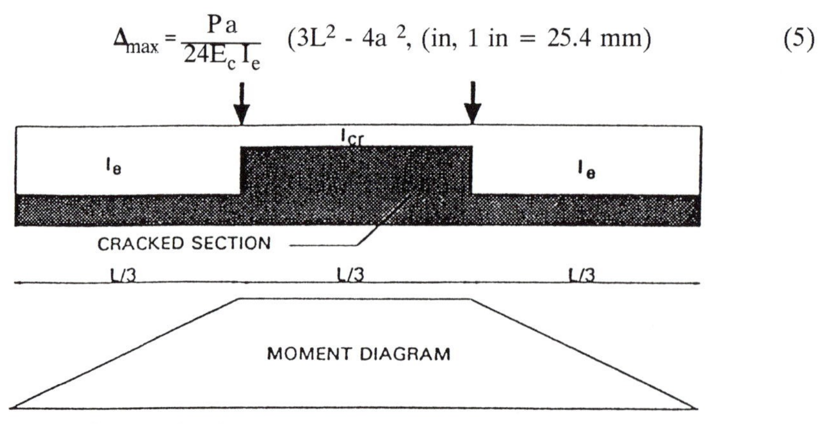

Figure 5 Load, Moment and Cracking Section of Loaded Beam

for a = L/3,

$$\Delta_{max} = \frac{23PL3}{648E_cI_e} \quad (6)$$

in which, E_c = Concrete modulus of elasticity (experimental values, psi)
I_e = Effective moment of inertia (in^4)
L = 108 in., the deflection expression can be written as

$$\Delta_{max} = \frac{44712\ P}{E_cI_e} \quad (7)$$

For evaluating I_e, the experimental cracking moment M_{cr}, observed from the tests was used.

It is seen from the load-deflection curves, Fig 3 and 4, that deflection by Eq.(6) overestimates the moment of inertia of the beam after the first crack. Thus, the calculated deflection values from Eq. (8) are lower than the observed values.

A better estimate of the moment of inertia is needed. In the following subsection, a new expression for the effective moment of inertia has been proposed by the researchers.

Modified Moment of Inertia

Due to the nature of crack pattern and propagation and the height of the neutral axis which is very small for FRP reinforced concrete beams, a new method in calculating the effective modulus of elasticity is introduced for FRP reinforced concrete beams. The new expression is based on the assumption that concrete section between the point loads is assumed to be fully cracked, while the end sections are assumed to be partially cracked (Fig. 5).

Therefore, expression for I_{cr} is used in the middle third section, and I_e is used in the end sections.

Using the moment area approach or other methods to calculate the maximum deflection at the centre of the beam, as shown in Fig. 5, would result in an expression for maximum deflection that incorporates both I_e and I_{cr} as shown in Equation (8)

$$\Delta_{max} = \frac{8PL^3EI_{cr} + 15PL^3EI_e}{648EI_{cr}EI_e}, \text{ in. } (1 \text{ in.} = 25.4) \quad (8)$$

$$= \frac{8PL^3I_{cr} + 15PL^3I_e}{648EI_{cr}I_e} \quad (9)$$

Rewriting the deflection expression in equation (9)

$$\Delta_{max} = \frac{23PL^3}{648E_cI_m} \quad \text{(in)} \quad (10)$$

in which,

$$1_m = \frac{23I_{cr}I_e}{8I_{cr} + 15I_e} \quad (11)$$

The resulting deflection equation (10) and the modified moment of inertia (11) which is valid for two concentrated point loads that are applied at the third points on the beams are plotted as shown in Fig. 3 and 4.

Using the same approach as in the case of two concentrated point loads, expressions for maximum deflection and modified moment of inertia are derived for a concentrated point load and for a uniform distributed load. However, no experimental information is available to check their validity.

For a concentrated point load applied at the centre of the beam the maximum deflection expression can be written as:

$$\Delta_{max} = \frac{PL^3}{48E_cI_m} \quad \text{(in) where} \quad 1_m = \frac{54I_{cr}I_e}{23I_{cr} + 45I_e} \quad (12)$$

For a uniform distributed load applied on the beam, the maximum deflection expression can be written as:

$$\Delta_{max} = \frac{5WL^4}{384E_cI_m} \quad \text{(in) where} \quad I_m = \frac{240I_{cr}I_e}{45I_{cr} + 202I_e} \quad (13)$$

Summary and Conclusions
Based on the mechanical properties of FRP rebars obtained by Wu (1991), twenty five concrete beams were designed and tested under bending. Test variables included concrete strengths (29 - 69 GPa), type of FRP rebar (smooth, ribbed, sand coated), and rebar size. The response of concrete beams reinforced with FRP rebars were investigated in terms of pre- and post-cracking load-deflection behaviour. The use of sand coated FRP rebars in addition to high strength concretes improved the overall behaviour of concrete beams in terms of the ultimate moment capacity, crack width and propagation, thus improving the load-deflection behaviour.

Theoretical correlations with experimental deflections were conducted using current provisions. The current design methodology for steel reinforced concrete beams cannot be applied directly to FRP reinforced concrete beams. New design equations for deflections similar to the ACI 318-89 building code provisions were established based on the experimental results outlined. Theoretical equations were established to compute modified moment of inertia.

Due to the nature of crack formation and propagation in FRP reinforced concrete beams and the low modulus of elasticity of FRP rebars, a modified effective moment of inertia equation is proposed herein to estimate deflection. The modified effective moment of inertia incorporates both the cracked moment of inertia as well as the current ACI code equation, and is valid for sand coated FRP rebars which exhibit a bond strength of at least 1500 psi [eq. (11)].

REFERENCES

1. ACI Building code requirements for reinforced concrete (ACI 318-89), American Concrete Institute, Detroit, MI, 1989
2. American Highway, Accelerating the search for innovation, Transportation Research Board
3. Faza S.S. PhD Dissertation, West Virginia University, Morgantown, West Virginia, 1991. 'Bending and bond behaviour and design of concrete beams reinforced with fibre reinforced plastic rebars'.
4. Faza S.S. and GangaRao H.V.S. Advanced composites materials, ASCE, Edited by Iyer S.S., 1991, pp 262. 'Bending response of beams reinforced with FRP rebars for varying concrete strengths.
5. Nawy E.G. and Neuwerth E.G. ASCE Journal of the Structural Division, Vol 103, No. ST2, Feb. 1977.
6. Nawy E.G. Reinforced concrete, A fundamental approach, 2nd edition, Prentice Hall International, 1990.
7. Rubinski I. and Rubinski A. Magazine of Concrete Research, (6), 71-78, 1954.
8. Wu W.P. 'Thermomechanical properties of fibre reinforced plastic bars'. PhD Dissertation, West Virginia University, 1991.

22 APPLICATION OF PALM TREE FRONDS IN BUILDING HEAT-INSULATING HOUSES

M.S. Abdel-Azim
Egyptian Atomic Energy Authority, Cairo, Egypt

The reasonable mechanical strength, the low thermal conductivity, resistance to corrosion and low cost of Palm Tree Fronds (PTF), were utilized in building heat-insulating and economic houses. Experiments were carried out to determine the mechanical strength and physical properties of PTF. A proposed vault structure was designed and tested applying different kinds of loads. The modulus of elasticity of PTF was found to be one tenth that of steel. Therefore certain process was recommended to build one storey building using PTF with concrete.

INTRODUCTION

The production within developing countries of building materials from indigenous agricultural residues and fibrous materials is necessitated by demands in those countries for efficient housing and low-cost shelter. Building programmes which utilize such materials will not only cut down on the import content of structures, but in most instances, their absolute cost as well.

The characteristics properties of Palm Tree Fronds (PTF) was too challenging a prospect to ignore as a qualified substitute for steel reinforcement and for conventional heat-insulating materials in building efficient and economic houses.

The date palm trees are available every where in Egypt. It constitutes one of the main agricultural product in most of the middle east countries. Pruning of those palm trees in Egypt, for example, gives annually more than 390,000 tons of fronds.

The main objective of the present work is the utilization of this huge amount of renewable natural resource to build heat- insulating and economic houses.

EXPERIMENTAL AND APPLICATION

The main characteristic properties of PTF that promote the idea of using it as a qualified substitute for steel reinforcement and for conventional heat-insulating materials in building efficient houses are its reasonable mechanical strength and its thermal conductivity.

The PTF consists of a 2m to 5m stalk that carried up to 800 mm long and 35 mm wide leaflets. Stalk and leaflets contain fibrovascular bundles which provide fluid transport and structural stability. They covered with a thick layer of cuticles to protect the frond from dehydration.

To avoid dimensional changes in the final product, only reasonably dried PTF may be used. Under normal conditions, in shaded open air, it was found that about 14 days are necessary for the fronds to reach the state of dimensional stability.

Experiments carried out on PTF, through visual observations and weight monitoring, showed that PTF have a high resistance to a wide spectrum of chemical solutions of PH from 1 to 13 at room temperature and up to 60° C.

The thermal conductivity of PTF was found to be identical to that of wood, 0.123 to 0.21 W/(m °C).

Experiments were also carried out to determine the mechanical properties of PTF stalks and leaflets. The ultimate tensile strength of the peripheral portion of the stalks was found to be 175 ± 25 N/mm^2, which was higher than the core Zone, 88 ± 10 N/mm^2. This was because the peripheral fibres in the stalk are more dense and congregated. The ultimate compression strength of PTF was found to be 60 ± 2 N/mm^2.

The experiments were carried out to determine the tensile strength of the leaflets, found that each bundle of PTF leaflets could carry tension in the range of 45 to 50 N. Considering the cross-sectional area of the bundle, the average tension is 1600 N/mm^2.

All the tested Specimens display a linear stress-strain relationship up to the point of rupture. The modulus of elasticity of PTF Stalks was found to be 20,000 ± 1500 N/mm^2.

BASIC PTF COMPONENTS AND APPLICATION

To produce a heat-insulating flat roof slabs, replacing the steel with PTF, the designer must determine the area of PTF stalks necessary to replace the steel bars, without changing the thickness of the slap and the deflection due to the load.

Since the deflection can be expressed as a function f of the modulus of elasticity of PTF,E, and the applied load W,

$$\text{The deflection} = f\,(W/E)$$

Since E of PTF is about one tenth that of steel, the area of stalk reinforcement should be increased by the ratio E_{Steal}/E_{Stalk} (which is about 10) in order to keep the deflection and the load carrying capacity constant.

To build one storey building using PTF, the proposed roofs are PTF membrane have the vault structure, as shown in Figure 1.

The vault structure has 1 m long repeatable module. The main carrying element consists of one layer of PTF, tied together to form the vault structure,

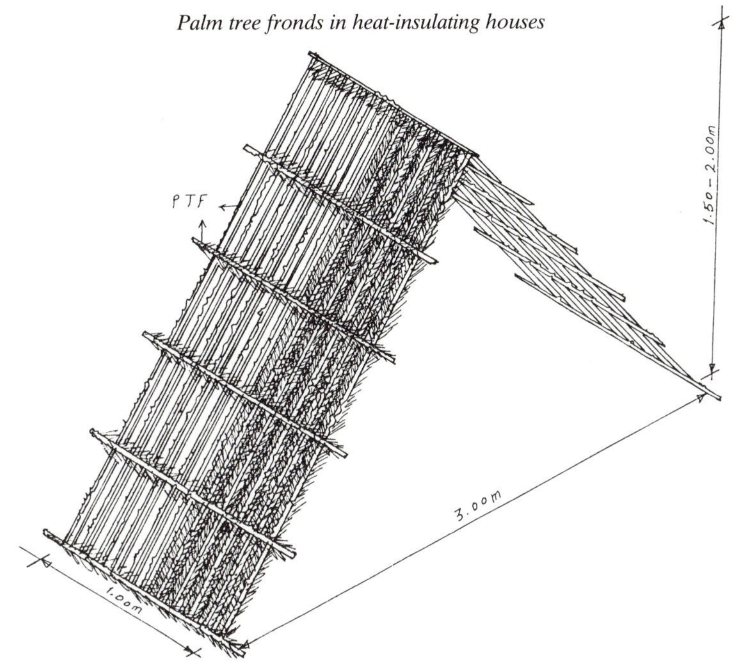

Figure 1. Basic unit of PTF vaulted roof membrane.

having opening span of 3 m and height of 2 m. The unit vault was built horizontally to a height of 1 m after defining its curvature on the floor. Fixed vertical posts were used to control the verticality and straightness of the walls. The vault was then hoisted and turned upright so that its lower edges were supported on the bearing walls. As previously mentioned, PTF stalks can resist high stresses, moreover, leaflets should be split so that PTF leaflet strips work as a matrix with PTF stalks for reinforcement the concrete. They are combined together to form a hard fibre-slab.

Leaflets should be split before tying the stalks together, in order to increase workability. Leaflet strips can be produced using special steel comb splitter designed for this purpose, without cutting the strong fibrovascular bundles.

Although the structure, at this stage, is strong and stable, it is not airtight, and not completely rigid. Therefore it was decided to test the vault after verifying these two conditions. Hence the vault was covered with a 50 mm thick layer of sand-cement mortar.

The vault was tested by applying vertical and horizontal loads successively. The deformation was followed during and after the loading process to determine the type of deformation. The vertical load consisted of a distributed load of 200 Kg/m^2 and concentrated load of 130 Kg applied at the top of the vault. The maximum vertical deformation took place at the top after 30 minutes; it was only 0.4 mm. The maximum horizontal deformation of 0.2 mm occurred at two-thirds of the total height.

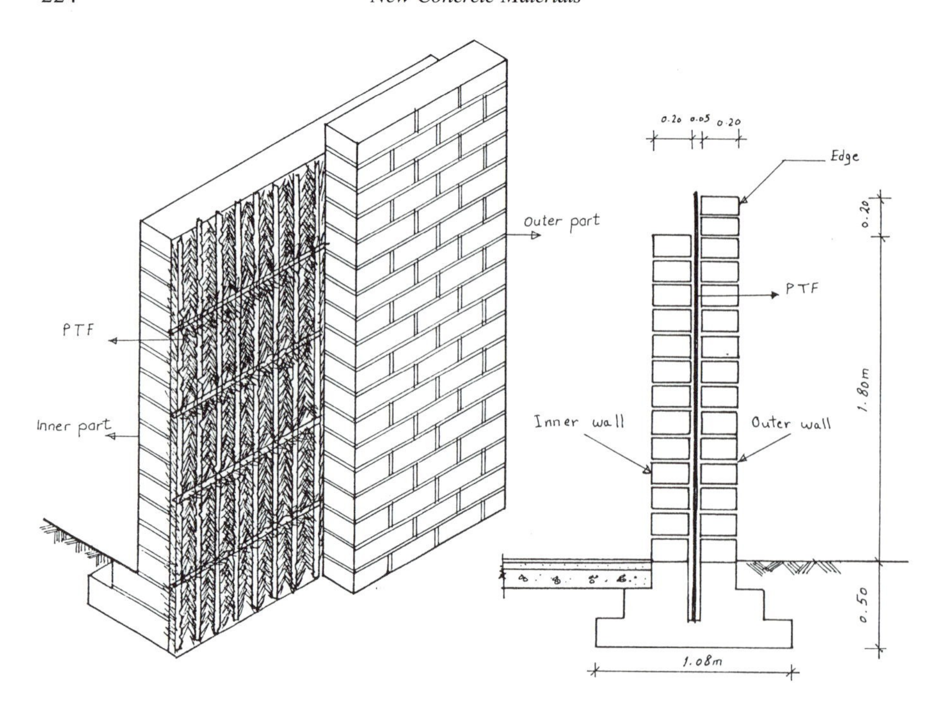

Figure 2. Section in outside bearing wall.

When the vault was unloaded all the displacement disappeared after 30 minutes indicating that the deformation was elastic. The horizontal load was 80 Kg/m^2, representing wind effect. The maximum horizontal deformation was 2.5 mm at the top of the vault. Again all the displacements were recovered, indicating elastic deformations.

To produce a heat insulating vaulted PTF roofs, 2 bearing walls should be first built. Each wall consists of two parts separated by PTF as shown in Figure 2. PTF should be attached to the first part of the wall and fixed to 50 cm in the ground with concrete. The second part of the bearing wall will cover completely the PTF and the first part of the wall, and it should have an edge 20 cm high as shown in Figure 2. The edge will prevent the PTF vault from sliding horizontally, moreover, it will hold the plain concrete when placed on the top of the vaulted roof.

Before placing the vaulted PTF unit of the roof, a mortar composed of sand, cement and water should be placed on form work having the same shape of the vaulted PTF roof, so that the PTF basic unit will sink in the mortar.
Figure 3, shows the general front view of completed one storey building.

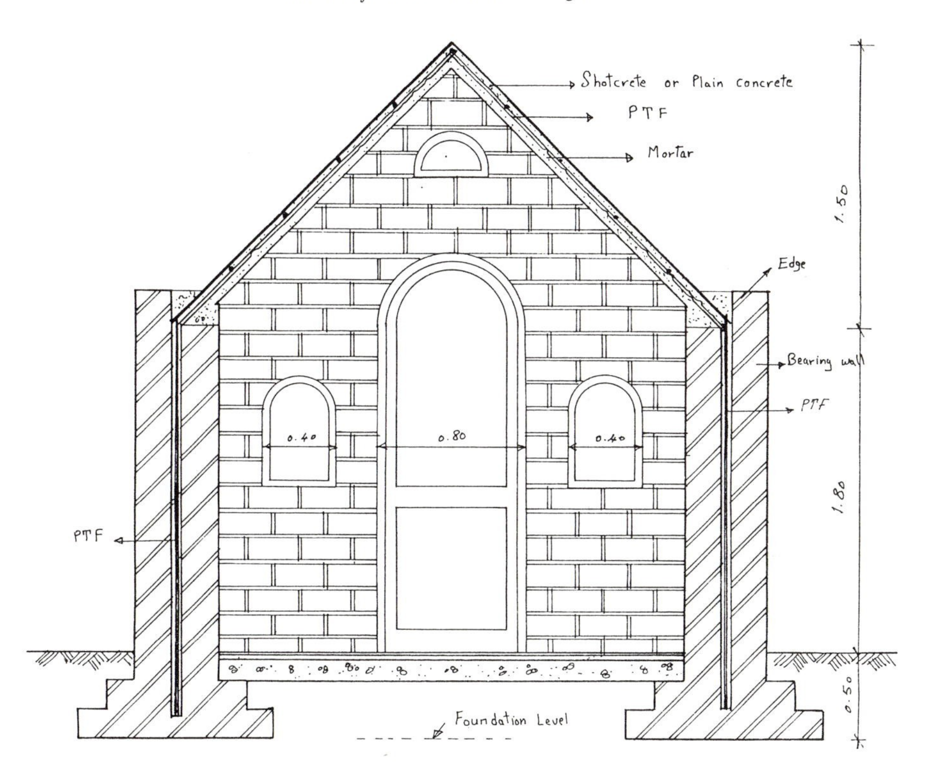

Figure 3. Front view of completed one storey building entrance

23 THE TESTING OF A MASS CONCRETE ARCH BRIDGE

C. Melbourne
Bolton Institute of Higher Education, UK

The paper describes the construction and testing of a mass concrete arch bridge which formed part of a contract to improve part of the A59 Samlesbury-Skipton Trunk Road.

INTRODUCTION

The arch has been used by engineers and architects in buildings and bridges for thousands of years. Its perennial appeal lies not only with its structural efficiency but also with its aesthetic appeal. The first arches and vaults were built using sun-dried mud bricks in mud mortar by the masons of ancient Egypt and Mesopotannia(1). These were probably a development of vaulted reed structures depicted in early Egyptian drawings.

To Europeans, arches are usually associated with the Romans who exploited this form of construction to its fullest not only in buildings of monumental proportions but also in their civil engineering works like bridges and aqueducts.

The majority of the world's stock of arch bridges are constructed of brick or stone and have stood for hundreds of years. They have shown themselves to be durable structures, capable of carrying loads very much greater than those envisaged by their designers and withstand significant relative deformation without loss of structural integrity. Consideration of the latter properties indicated that it was the particulate nature of the structure which allowed such movement to take place whilst maintaining a load path through the masonry block interfaces.

If this could be replicated in a mass concrete structure and thus produce the same articulation then a structure of comparable structural efficiency and potential longevity would have been produced. This can be achieved using plastic sheets fabricated to produce a grillage which when cast into the mass concrete produces equivalent voussoirs (wedge shaped blocks) similar to the historical stone arch structures with which we are all familiar. The entire barrel does not have to be treated in this way; only sufficient to ensure that the barrel articulation is assured.

Details of the technique are presented elsewhere[2]. This paper concentrates on the construction and test loading of a mass concrete arch bridge constructed by Mowlem Regional Construction Ltd for the Lancashire County Council on behalf of the Department of Transport.

DESCRIPTION OF THE WORKS

The Works were located on the A59 trunk road 3 km east of Gisburn Lancashire and were designed to improve the existing road, traffic movements at the junction with the B5251, and to provide an overtaking facility on the up hill section east of Monk Bridge.

Monk New Bridge carries the re-alignment of the A59 Samlesbury-Skipton Trunk Road over Stock Beck. The existing Monk Bridge, a single span masonry bridge of 10.98 m skew span, was demolished as part of the contract.

The new bridge was designed by the Lancashire County Council and is a single span in-situ mass concrete voussoir arch, 725 mm thick, with a clear square span between springings of 9.9 m (11.32 m skew), a rise above the springings of 2.2 m, a square width over parapets of 13.577 m and a skew angle of 29° (Figure 1).

The arch is carried on reinforced concrete abutments which are connected at foundation level by precast reinforced concrete beams. The wing walls are in reinforced concrete and both abutments and wing walls are supported by bored cast-in-place continuous flight auger piles founded on dense to very dense gravel. The parapets over the arch are independent walls with spread footings, the footings being tied by reinforced concrete ties. The exposed faces of the wing walls, spandrels and parapets are masonry faced.

CONSTRUCTION AND INSTRUMENTATION OF THE BRIDGE

Construction started during the summer of 1990. The piling sub-contractor experienced some difficulties during the installation of the piles which resulted in delays. Consequently, manufacture and installation of the ground beams was delayed until April 1991.

The arch was formed on an RMD triangulated scaffolding frame onto which timber joists and plywood were attached. The 6 mm thick UPVC crack inducers were positioned against the springings. Side shutters were erected and sections cast up to the position of the quarter span grid of the crack inducers. The extrados of this section of the arch was formed by using an Expamet flyrib system. It can be seen from Figure 2 that the prior casting of the sections of the cementitious 'backfill' provided a substantial support for the temporary works as well as allowing early backfilling with granular material in preparation for the road surfacing.

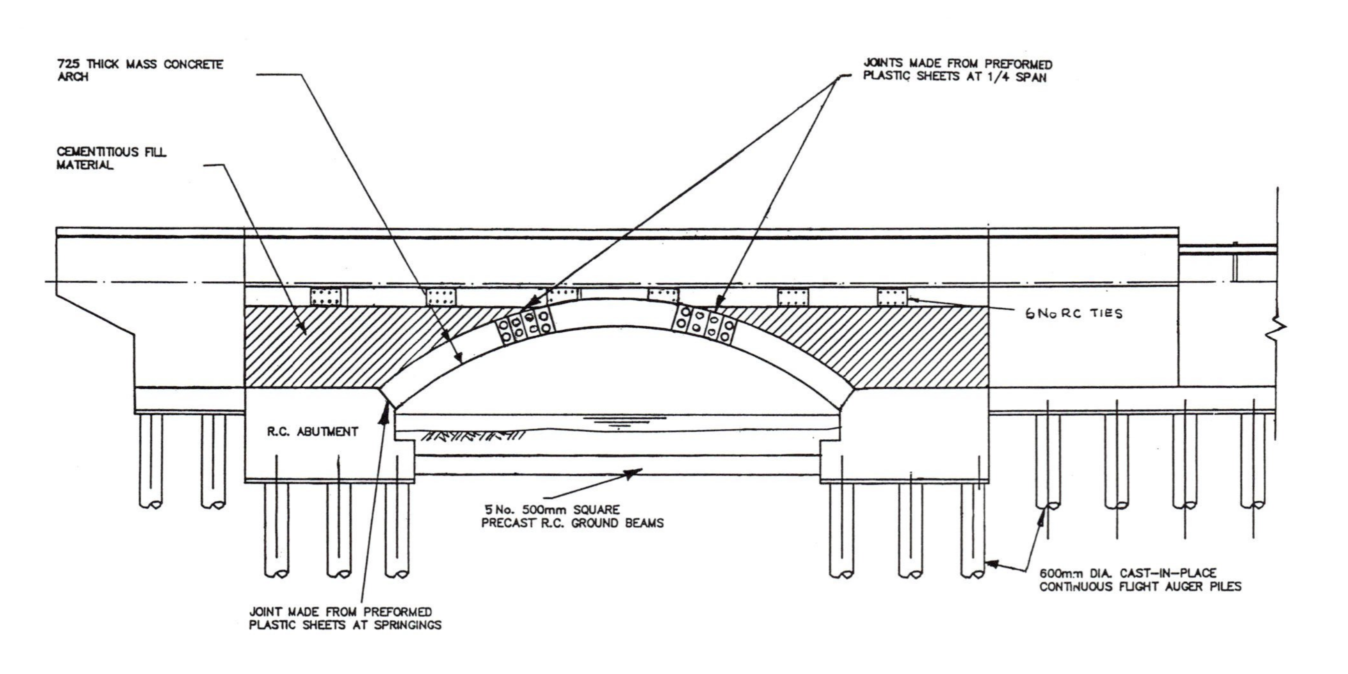
725 THICK MASS CONCRETE ARCH
CEMENTITIOUS FILL MATERIAL
JOINTS MADE FROM PREFORMED PLASTIC SHEETS AT 1/4 SPAN
6 No RC TIES
R.C. ABUTMENT
5 No. 500mm SQUARE PRECAST R.C. GROUND BEAMS
JOINT MADE FROM PREFORMED PLASTIC SHEETS AT SPRINGINGS
600mm DIA. CAST-IN-PLACE CONTINUOUS FLIGHT AUGER PILES

Figure 1

Figure 2

Figure 3

Marrying the 6 mm thick UPVC grillages up to the lower section of the arch proved to be difficult and resulted in cutting the plastic units and re-jointing them once geometrical compatibility had been satisfied.

Vibrating wire strain gauges were used to monitor internal strains. They were held in position by prior casting into compatible concrete prisms which had sockets cast into them to allow bolting to the grillage, Figure 3. Both grillages were cast at the same time and for the full width of the bridge.

The strain gauges at the crown were precast into compatible concrete slabs which were positioned and concreted into the arch. This section was cast in two sections. Once the arch was completed, the surface mounted vibrating wire strain gauges were attached to the extrados in the same positions as the embedment gauges.

Side shutters were fixed and the cementitious backfill was placed over the arch. This was brought up in lifts until the gauges were covered and the underside level of the ground slab connecting the parapets had been reached. At this stage the reinforcement for the parapets and tying ground slab was installed, instrumented and concreted ready to receive the artificial and natural stone facing. The bridge received its kerb line and black top surfacing and was opened at the end of 1991.

TEST ARRANGEMENT

Loading

The bridge was subjected to five separate load tests using an 'HB' trailer with three increments of 25T, 35T and 45T (nominal) load. For 25T and 35T (nominal) loads the trailer was positioned to produce centreline influence lines only. For 45T (nominal) load influence lines were produced for the centreline, eastbound lane and westbound lane respectively. The influence lines were produced by positioning the trailer at pre-determined longitudinal locations at approximately one-eighth span increments across the bridge.

Instrumentation

Deflections were measured using a total of twenty potentiometer and LVDT type linear displacement transducers positioned beneath the intrados of the arch barrel and attached to an independent scaffolding frame secured in the stream bed. The resolution of each type of gauge was 0.11 mm and 0.01 mm respectively.

The backfill pressures between the cementitious backfill and the granular backfill were measured using six 'Gage Technique' BSP type vibrating wire earth pressure cells which were incorporated into the cementitious backfill drainage layer at the west side of the bridge.

In order to evaluate the relative contribution of each structural element within the bridge, a representative distribution of embedment vibrating wire strain gauges were installed in the ground beams, tie beams and arch barrel. In the beams, the gauges were cast directly into the element with an adjacent 'dummy' beam. The barrel presented a special problem as there was no steel reinforcement to which the gauges could be attached. The instrumentation was held in position by prior casting of the gauges into compatible concrete prisms which had sockets cast into them to allow for bolting to the plastic grillage. The gauges at the crown and springings were precast in compatible concrete slabs which were positioned and concreted into the arch.

DISCUSSION OF RESULTS

Ground Beams

The gauges indicated that the tie beams played an active role during the loading cycle of the bridge. Those in tie beams recorded tensile strains which increased to a maximum of 5 μ strain when the 45T axle was at the crown. This corresponded to a load of about 47 kN, assuming a modulus of elasticity for the concrete of 35 kN/mm^2.

Deflections

The deflections recorded during the test were within the stated accuracy of the gauges. Although the scaffolding was braced and secure, the stream had been in flood and the flow was still substantial, this would inevitably cause some vibration and hence movement of the scaffolding and shear transducers. None of the gauges gave sufficiently consistent readings to warrant detailed analysis. However, within the accuracy of the above it can be recorded that the bridge is unlikely to have experienced any deflections greater than 0.10 mm.

Vibrating Wire Strain Gauges

The vibrating wire strain gauges were used to determine:

- the strain profile at various positions in the arch barrel
- whether or not the backfill concrete acted compositely with the arch barrel
- the degree of load distribution
- the effect of skew

Springing Gauges

Nine strain gauges were installed at each springing. The maximum change in strain was 2 μ strain; this represented a stress change of the order of 0.07 N/mm^2. The gauges on the centreline and obtuse corners of the bridge were more active - indicating that the bridge was trying to span square to the abutments. Only when the axle was immediately over the acute corners was there a comparable response.

The bottom gauges were generally more active than the mid-depth and top gauges. This suggests that the arch was acting compositely with the concrete backfill.

Quarter Span Gauges
At the quarter span; longitudinal and transverse embedment strain gauges were installed with intrados and extrados gauges fixed to the surface to be correlated with the internal gauges. The embedment gauges recorded low strain changes of up to 4 μ strain. The gauges at each level within the arch barrel tended to follow the same trend of stress either tensile or compressive - this suggested that the concrete backfill was behaving compositely with the barrel.

The influence lines for the strains are compatible with the notion of the arch 'swaying' as the axle moved from one side of the bridge to the other. The intrados and extrados gauges confirmed this observed behaviour. However, the magnitudes of the recorded intrados strains were large compared to the corresponding internal gauge strains. The most likely explanation is that the backfill concrete encouraged a 'mass concrete' response of the extrados gauges whilst the intrados gauges which bridged the plastic crack inducers were 'free' to respond to the imposed stress, as the modulus of elasticity of the plastic is very much lower than that of the concrete. If the 'E' for the UPVC is 2 kN/mm^2, ie a twentieth of that of the concrete then a plastic would have produced twenty times more strain.

It was also noted that the mid-depth internal gauges attracted more strain than those near the intrados and extrados. This raised an interesting question, if the arch was subjected to flexural loading and plane sections remained plane then the strain gauges should display this by a linear distribution - this was not the case. It is likely that the plastic acted as a stress reliever to the adjacent concrete thus encouraging the thrust to flow through the 'cut-outs' in the plastic. The net effect is to create a 'diffused hinge' at the quarter points the effective depth of which is reduced in the elastic range thus making the 'hinge' more flexible but which will not affect the ultimate condition.

Crown Gauges
The embedment gauges at the crown of the bridge gave the most predictable results with maximum strains being recorded when the axle was over the crown gauges. If it is assumed that plane sections remain plane the relative range of the strain measurement for any load case is indicative of the induced bending moment. It follows that the bending moment on the centreline, when the axle was on the centreline crown, was of the order of two to three times that recorded in the corresponding kerb line gauges. Additionally, the kerb line gauges showed a significant increase in bending moment when the loads were applied adjacent to them and the implied bending moments were greater than the corresponding maximum for the centreline gauges with centreline loading.

Transverse Gauges
The transverse gauges were installed to monitor the transverse behaviour of the arch barrel. Generally, through the passage of the axle the variation in recorded strain was ± 1μ strain when allowance for the drift was made. Only in the cases of asymmetric loading near the crown did the gauges adjacent to the load show significant changes in strain compatible with local hogging. The strains were

approximately half the value of the local longitudinal sagging strains recorded under the same loading conditions.

CONCLUSIONS

The following conclusions may be made:

1) the ground beams played an active role in restraining the spread of the abutments;
2) the concrete backfill acted compositely with the concrete arch barrel;
3) the plastic crack inducers acted as a local stress reliever thus encouraging the thrust surface to flow through the cut outs in the plastic;
4) only a limited amount of transverse distribution took place;
5) the carriageway tie beams were unaffected by the test load;
6) horizontal soil pressures increased significantly as the axle approached the bridge but were not affected by the axle once it was on the bridge;
7) the influence lines for the intrados gauges implied that the barrel 'swayed'.
8) the barrel tended to span square to the abutments.

REFERENCES

1. Van Beck G.W. 'Arches and Vaults in the Ancient Near East' Scientific America (July 87).
2. Melbourne C. 'A New Arch Construction Technique' Structural Engineering Review, 1, 91-95 (1988)

ACKNOWLEDGEMENTS

The author wishes to acknowledge the support and help given by the Lancashire County Council, NW Regional office of the Department of Transport, the TRL (especially John Page and Colin Beales), Mowlem Regional Construction Ltd and my colleagues at the Bolton Institute of Higher Education.

24 THE CONCEPT OF CONCRETE COVER DESIGN IN REINFORCED CONCRETE COLUMNS

W.H. Chen
National Central University, Taiwan, R.O.C.

To investigate the cracking, cleavage and spalling behaviour of the concrete cover of the reinforced concrete columns, the 27 series of 81 axially loaded specimens with uniaxial small eccentric compressive loading were tested. Factors considered were the thickness of the cover, the spacing of the transverse steel and the longitudinal bar diameter. It was concluded that 51% of cracks have the internal cleavage failure between the transverse steel and concrete cover at 97% of its average ultimate strength. The cleavage length of transverse steel increased with decreasing the cover thickness, and decreasing the spacing of transverse steel.

INTRODUCTION

The internal cleavage failure between the concrete cover and longitudinal/transverse steel on reinforced concrete columns in the relatively new buildings due to the moderate earthquake motion is almost unavoidable in the active earthquake areas such as in Taiwan. When a crack opens up from surface of concrete to steel and provides access to external corrosive agents in the vicinity of cleavage, the function of concrete cover would be lost and the corrosion of steel would arise(1)(2)(3). As a result, the products of steel corrosion occupy a volume several times greater than the steel from which it is produced(4), this volumetric changes lead to a bursting effect to the surrounding concrete, it also results in cracking and spalling of concrete cover, once the cover has been spalled off, rapid corrosion of the exposed steel will take place. Therefore from the stand point of corrosion protection of reinforcing steel and durability of structures, there is a need for more research in the cracking, cleavage and spalling behaviour of the concrete cover of the reinforced concrete compress member for concrete cover design.

In spite of considerable experimental research, design philosophy of concrete cover for columns is still not fully understood. Generally, the approximations introduced are: (a) using gross area of concrete instead of core area and the spalling of the cover concrete at high strains was ignored for strength calculations, (b) using the core area of concrete instead of gross area and the cover was considered to be ineffective for ductility analysis, (c) considering corrosion and fire resistance an adequate concrete cover thickness over the embedded steel should be provided for durability. It is difficult to determine the strain at which cleavage of the cover concrete commences because the cleavage process occurs gradually. However, cleavage failure need not consider to repair from the durability viewpoint. It was the purpose of the experimental program described in this paper to examine the various parameter on contribution of concrete cover for strength, ductility and durability.

TEST PROGRAM

The eighty one short column test specimens were cast in 27 series of 3 columns each. The variables considered were the thickness of the clear concrete cover Dn, the spacing of the transverse steel S and the longitudinal bar diameter Db. The specimens were 16 cm, 18 cm and 20 cm square cross section and 70 cm high. The core perimeter measured from the out side to out side of square ties were kept constant at 12 cm for all specimens. Clear concrete cover thickness Dn of 2 cm, 3 cm and 4 cm were chosen for investigation. The tie bars were manufactured from 10 mm (#3) diameter deformed bars and had a spacing S of 4 cm, 8 cm and 16 cm. The longitudinal bar diameter Db of 13 mm (#4), 16 mm (#5) and 19 mm (#6) were used and placed in the four corners of square ties.

The concrete used in the specimens was made of ordinary Portland Cement (Taiwan Suini). Tou Zhian River sand with a maximum size of 5 mm and Dah Hann River gravel with a maximum size of 15 mm were used in all of the mixes. A water-cement ratio of 0.73 was used. Slump for all the mixes varied between 16 cm and 18 cm . The specimens were cast with the longitudinal bars in a vertical position. The average compressive strength fc', obtained from 150 300 mm standard cylinders at 28 days was 19.3 N/mm^2, the average 300150 mm diameter cylinder splitting strength was 1.4 N/mm^2, the average modules of rupture was 2.5 N/mm^2. The yield strengths, determined by the 0.2 percent offset method, were 358.9 N/mm^2 for 10 mm reinforcing bars, 379.5 N/mm^2 for 13 mm reinforcing bars, 386.4 N/mm^2 for 16 mm reinforcing bars and 361.9 N/mm for 19 mm reinforcing bars.

TEST PROCEDURE

The specimens were tested in a 100 ton universal hydraulic testing machine. Loads were applied to the specimens with the minimum eccentricity of 0.10 h specified by the ACI code 318-71. Crack widths, obtained perpendicular to direction of a particular crack at the position of the reinforcement were measured with a 100-power microscope with a least count of 0.01 mm.

The crack width measurements identified by the Wc and Wo are the measure crack width on the concrete surface under loading and unloading, respectively. The cleavage length measurements identified by the Ws are the cleaved length of the concrete cover measured at the steel-concrete interface under loading. The major steps of measuring these crack widths are as follows:

a) As a first load cycle, the specimen was loaded to formation of a crack at a rate of about 4000 N per minute. Both the load Pcr1 that produced the first crack and the crack width of the first crack Wc1 were measured. After that, the 1% phenolphthalein alcohol solution was injected from the opening of crack into the concrete, and then the load was released.
b) After the load had reached a value of zero, the residual crack width of first crack Wo1 was measure.
c) The second cycle of load was applied to the specimen until the second crack occurred, the Wc1 of the first crack, Wc2 of the second crack and the second cracking load Pcr2 were measured respectively, the phenolphthalein alcohol solution was injected and then the load was released.
d) After the specimen was unloaded, the Wo1 and Wo2 of first and second crack widths were measured respectively.
e) The third cycle of load was applied until the third crack occurred, the same procedure, as described above, was repeated, the Wc1, Wc2, Wc3 and Pcr3 were measured, the phenolphthalein alcohol solution was injected, Wo1, Wo2 and Wo3, corresponding to Pcr3 were measured.
f) If the specimen did not fail during the three load cycles, the specimen was then loaded to failure under the fourth load cycle. After failure, a careful visual inspection of the specimen was made and the length to be dyed red with the phenolphthalein alkaline reaction on the plane at the steel surface was measured as the length of Ws1, Ws2 and Ws3. Total testing time for a single specimen varied between 2 and 3 hours.

TEST RESULTS & ANALYSIS

The results of experiments to study the cleavage behaviour of concrete cover in reinforced concrete columns indicate that:

a) 51% of cracks have the internal cleavage failure between the transverse steel and concrete cover and 24% of cracks have the internal cleavage failure between the longitudinal steel and concrete cover at 97% of its average ultimate strength, as shown in Table 1.
b) For a given crack width on the concrete surface under loading Wc, the average cleavage length under loading along the longitudinal/ transverse steel Ws,increased with decreasing of clear concrete cover thickness Dn, as shown in Figs. 1 and 7. The ratios of Ws/Wc and Ws/Wo for various parameters were shown in Tables 2, 3, 4 and 5.
c) For a given crack width Wc, the Ws along the longitudinal steel was found to increase with the increase of the spacing of transverse steel S as shown in Fig.2.
d) The Ws along the longitudinal steel was found to increase with the decrease of clean concrete cover thickness Dn and with the increase of the spacing of the transverse steel S, as shown in Figs. 3 and 4.

e) For a given Db, the Ws along the longitudinal steel was found to increase with the decrease of the Dn, and increase of the S as shown in Figs 5 and 6.
f) For a given crack width Wc, the Ws along the transverse steel was found to increase with the decrease of the S as shown in Fig.8.
g) The Ws along the transverse steel was found to increase with decrease of the Dn and the S as shown in Figs. 9 and 10.
h) For a given Db, the Ws along the transverse steel was found to increase with the decrease of the Dn and the S as shown in Figs. 11 and 12.
i) The importance of the factors affecting the cleavage of concrete cover was the concrete cover thickness and the pitch of transverse steel, and the influence of the longitudinal bar size was not significant.

Table 1. Percent of cleavage for longitudinal and transverse steel

Parameters		Longitudinal steel		Transverse steel	
		Number	%	Number	%
Clear cover thickness (mm)	20	31	38.3	53	65.4
	30	23	28.4	47	58.0
	40	5	6.2	23	28.4
Transverse steel spacing (cm)	4	8	9.9	54	66.7
	8	21	25.9	42	51.9
	16	30	37.0	27	33.3
Longitudinal steel dia. (mm)	13	16	19.8	38	46.9
	16	21	25.9	45	55.6
	19	22	27.2	40	49.4
Total samples = 81x3		59	24.3	123	50.6

Table 2. Ws/Wc and Ws/Wo for Wc>0.41 mm

Type of steel	Cover cleavage Number %	Ave. Ws mm	Ave. Ws/Wc	Ave. Ws/Wo	Ave. Pcr3/Pu %	Ave. Pcrl/Pu %
Longitudinal	59 39.1	9.213	10.601	11.205	97.2	93.9
Transverse	110 72.8	18.414	20.896	21.629	97.2	

Note: Based on 151 samples (Wc=0.41 mm, Z=30 MN/m, AC1318M-89, Eq.10-4)

Table 3. Ws/Wc and Ws/Wo for different thickness of clear concrete cover

Type of steel	Clear cover mm	Wc > 0.41 mm Number	Cover cleavage		Ave. Ws mm	Ave. Ws/Wc mm/mm	Ave. Ws/Wo mm/mm	Ave. Pcr3/Pu %	Ave. Pcrl/Pu %
			Number	%					
Longi-tudinal	20	50	31	62.0	17.698	23.188	23.032	97.5	94.4
	30	51	23	45.1	8.609	7.983	8.806	96.5	93.5
	40	50	5	10.0	1.344	0.685	0.824	97.4	94.4
Trans-verse	20	50	43	86.0	23.860	33.337	34.377	97.5	94.4
	30	51	44	86.3	23.580	22.991	23.728	96.5	93.5
	40	50	23	46.0	7.698	6.318	6.741	97.4	94.4

Table 4. Ws/Wc/ and Ws/Wo for different spacing of transverse steel (S)

Type of steel	S mm	Wc > 0.41 mm Number	Cover cleavage		Ave. Ws mm	Ave. Ws/Wc mm/mm	Ave. Ws/Wo mm/mm	Ave. Pcr3/Pu %	Ave. Pcrl/Pu %
			Number	%					
Longi-tudinal	4	54	8	14.8	1.476	1.646	1.690	96.7	93.4
	8	51	21	41.2	8.318	9.802	10.065	96.8	93.5
	16	46	30	65.2	19.288	22.001	23.637	97.9	95.4
Trans-verse	4	54	48	88.9	27.128	29.596	30.462	96.7	93.4
	8	51	40	78.4	17.134	20.930	21.652	96.8	93.5
	16	46	22	47.8	9.539	10.645	11.234	97.9	95.4

Table 5. Ws/Wc and Ws/Wo for different diameter of longitudinal steel (Db)

Type of steel	Db mm	Wc > 0.41 mm Number	Cover cleavage		Ave. Ws mm	Ave. Ws/Wc mm/mm	Ave. Ws/Wo mm/mm	Ave. Pcr3/Pu %	Ave. Pcrl/Pu %
			Number	%					
Longi-tudinal	13	49	16	32.7	6.187	8.422	9.225	97.5	94.6
	16	53	21	39.6	8.825	9.997	10.388	97.0	93.9
	19	49	22	44.9	12.659	13.456	14.068	96.9	93.9
Trans-verse	13	49	34	69.4	17.058	19.947	20.724	97.5	94.6
	16	53	41	77.4	19.089	22.413	23.240	97.0	93.9
	19	49	35	71.4	19.040	20.204	20.793	96.9	93.9

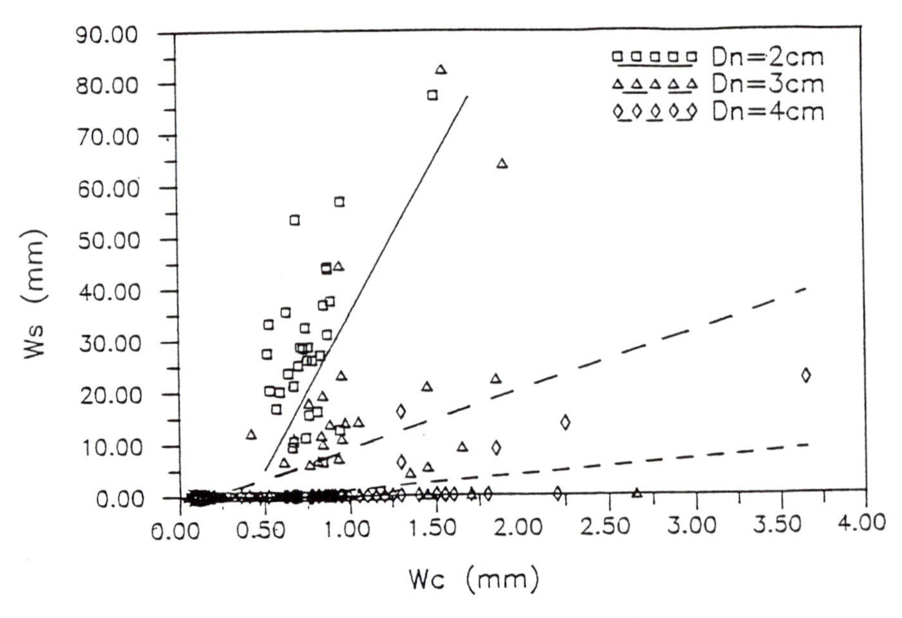

Figure 1. Ws-Wc-Dn relationships of longitudinal steel

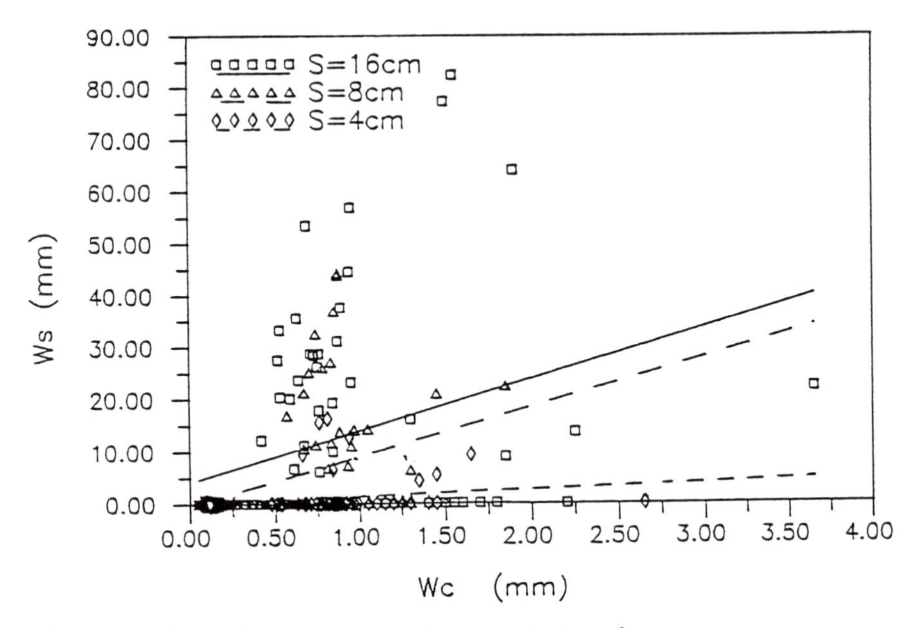

Figure 2. Ws-Wc-S relationships of longitudinal steel

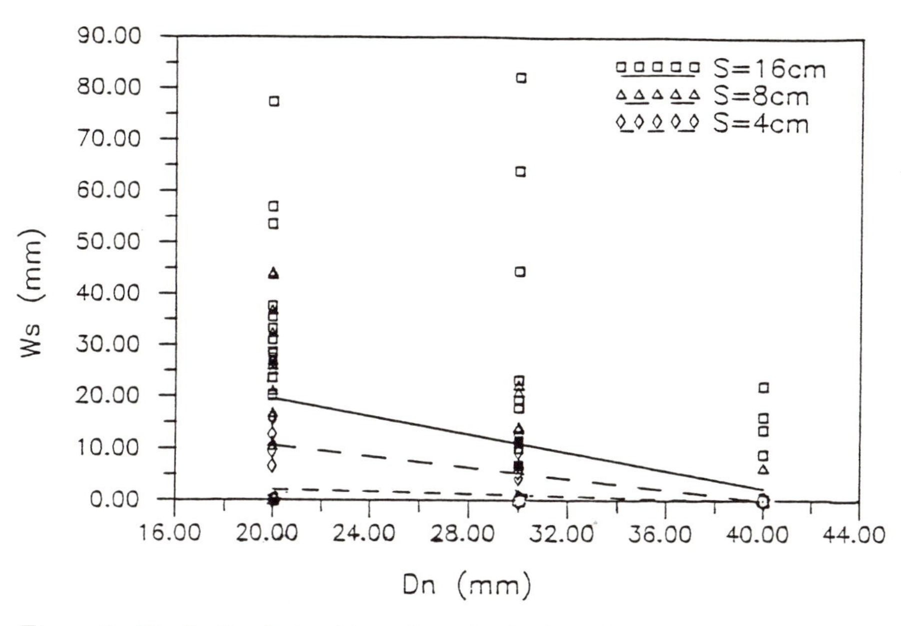

Figure 3. Ws-Dn-S relationships of longitudinal steel

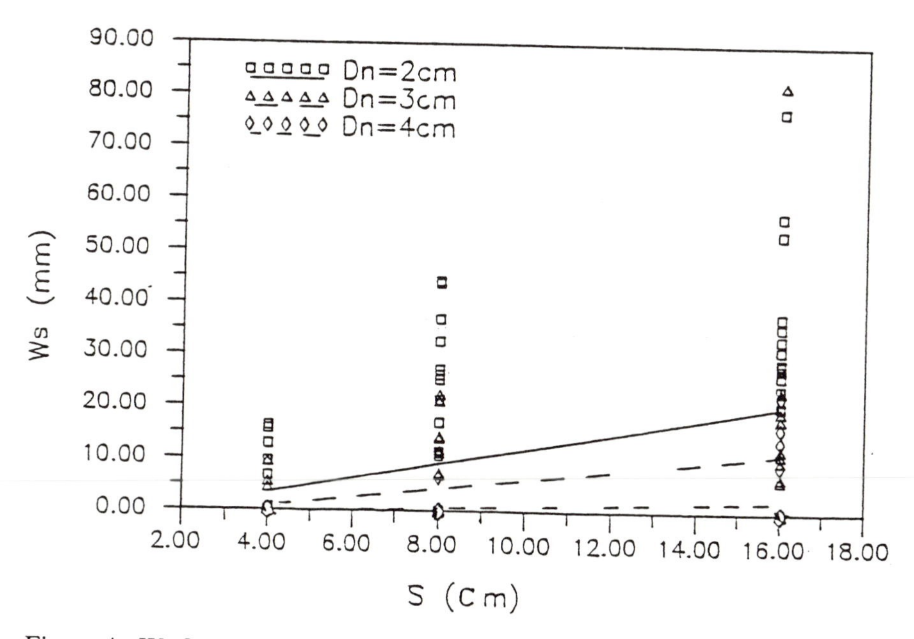

Figure 4. Ws-S-Dn relationships of longitudinal steel

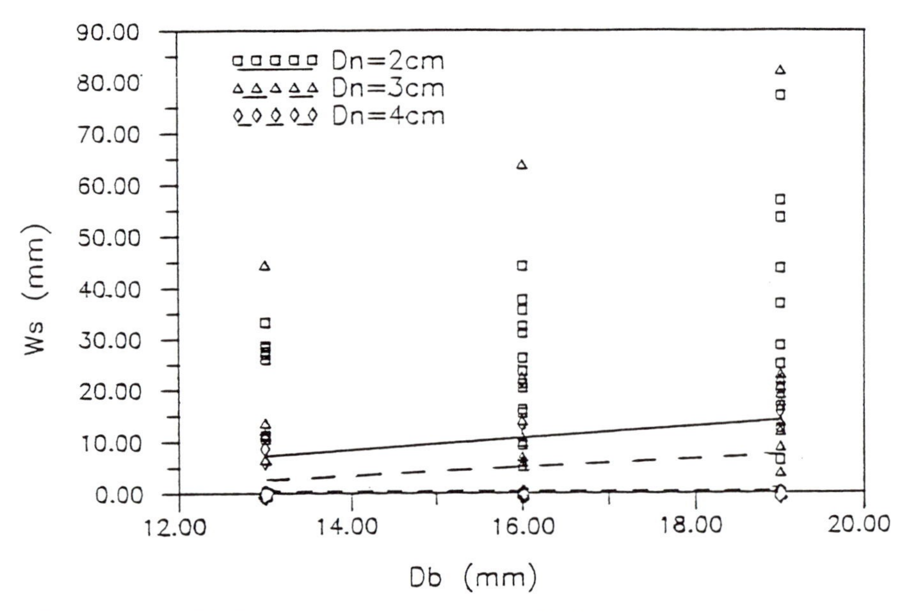

Figure 5. Ws-Db-Dn relationships of longitudinal steel

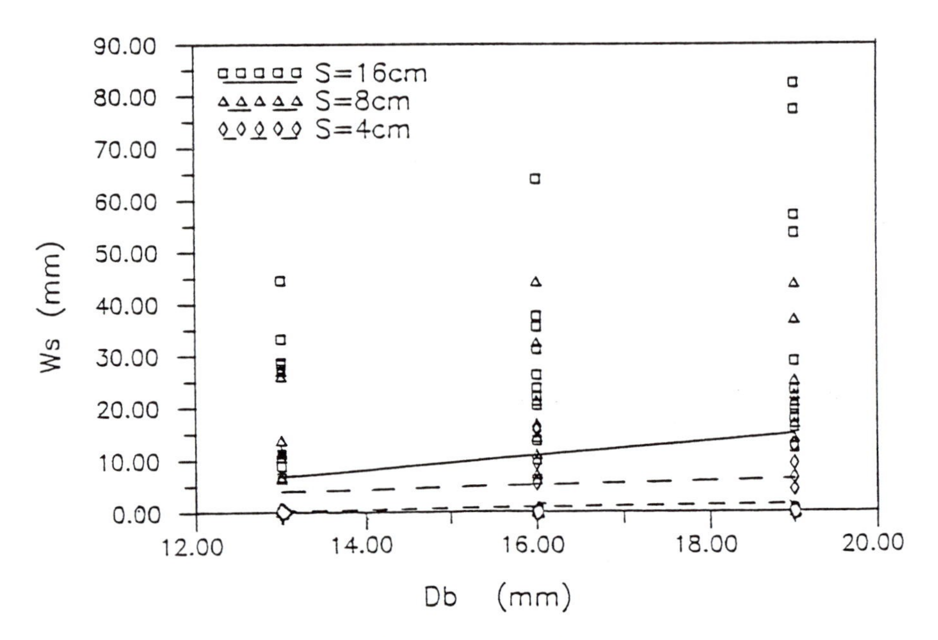

Figure 6. Ws-Db-S relationships of longitudinal steel

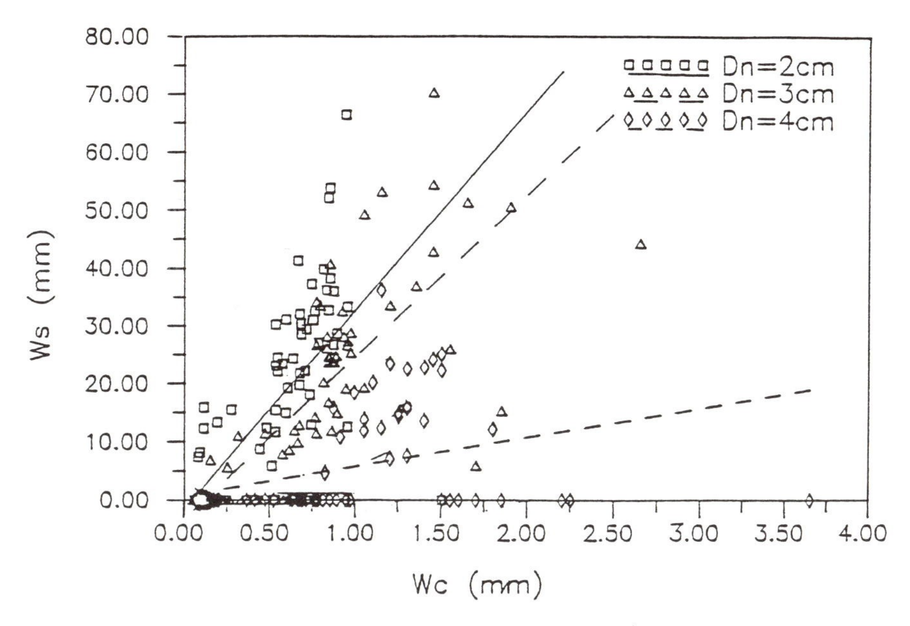

Figure 7. Ws-Wc-Dn relationships of transverse steel

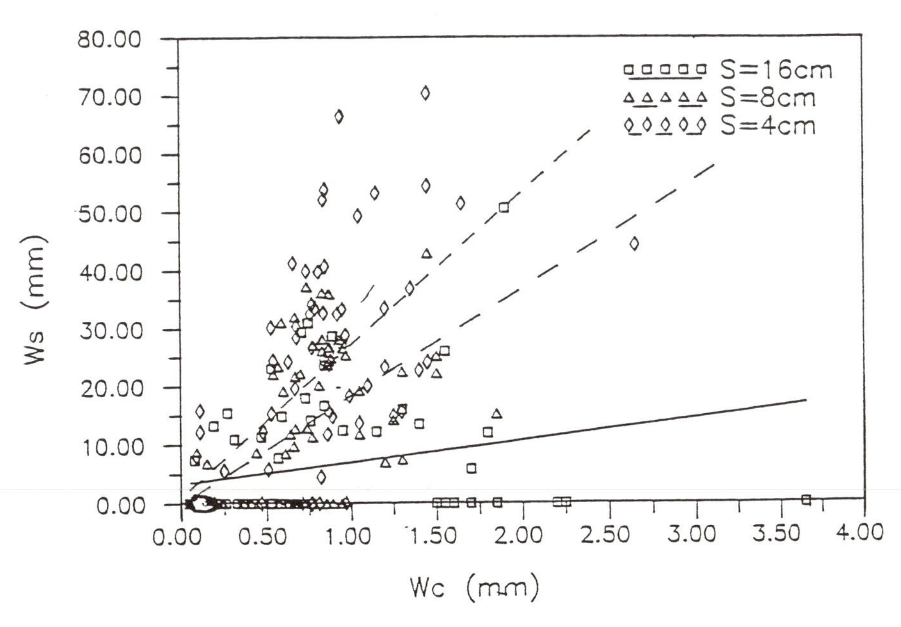

Figure 8. Ws-Wc-S relationships of transverse steel

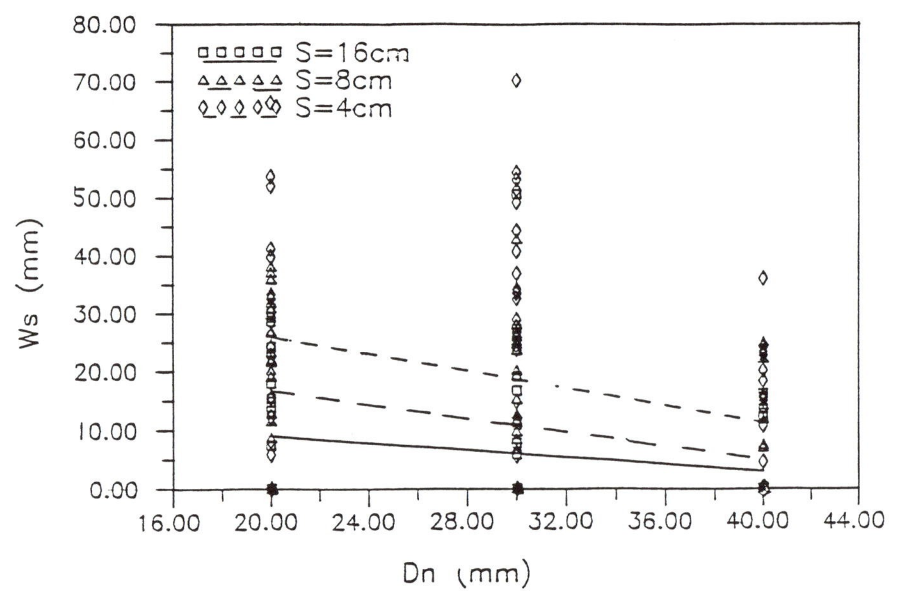

Figure 9. Ws-Dn-S relationships of transverse steel

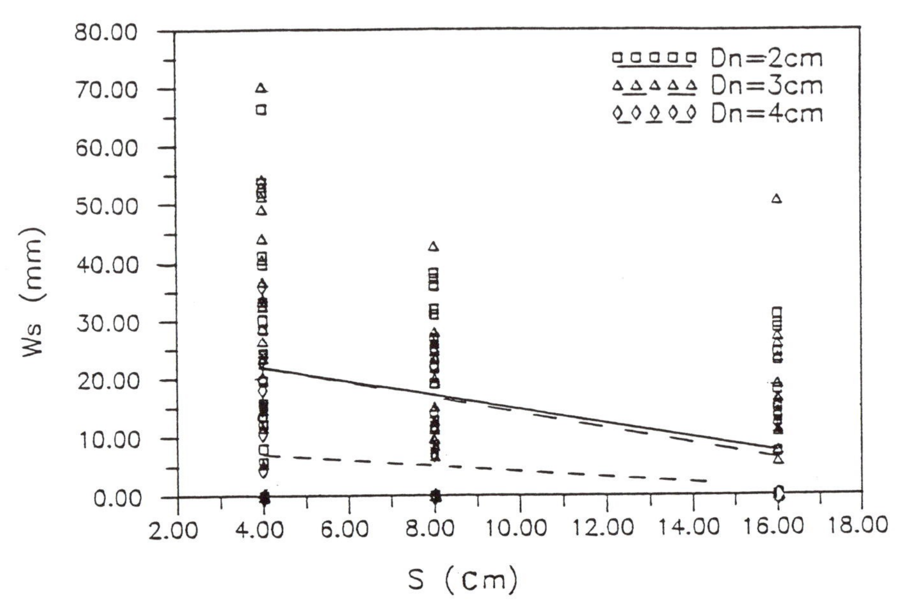

Figure 10. Ws-S-Dn relationships of transverse steel

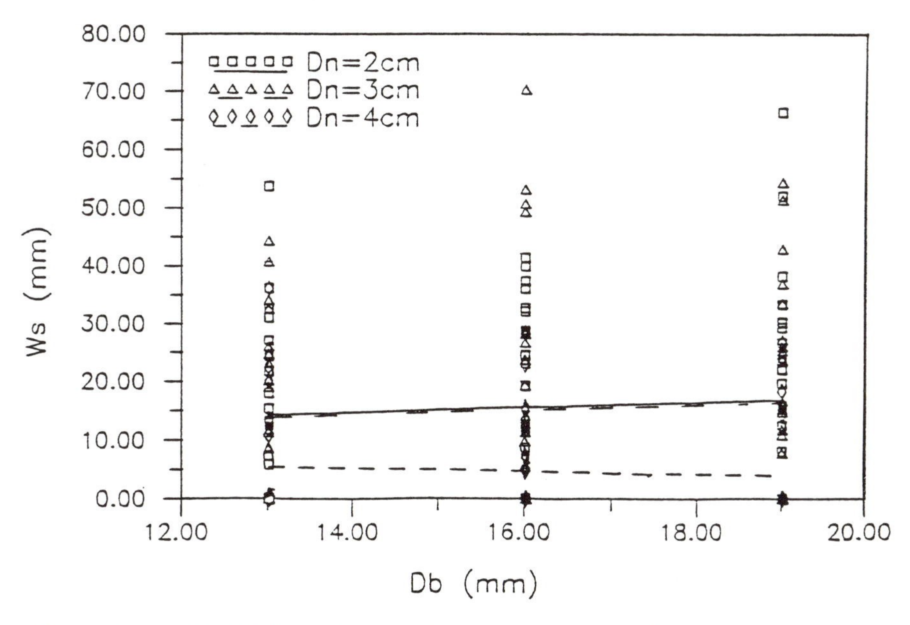

Figure 11. Ws-Db-Dn relationships of transverse steel

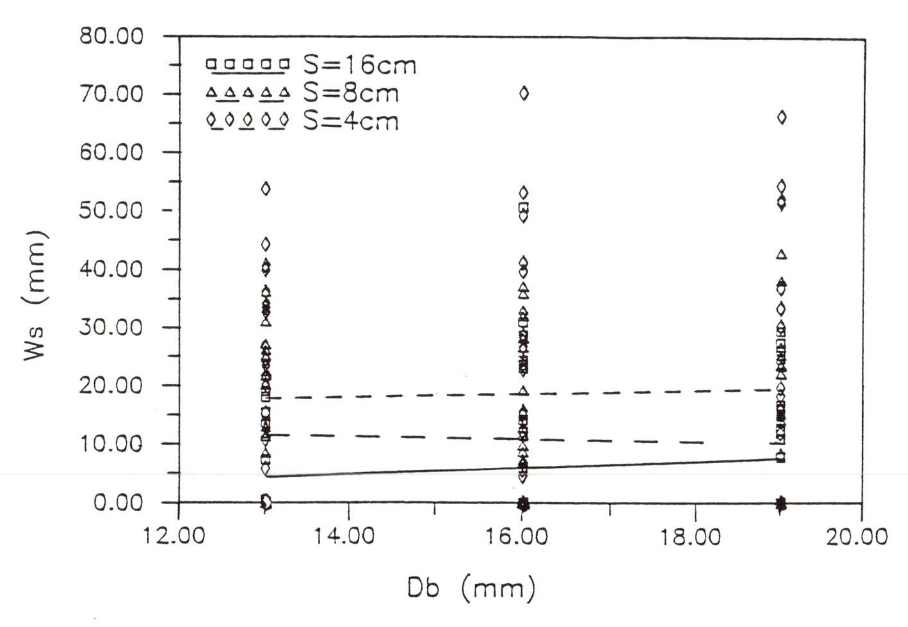

Figure 12. Ws-Db-S relationships of transverse steel

DISCUSSION AND CONCLUSIONS

From the results obtained the following generalizations can be made : (a) 51% of cracks have the internal cleavage failure between the concrete cover and the transverse steel, this phenomena implies that the ability of the concrete cover to protect the rebar from corrosion after cracking is doubtful, (b) the ACI code for the determination of transverse steel is based on a philosophy of maintaining the axial load strength of the column after spalling of the cover concrete, this implies the contribution of concrete cover for column strength is unreliable for seismic design of ductile frames, and (c) to repair the internal cleavage failure is a complicate work, it requires proper supervision, skilled workmen and good workmanship. Repairs should preferably be performed with complete removal of the existing concrete cover and replacing with new concrete.Repairs that simple sealing of cracks with low viscosity materials require much greater care and have less change of being satisfactory.

From the foregoing investigations it may be suggested that a more realistic design approach is to reduce concrete cover thickness as much as possible, while the prime role of concrete cover: (a) to protect rebar from corrosion, (b) to protect the core concrete from the deterioration, (c) to resist the column from the mechanical abrasion, (d) to protect the column from the effects of fire, and (e) to improve the aesthetics of the structures, may be considered to achieve by the other means.

ACKNOWLEDGMENTS

The author wishes to express his appreciation to the National Science Council of R.O.C. for the financial support for this investigation. Thanks are due to Ing-Hsiao Jun for his assistance during the experimental program.

REFERENCES

1. American Concrete Institute Committee 222. 'Corrosion of metals in concrete'. Journal of the American Concrete Institute, Volume 82, Number 1, January/February 1985.
2. American Concrete Institute Committee 318. 'Building Code requirements for reinforced concrete' (ACI 318-89). American Concrete Institute, 1983.
3. American Concrete Institute Committee 224. Concrete of cracking in concrete structures. Journal of the American Concrete Institute, V.69, No.12, December 1972.
4. Kumar Mehta P. Concrete, Prentice-Hall, Inc. Englewood Cliffs, New Jersey, 1986, pp 153.

PART FIVE

NEW CONCRETE TECHNIQUES AND MASONRY STRUCTURES

25 UTILIZATION OF AN INDUSTRIAL WASTE PRODUCT TO DEVELOP A CORROSION PROTECTIVE COATING TO STEEL IN CONCRETE

H.A. El Sayed and B.A. El-Sabbagh
Building Research Institute, Egypt

The present investigation utilized black liquor, a commercial by-product in the pulp and paper industry in Egypt, to develop a new protective coating scheme to steel reinforcement in concrete. The efficiency of this coating - compared to some other known coatings - has been evaluated using anodic polarization of steel in concrete when aggressive ions were mixed with the concrete and when they diffused from outer media. also, the steel/concrete bonding as affected by the applied coating has been evaluated.

INTRODUCTION

A considerable amount of research has been carried out, and is still continuing, with the object of finding concrete admixtures that reduce reinforcement vulnerability to attack by aggressive agents. Unfortunately, the adverse long-term effects of several corrosion inhibitors on the mechanical properties of concrete made their suitability to be used as admixtures in concrete questionable.

An alternative method of corrosion prevention is to screen the steel itself from the concrete and thus from the aggressive medium. This can be achieved by different techniques such as employing non-reactive metal or resin coating, painting or applying a coating of cement slurry that may incorporate anodic corrosion inhibitors.

In a previous study(1), black liquor (or lye), which is a by-product in the pulp and paper industry in Egypt, proved to be a useful material as a plasticizer or a water-reducer in concrete besides, behaving as a good anodic corrosion inhibitor. In the present study, black liquor has been tried as an admixture to cement slurry to develop a new protective coating to steel reinforcement. The corrosion protection effectiveness of this coating has been compared with that of plain cement slurry coating and that incorporating some efficient anodic corrosion inhibitors, namely, 1% sodium nitrite + 1% sodium benzoate and 0.2% calcium stearate. Also, such cement slurry-based coating has been evaluated relative to some commonly known industrial protective coatings, namely, wash primer (zinc tetroxy chromate) and epoxy-resin as well as to coating-free steel.

The corrosion prevention efficiency of the different coatings has been evaluated using galvanostatic anodic polarisation of the steel in concrete under the effect of various severe environments that could confront it in practice. The conditions investigated were: (a) concrete mixed with tap water, 2% Na_2SO_4, 2% NaCl or 1%NaCl+1% Na_2SO_4 and (b) concrete immersed in sea water, 10% NaOH or 10% $(NH_4)_2 SO_4$ solution.

The steel /concrete bonding characteristics as affected by the different coatings have been also compared.

EXPERIMENTAL

The reinforcing steel used was mild steel bars 6 mm in diameter. The steel coating scheme was as follows:

a) De-rusting of the steel surface: The steel rods were dipped into 5% H_2SO_4 acid-to which 0.1% thiourea corrosion inhibitor has been added - for about 20 minutes. The rods were removed from solution as soon as the oxides next to the metal surface were removed satisfactorily and bright surfaces were obtained.
b) Acid neutralization: The rods were cleaned with a wet waste cloth carrying sodium carbonate so as to neutralize any acid that may be lurking in crevices.
c) Phosphating: It is known that when steel is treated with phosphoric acid, alone or containing certain metal phosphates in solution, an adherent phosphate film is produced on the metal surface. This type of coating does not suffice by itself to prevent the rusting of steel for a long time. The advantage of such phosphate film lies in the acid attack of the steel and the production of a network of porous iron phosphate crystals firmly bonded to the steel surface. This will ensure a good adherence of subsequent coating and decrease the tendency for corrosion to undercoat the coat film at scratches or other defects in the coat at which corrosion could initiate. Phosphating (sometimes called phosphatizing or packerizing or bonderizing) was applied by brushing the clean surface of steel by a cold 3% zinc orthophosphate to which orthophosphoric acid has been added till acidity then leaving the phosphating solution on the steel surface for 45-60 minutes. The thickness of the phosphate film thus produced was about 3 μm.
d) Steel coating: For the coatings based on cement slurry, a slurry has been prepared by stirring 1000 gm of Portland cement with 500 ml of distilled water for 2 hours. 0.2% black liquor (or the used inhibitors), by weight of cement in the slurry, was added to the water before mixing. The cement slurry coating was applied by brushing a first coat to the steel in the same day of performing the steps of de-rusting, neutralizing and phosphating. After 24 hours of air drying, brushing with a sealing solution (80% $K_2 Cr_2 O_7$ solution)) was carried out then applying the second coat of cement slurry. Then, after 24 hours of air drying, the coat was brushed with the sealing solution and after 4 hours of air drying again the sealing solution was applied. The entire procedure needed about 3 days.

For the coatings not based on cement slurry, ie wash primer and epoxy-resin, these were applied to the steel immediately after the de-rusting operation.

The ingredients of the concrete mix used were ordinary Portland cement, sand and gravel at a mix proportions of 1:2:4 by weight. The water/cement ratio was 0.6 by weight. In the case of concrete samples admixed with aggressive salts, the required amount of salt was dissolved in the mixing water.

For carrying out the corrosion behaviour measurements, the reinforcing steel rods were centrally placed in cylindrical 5x10 cm steel moulds then the rods were covered by the prepared concrete mix. The samples were cured for 24 hours after casting then demoulded and continuously cured in the test medium for 28 days. To avoid salt diffusion out of the solid during the curing duration, the steel-in-concrete electrodes were cured in the same mixing solutions. After curing, the steel-in-concrete electrode was made as an anode in a circuit using an auxiliary platinum electrode as cathode. A constant current density of 10/μA cm^{-2} was applied to the steel electrode and the corresponding potential was recorded as a function of time.

For carrying out the bond strength measurements, the concrete mix was cast into standard 10x20 cm cylindrical moulds in which steel rods were centred. The concrete samples were demoulded 24 hours after casting and then cured for 6 months in the test media. The bond strength was then determined by carrying out the pull-out tests for the steel embedded in the concrete cylinders and recording the respective loads at which initial and then ultimate slips occurred. A set of 3 tests were carried out for each measurement and the mean bond strength value was calculated.

RESULTS AND DISCUSSION

Anodic polarization measurements

Anodic polarization behaviour of coated reinforcing steel in concrete mixed with different aggressive ions:

The anodic polarization behaviour of reinforcing steel coated with cement slurry-based coatings ie plain cement slurry (C.L), C.L+1% sodium benzoate+1% sodium nitrite, C.L.+0.2% calcium stearate, C.L.+0.2% black liquor, C.L.+ 0.5% black liquor and coatings not based on cement slurry ie wash primer and epoxy resin embedded in concrete mixed with 2% NaCl is shown in Figure 1 as an example of the results obtained using concrete mixed with the different aggressive ions investigated. The results pertaining to uncoated steel are also given for comparison.

It is of interest to notice that, the potential of the uncoated steel rises sharply in noble direction reaching the oxygen evolution potential, but, it cannot remain there and after the elapse of 8 minutes starts to drift with time towards the active direction. Such behaviour indicates that 2% NaCl in concrete represents the threshold concentration at which breakdown of steel passivity starts to occur.

Figure 1 shows that, coating the steel reinforcement with cement slurry alone improves its corrosion resistance, hence the potential could reach the oxygen evolution potential after the elapse of 11 minutes. Such protection given by the cement slurry may be attributed to the fact that, it does not merely hinder admission of aggressive ions but primarily because it raises the alkalinity around the steel reinforcement and consequently high concentrations of aggressive ions, such as 2% NaCl, could be tolerated.

When the cement slurry contained the anodic corrosion inhibitors, 1% sodium benzoate + 1% sodium nitrite or 0.2% calcium stearate, the corrosion resistance of steel has improved where passivation has reached after 9 minutes for both media. According to Brasher, et al[2], the function of the inhibitive ion when present in sufficient concentration is to keep the proportion of sites on the surface of the oxide film which are occupied by aggressive ions, below a certain critical level. It has also a secondary function, namely, that of suppressing incipient dissolution of Fe^{++} ions at points where the metal itself is in contact with the solution, in other words, at points of breakdown in the oxide film. This enables oxygen or other oxidising agents in the medium to complete the oxide film at these points by the anodic process of oxide formation.

Figure 1 shows, also, that cement slurry incorporating black liquor behaves as an efficient corrosion inhibitor where it brings about the shortest time to reach passivation (7 min.) upon using 0.2% black liquor. In case of 0.5% black liquor concentration in the cement slurry, the time to reach passivation is slightly longer (8 min.). Some investigations[3] recommend the addition of a plasticizer to the cement slurry to avoid cracking of the applied steel coating. Fortunately, black liquor proved to be an efficient plasticizer[1]. Hence, utilization of black liquor in the cement slurry will satisfy such coating requirement.

Figure 1 clarifies also that wash primer offers an efficient coating against the aggressive action of chloride ions where the time for passivation has been reached after only 8 minutes.

It is worth mentioning that, extraordinary noble potential values attained after few seconds have been recorded for the steel coated with epoxy resin. Such behaviour illustrates the extremely high electrical resistance of the epoxy-coated steel.

Figure 2 compares the times to reach passivation for reinforcing steel coated with six different coatings investigated compared to uncoated steel embedded in concrete admixed with tap water, 2% $Na_2 SO_4$, 2% NaCl and 1% NaCl + 1% Na_2SO_4. From such comparison, it is evident that a cement slurry coat incorporating 0.2% black liquor exhibits an outstanding protection towards all aggressive ions incorporated in concrete.

Anodic polarization behaviour of coated reinforcing steel embedded in concrete immersed in different aggressive media

Figure 3 compares the times to reach passivation for reinforcing steel coated with the six different coatings relative to uncoated steel when embedded in concrete immersed in sea water, 10% NaOH and 10% $(NH_4)_4$ solutions. It can be seen that, cement slurry mixed with 0.2% black liquor still presents a versatile cement

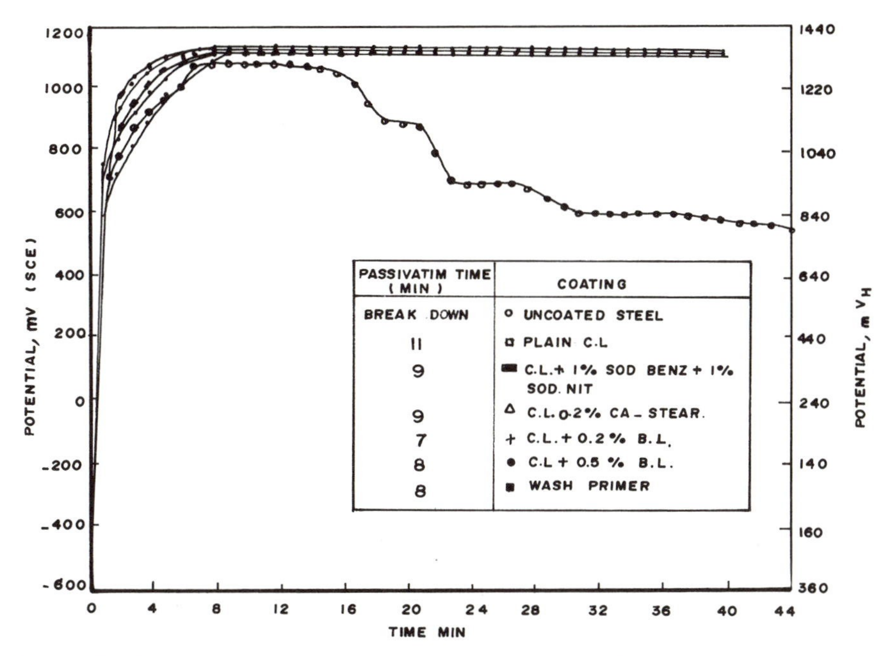

Figure 1. Anodic polarization of steel coated with six different coatings relative to uncoated steel, embedded in concrete admixed with and tested in 2% NaCL at a current density of 10μA cm^2

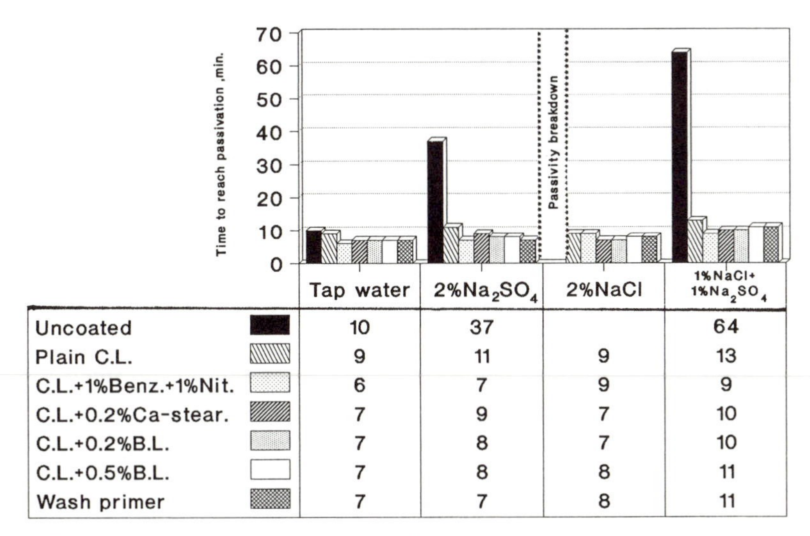

	Tap water	2%Na_2SO_4	2%NaCl	1%NaCl+ 1%Na_2SO_4
Uncoated	10	37		64
Plain C.L.	9	11	9	13
C.L.+1%Benz.+1%Nit.	6	7	9	9
C.L.+0.2%Ca-stear.	7	9	7	10
C.L.+0.2%B.L.	7	8	7	10
C.L.+0.5%B.L.	7	8	8	11
Wash primer	7	7	8	11

Figure 2. A presentation comparing the time to reach passivation for the differently coated reinforcing steel embedded in concrete admixed with different salts.

slurry-based coating highly efficient in protecting steel reinforcement against the three hostile surrounding media.

Bond strength measurements
Previous investigations[4][5] reported that, treatments of steel surfaces might involve a loss of the steel/concrete bond strength either before or after some slight corrosion has occurred. The bond between concrete and coated steel should not be significantly less than that between concrete and uncoated steel.

This part of the study has been carried out to assess the effects of the applied steel treatments on the concrete/steel bond strength relative to the bond between concrete and uncoated steel.

Bond strength measurements for coated reinforcing steel in concrete mixed with different aggressive ions:
Example of the results obtained are those for coated and uncoated steel in concrete mixed with and cured in 2% NaCl solution which are illustrated in Figure 4. It is clear that, coating the steel with cement slurry improves appreciably the bond strength with the surrounding concrete. Such improvement reaches high proportions on incorporating anodic inhibitors, particularly, 1% sodium nitrite + 1% sodium benzoate in the cement slurry. Such a coating as well as that incorporating 0.2% black liquor manifest the highest bond strength values among all the cent slurry-based coatings.

It can be seen that the bond between steel and the surrounding concrete is almost unaffected by applying wash primer as a steel coating. An interesting feature is that, epoxy coating slightly reduces the initial bond strength, relative to uncoated steel, but, increases the ultimate strength by about 4%. This may be due to the fact that, epoxy coating gives smooth surfaces, thus, causing an early slipping and reduction in the initial bond values. Then, during the course of the pull-out test, increased surface roughening occurs as a result of the friction between concrete and the coated steel, thus, yielding high ultimate bond strength values.

Figure 5 has been prepared summarizing the bond behaviour of the differently coated steels towards the encasing concrete. The values for the initial bond strengths, at which slipping starts to occur, have been used for comparison.

It is evident that, among all treatments applied, cement slurry incorporating 0.2% black liquor exhibits the highest bond strength values.

Bond strength measurements for coated reinforcing steel in concrete immersed in different aggressive media
Figure 6 summarizes the bond behaviour of the differently coated steels towards the encasing concrete in the three immersion media. It can be seen that, a cement slurry coating incorporating 0.2% black liquor still develops the highest bond strength values.

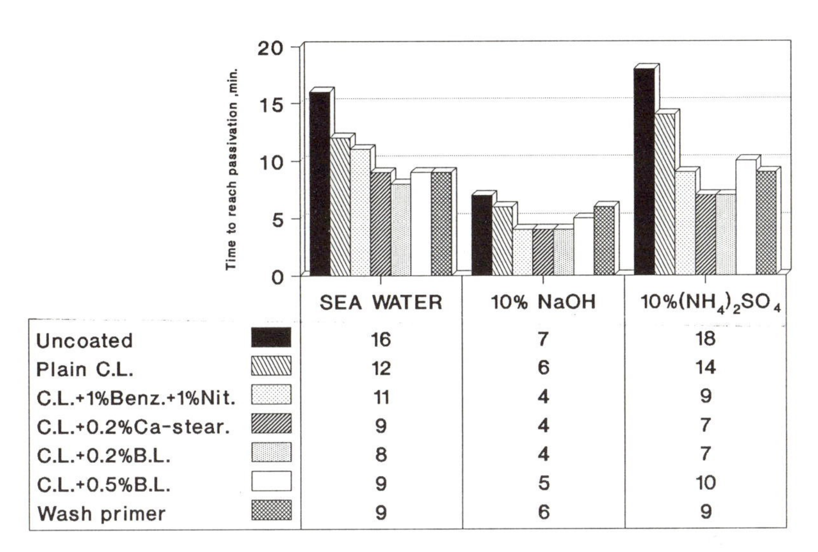

	SEA WATER	10% NaOH	10%$(NH_4)_2SO_4$
Uncoated	16	7	18
Plain C.L.	12	6	14
C.L.+1%Benz.+1%Nit.	11	4	9
C.L.+0.2%Ca-stear.	9	4	7
C.L.+0.2%B.L.	8	4	7
C.L.+0.5%B.L.	9	5	10
Wash primer	9	6	9

Figure 3. A presentation comparing the times to reach passivation for the differently coated reinforcing steel embedded in concrete immersed in different media.

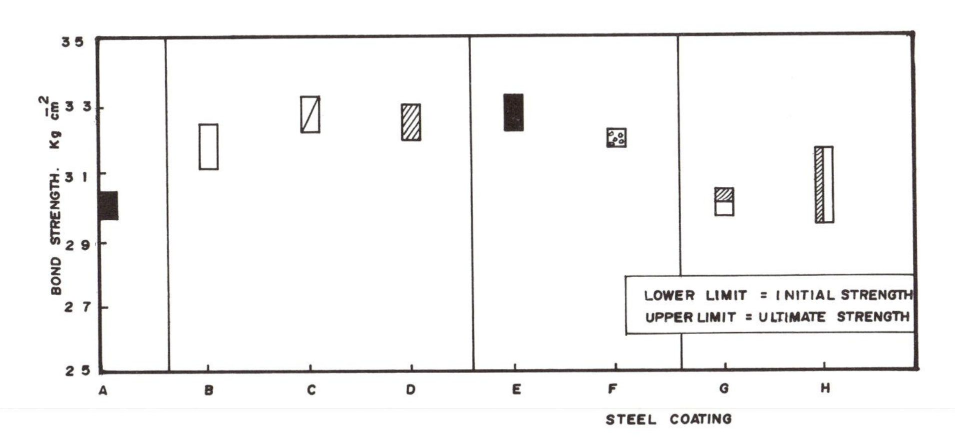

Figure 4. Bond strength for uncoated and coated reinforcing steel in concrete mixed with and cured in 2% NaCL solution for 6 months
A = uncoated steel, B = C.L., C = C.L. + 1% sodium nitrite + 1% sodium benzoate, D = C.L. + 0.2% Ca-stearate, E = C.L. + 0.2% black liquor, F = C.L.+ 0.5% black liquor, G = wash primer and H = epoxy resin

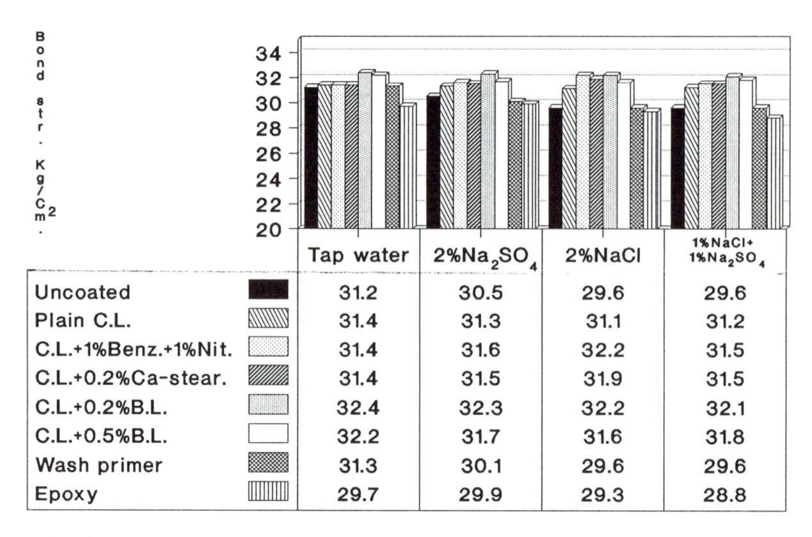

	Tap water	2%Na_2SO_4	2%NaCl	1%NaCl+ 1%Na_2SO_4
Uncoated	31.2	30.5	29.6	29.6
Plain C.L.	31.4	31.3	31.1	31.2
C.L.+1%Benz.+1%Nit.	31.4	31.6	32.2	31.5
C.L.+0.2%Ca-stear.	31.4	31.5	31.9	31.5
C.L.+0.2%B.L.	32.4	32.3	32.2	32.1
C.L.+0.5%B.L.	32.2	31.7	31.6	31.8
Wash primer	31.3	30.1	29.6	29.6
Epoxy	29.7	29.9	29.3	28.8

Figure 5. A presentation comparing the initial bond strength for uncoated and coated reinforcing steel in concrete admixed with different salts and cured in the same media for 6 months

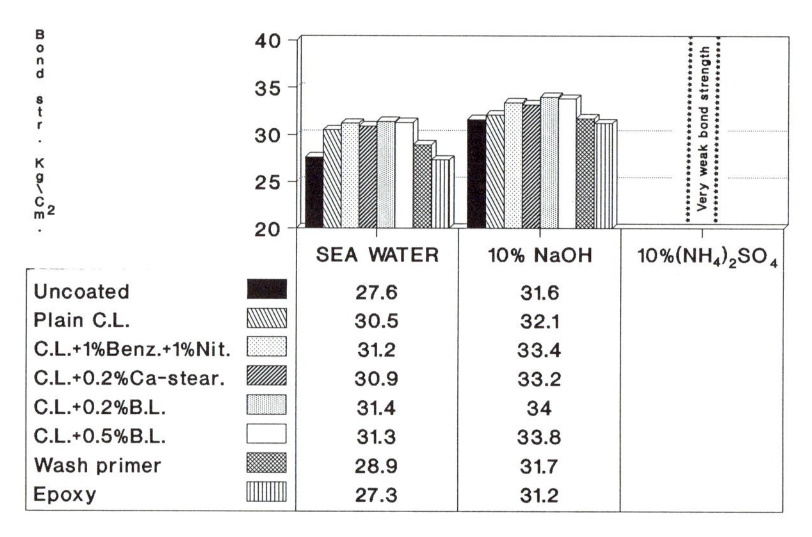

	SEA WATER	10% NaOH	10%$(NH_4)_2SO_4$
Uncoated	27.6	31.6	
Plain C.L.	30.5	32.1	
C.L.+1%Benz.+1%Nit.	31.2	33.4	
C.L.+0.2%Ca-stear.	30.9	33.2	
C.L.+0.2%B.L.	31.4	34	
C.L.+0.5%B.L.	31.3	33.8	
Wash primer	28.9	31.7	
Epoxy	27.3	31.2	

Figure 6. A presentation comparing the initial bond strength for uncoated and coated steel in concrete immersed in the different media for 6 months

It is of interest to record that, curing the differently coated steels in concrete for 6 months in 10% $(NH_4)_2$ SO_4 solution reduced considerably the binding properties of the concrete and consequently the bond strength towards the embedded reinforcement. The concrete samples appeared cracked and having a loose texture full of voids. However, the integrity of the steel coatings could not be affected and appeared intact.

CONCLUSIONS

A cement slurry admixed with 0.2% black liquor undoubtedly forms an outstanding corrosion resistant coating that efficiently suppresses the corrosion of steel reinforcement that can ensue due to the presence of aggressive ions admixed with the concrete or penetrating from outer media without any adverse effects on the strength of the concrete-steel bond, but, on the contrary improves it appreciably, besides being economic and easily applied.

REFERENCES

1. El-Sayed H.A. 2nd Inter.Symp. on Vegetable plants and their fibres as building materials, Nucleo De Servicos Tecnologicos, Salvador-Ba-Brazil (1990).
2. Brasher D.M. Reinchenberg D. and Mercer A.D. BR.Corros.j., 3, 144 (1968).
3. Ellyin F. and Matta R. ACI.J., 79, 366 (1982).
4. Lewis D.A. 1st Inter. Congr. on Metall. Corros. Butterwort, London (1962).
5. Clifton J.R. Mater Perf. 15, 14 (1976)

26 HYRIB: LOOKS TO REDUCE PRESSURE ON FORMWORK

P.S. Chana
British Cement Association, UK
V. Camble
Expamet Building Products, UK

This paper describes the development of Hyrib expanded metal as a permanent formwork material in situations where a fair-faced finish is not required or where external cladding or rendering is to be applied. Results from six full scale trials to estimate the reduction in formwork pressures compared to conventional plywood shuttering are presented. It is shown that the pressures are reduced to less than half as the pore water pressure is dissipated. Core tests also indicate that the quality of concrete, as measured by density and strength, is comparable to that obtained using a conventional plywood shutter.

INTRODUCTION

Hyrib expanded metal has a long track record in the UK for forming construction joints. It is manufactured from pre-coated hot dipped zinc coated steel sheets. The mesh and ribs are formed on machines which first cut and press the sheet and then roll form the ribs. The sheet is stretched to form a standard width of 445 mm expanded mesh. Hyrib is available in three grades; 2411, 2611 and 2811 with decreasing weight density and stiffness.

The traditional use of Hyrib is now being extended to cover the area of permanent formwork. Formwork accounts for a significant proportion of the cost of in-situ concrete structures. The main advantage of using Hyrib is that the high formwork pressures associated with traditional plywood formwork are significantly reduced as the pore water pressure is dissipated. Hence, the formwork's supporting system is considerably simplified with fewer ties required. Hyrib can therefore be an economical solution for back faces of retaining walls or, more generally, it can be used as formwork where external cladding or rendering is to be applied.

Full scale trials were carried out at the British Cement Association to investigate the use of Hyrib as permanent formwork. The first three trials had the prime aim of providing some technical data on tie bolt forces and formwork pressures when Hyrib is used in this application. In a limited programme of tests it is not possible to generate sufficient data to provide formwork pressure design charts, but by testing at extreme conditions (i.e. very rapid rates of placing and high workability concrete), these data will provide some guidance to a designer.

Some limited further testing was done in Trials 4 to 6 to assess the quality of concrete on the Hyrib face.

TEST DETAILS

In the first phase, three short walls were cast. Each wall was 2 m long by 0.5 m wide and 5 m high. Two of the walls were formed with opposite faces containing different grades of Hyrib and the third wall was constructed using Hyrib on one face and conventional plywood formwork on the other face.

The formwork for the first wall comprised mesh 2411 on one face and mesh 2611 on the opposite face. The mesh was laid with the ribs in the horizontal direction. This was supported by soldiers at variable centres ranging from 350 mm to 450 mm. In turn the soldiers were supported by steel walings at 600 mm centres. Pairs of walings were tied together with 15 mm Dividag ties. The scheme, together with the positions of the load cells, is given in Figure 1. The load cell readings are converted directly into tie bolt forces and indirectly into formwork pressures.

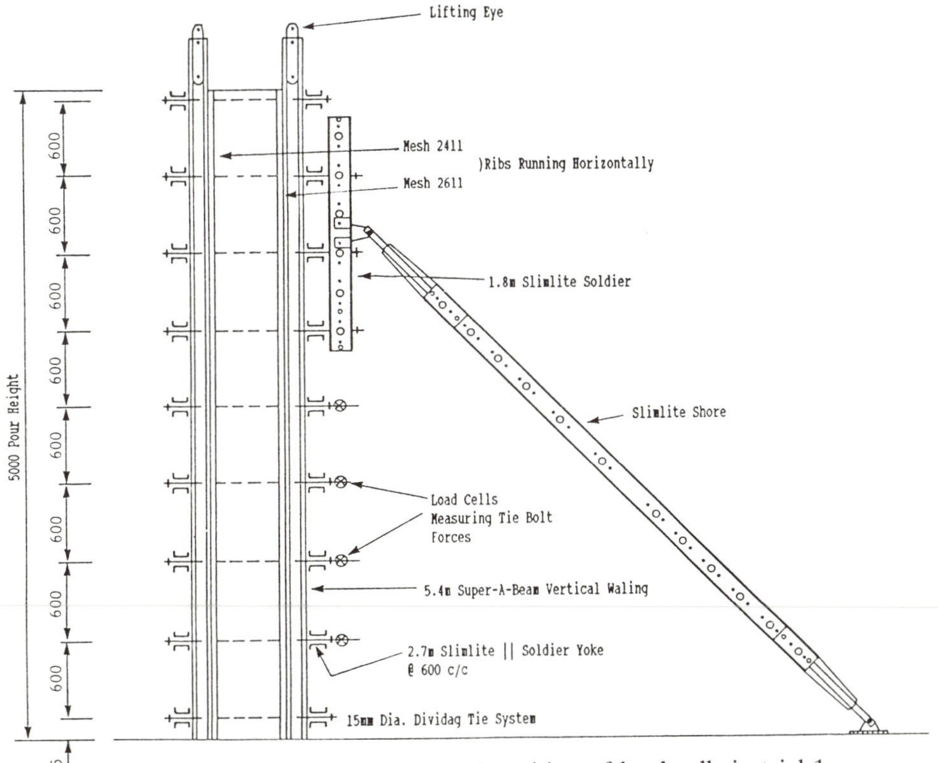

Figure 1. Column section and position of load cells in trial 1

Based on the results at the first test, the formwork design was modified for the second and third tests. The Hyrib was placed with the ribs running vertically (with the fingers pointing in the downward direction) and this was supported by walings at 600 mm centres. The walings were supported by two pairs of soldiers and linked with 4, 15 mm Dividag ties at 1200 mm centres.

The main difference between trials 2 and 3 was that trial 2 had one face of conventional plywood and the other face was 2411 mesh whilst trial 3 had mesh 2611 on one face and mesh 2411 on the other face.

In all the trials the formwork to the stopends was plywood. The system was designed so that the formwork system for the stopends was independent of the formwork system for faces.

From theoretical considerations of the pore-water pressure theory(1), it was deduced that the pressure should be less than normal(2) and therefore the form design was based on 50 kN/m^2. As tie bolt forces were being continuously measured, this provided a safety system to forewarn the team if pressures were higher than expected.

CONCRETE AND PLACING

In order to test at extreme conditions, high workability, pumped OPC/ggbs mixes were selected. The details of the mixes are given in Table 1.

The mixes were pumped and internally vibrated. The normal recommendation for Hyrib is not to move the vibrator closer than 450 mm, but in a 500 mm wide wall such guidance could not be followed. The selected high workability gave concrete that was self compacting, but to simulate normal site practice internal vibration was applied.

The method of placing was to place a first layer about 600 mm deep. The pump was stopped and this layer was vibrated. Once vibrated, the rest of the wall was pumped whilst simultaneously the concrete was vibrated as necessary. Due to the high workability continuous vibration was not necessary.

Figures 2 and 3 show the concrete being placed and the formwork scheme for trials 2 and 3.

METHODS OF ESTIMATING AND FORMWORK PRESSURE

Two methods were used to estimate the formwork pressure. The first method simply divided the tie bolt force by the effective area. The second method assumed that the concrete exerted a uniform pressure on the form and this pressure was calculated from the tie bolt forces using normal structural design.

Table 1. Concrete mix design details

MIX	TRIAL 1	TRIAL 2	TRIAL 3
Water kg/m^3	179	163	163
Cormix kg/m^3	1 (P4)	.68 ltrs. (P10)	.68 ltrs. (P10)
Cement, OPC kg/m^3	220	205	205
ggbs kg/m^3	145	135	135
sand kg/m^3 S.S.D.	782	813	813
10mm agg. kg/m^3 S.S.D.	262	260	260
20mm agg. kg/m^3 S.S.D.	775	768	768
Air kg/m^3 measured	2.25	2.8	3.0
Total, i.e. density, kN/m^2			
Aimed	2364	2345	2345
measured fresh	2327	2315	2307
measured demoulded	2361	2329	2359
Temp. °C:			
Air start	7.6	4.0	4.6
end	7.4	4.8	5.0
Concrete start	12.6	8.6	10.1
end	12.8	8.6	10.1
Slump, mm			
initial	105	60	55
final	155	100	80

An example of the calculation is as follows:

Level (m)	.17	.17	1.37	1.37	2.57	2.57	4.07	4.07
1(m)	.765	.765	1.20	1.20	1.35	1.35	1.685	1.685
Effective area (m^2)	.765	.765	1.20	1.20	1.35	1.35	1.685	1.685
α	.8	.8	1.033	1.033	1.033	1.033	.8	.8
Tie force kN	8.45	5.82	11.57	18.51	27.99	25.28	23.59	21.52
Est'd.Press. (kN/m^2) (1)	11.05	7.61	9.64	15.43	20.73	18.73	14.00	12.77
Est'd.Press. (kN/m^2) (2)	13.81	9.51	9.33	14.93	20.07	18.13	17.50	15.96

Figure 2. Concrete placing

Figure 3. Formwork schemes for trials 2 and 3

RESULTS

Trial 1

The key results are as follows:

Rate of placing = 20 m/hour
Maximum estimated formwork pressure
Method 1 = 36 kN/m^2
Method 2 = 42 kN/m^2
CIRIA Report 108[(2)] design pressure for the actual placing conditions = 115 kN/m^2.

Trial 2

The key test results are as follows:

Rate of placing = 38 m/hour
Maximum estimated formwork pressure
Method 1 = 44 kN/m^2
Method 2 = 43 kN/m^2
CIRIA Report 108 design pressure for the actual placing conditions = 124 kN/m^2.

Trial 3

The key test results are as follows:

Rate of placing = 43 m/hour
maximum estimated formwork pressure
Method 1 = 22 kN/m^2
Method 2 = 21 kN/m^2
CIRIA Report 108 design pressure for the actual placing conditions = 124 kN/m^2.

The measured deflections in the Hyrib and plywood are given in Table 2.

DISCUSSION OF RESULTS

Hyrib by its nature has voids of sufficient size to pass the fine fraction of concrete and therefore the normal guidance is to keep the vibrator 450 mm from the face. This is sufficient to ensure compaction and minimise the loss of fines. In walls 500 mm wide it is not possible to follow this guidance.

In trial 1 which had a concrete with 150 mm slump, vibration was relatively heavy and consequently the loss of fines was high. The concrete continued to lose excess pore-water after vibration, but on inspection the following day the loss of fines and pore-water had not led to honeycombing. It was also easy to remove the fresh paste from the forms with a water jet.

The 600 mm base layer in trial 2 was also heavily vibrated with some loss of fines and high deflections (15 mm on the mesh). For the rest of the trials,

Table 2. Measured deflections

Side	Level (m)	Clear spans(mm)	Deflection (mm)	Deflection/ clear span
TRIAL 1				
Mesh 2411	0.9	400	1.39	1/ 288
	0.9	400	1.29	1/ 310
	0.9	350	2.87	1/ 122
	0.9	300	1.33	1/ 226
	0.9	300	0.26	1/1154
Mesh 2611	0.9	400	0.81	1/ 494
	0.9	400	1.57	1/ 255
	0.9	350	0.71	1/ 423
	0.9	300	2.92	1/ 103
	0.9	300	0.78	1/ 385
Formwork, Side 2611	0.6	2600	4.92	
	1.2	2600	6.36	
Formwork, Side 2411	0.6	2600	7.23	
	1.2	2600	6.80	
TRIAL 2				
Plywood face Mesh 2411	0.47	420	15.01	1/ 28
	1.07	420	6.92	1/ 61
	1.67	420	4.45	1/ 94
	2.27	420	1.23	1/ 341
Formwork, Plywood face	0.77	2600	11.76	
	1.37	2600	14.35	
Formwork, Side 2411	0.77	2600	9.11	
	1.37	2600	10.73	

vibration was applied in short bursts of 5 to 10 seconds. This gave adequate compaction and kept the loss of fines to a reasonable level.

The differences in the bolt forces between the two ties at the same level were small and this indicates that variation in pressure at one level and variations in the support system are acceptably small.

The methods used to estimate the formwork pressures were simple, but adequate given the variability of formwork pressures. Method 1 simply divided the tie bolt force by the effective area. In reality when concrete is placed in a form,

Table 2 (continued). Measured deflections

Side	Level (m)	Clear spans(mm)	Deflection (mm)	Deflection/ clear span
TRIAL 3				
Mesh 2611	0.47	420	4.53	1/ 93
	1.07	420	2.40	1/ 175
	1.67	420	1.37	1/ 307
	2.27	420	1.68	1/ 250
Mesh 2411	0.47	420	2.89	1/ 145
	1.07	420	1.30	1/ 323
	1.67	420	1.43	1/ 294
	2.27	420	0.81	1/ 518
Formwork, Side 2611	0.77	2600	6.45	
	1.37	2600	5.01	
Formwork, Side 2411	0.77	2600	4.23	
	1.37	2600	4.25	

the adjacent tie bolt takes most of the load but some is distributed to the other ties.

The second method of estimating the pressure assumes that the concrete exerts a uniform distributed load (UDL) and this load is calculated from the tie bolt forces using normal structural mechanics. Towards the top of the wall, the pressure must reduce as it is unlikely to exceed the fluid pressure and therefore the assumption of a UDL is not strictly correct. However the errors induced by the assumption are likely to be small and acceptable. The two systems give similar estimates of formwork pressure.

The pore-water pressure theory shows that two factors are key in controlling the maximum formwork pressure. These are the length of the drainage path and the extent of vibration.

In trials 1 and 3 the maximum length of drainage path is 250 mm (half the width of the section) whilst in trial 2 it is 500 mm (this assumes that the plywood has no permeability). From theory trial 2 should give higher pressures than trials 1 and 3 and the measured data confirm this prediction. By method 1 the estimated maximum formwork pressure on trial 2 was 44 kN/m^2 and for trials 1 and 3, 36 and 22 kN/m^2 respectively.

Whilst the concrete is subjected to vibration, a particle structure cannot form and consequently the vertical load cannot be transferred from the pore-water to the particle structure and thereby reduce the pressure. In trial 3 the rate of placing was twice that of trial 1 and for other factors being equal, the formwork pressure in trial 3 should have been higher than trial 1.

However the 'other factors' were not equal and in particular trial 1 had more vibration than trial 3 and the workability was higher. At the very high rates of placing achieved, the differences due to placing rate tend to reduce and the dominant factor is likely to have been the level of vibration. This is the probable reason why the estimated pressures were higher in trial 1 than in trial 3.

When the estimated formwork pressures are compared with those calculated from the CIRIA Report 108 formwork pressure design charts for the actual placing conditions, the pressures are substantially lower than the design values.

FURTHER TESTING

Following the completion of the investigation in Trials 1,2 and 3, three further walls were cast using Hyrib formwork. These are referred to as Trials 4,5 and 6.

In Trial 4, a single wall was cast, the formwork being similar in all respects to Trial 3 except that Mesh Grade 2811 was used to form both faces.

Trials 5 and 6 were primarily carried out as a demonstration of the use of Hyrib formwork. Trial 5 was similar to Trial 3 but with Mesh Grade 2411 used for both formwork faces. Trial 6 was similar to Trial 2, but using Mesh Grade 2611; no loadcell readings were taken during Trial 6.

After casting the Trials 5 and 6 walls, 12 cores were taken (3 from the face of each wall) for the purpose of comparing strength and densities between the different face concretes.

The 100 mm diameter, 210-280 mm length cores were taken at a level 110 mm up from the ground. These cores were then sawed, where appropriate, to remove the outer 20-25 mm containing the Hyrib and to give an 'outer' test core of approximately 120 mm length. For three of these 'outer' cores, 'inner' cores were also sawn from the original core to compare the strength and density of the 'outer' and 'inner' regions.

The results are summarised below:

Comparison of the densities of the 'outer' cores and the 'inner' cores/remainders indicates that the Hyrib Formwork had little adverse effect upon the density of the placed material.

Comparison of the strengths derived for the 'outer' cores with those of the three 'inner' cores and also, the low standard deviation of all the test core strengths also indicate no adverse influence.

Comparison of the strength obtained from the plywood face with the Hyrib face in Trial 6 also indicates satisfactory strength performance of the Hyrib face concrete.

CONCLUSIONS

1. The use of Hyrib formwork gave substantially lower pressures than the design values obtained from CIRIA Report 108 in 5 m high walls which were placed rapidly, using OPC/ggbs high workability concretes.

2. The estimated pressures (method 1) were 36, 44 and 22 kN/m^2 for trials 1 to 3 compared with design pressures of 115, 124 and 124 kN/m^2.
3. The loss of fines on vibration did not lead to honeycombing and because the concrete can be seen it is possible to achieve adequate compaction with reduced vibration.
4. The use of Hyrib formwork did not lead to any loss in strength on density of concrete cores extracted from the face.

ACKNOWLEDGEMENTS

The authors would like to thank Dr T.A. Harrison, Technical Director, BRMCA, (formerly of the British Cement Association) for his helpful comments during this project.

REFERENCES

1. Harrison T.A. 'The pressure on vertical formwork when concrete is placed in wide sections'. Cement and Concrete Association, Research Report, (22 March 1983).
2. Clear C.A. and Harrison T.A. 'Concrete pressure on formwork', CIRIA Report 108, (1985).

27 STRUCTURAL STEEL SHEARHEADS FOR CONCRETE SLABS

P.S. Chana
British Cement Association, UK

Shearheads are structural steel section assemblies normally placed within the slab on top of a supporting column to enhance the punching shear resistance. They are an effective solution to the problem of providing large service holes or openings next to the column. This paper describes test work carried out in order to establish the most efficient shearhead configuration and to develop a method of design. The paper also discusses some of the advantages of structural steel shearheads to the developer, designer and contractor.

INTRODUCTION

Flat slabs are a popular form of construction for a whole range of buildings such as office blocks, hospitals, warehouses and car parks with large uninterrupted concrete floors and minimum construction depths. A major design consideration in this type of construction is punching shear around the column support. Also, services are becoming increasingly important in structural design and the problem of punching shear is exacerbated when large service holes are required adjacent to columns. The shearhead development project was undertaken to provide an effective practical solution to this problem.

The project was undertaken at the BCA and supported by Square Grip Ltd. Practical design advice was provided by a number of consultants including Ove Arup and Partners, Kenchington Forde plc and Bunyan Meyer and Partners. In addition, the project benefitted from a Support for Innovation grant from the Department of Trade and Industry. There were two main aspects of the programme:

a) the development of prefabricated reinforcement assemblies to act as shear reinforcement.
b) the development of structural steel shearheads for use in reinforced and post-tensioned slabs.

The prefabricated reinforcement solution was called Shearhoops(1) and launched in 1991. This programme won the Construction News/Amec R&D Award for 1991. This paper is solely concerned with structural steel shearheads.

A shearhead is a fabricated structural steel assembly placed within the slab depth on the column support. Shearheads have been used widely in the United States and on the Continent for twenty years but their use has been very limited in the United Kingdom. Shearheads used in the States are generally of a cruciform arrangement with the steel members passing through the columns. Generally the top reinforcement passes over the shearhead. The bottom reinforcement may either pass under the shearhead or be curtailed and supported on the bottom flange or welded to the flange. The protection of the steel in the heads against fire and corrosion needs to be considered carefully.

Rules for design of the cruciform type of shearhead are given in the ACI Code(2). These rules are based on work carried out at the Portland Cement Association and reported by Corley and Hawkins(3). Three design criteria are considered: (i) shearheads increase the punching shear perimeter; (ii) the flexural strength of the shearhead arm needs to be adequate; (iii) the slab reinforcement for the negative moments over the columns can be reduced.

The ACI type shearheads were taken as the starting point of this investigation. This paper describes the test work carried out in order to establish a more efficient configuration and a suitable method of design.

TEST WORK

Three series of tests were carried out:

a) Model tests on internal slab-column connections; slabs were 120 mm deep.
b) Full scale tests on internal slab column connections; slabs were 250 mm deep.
c) Large scale model tests on edge slab column connections; slabs were 150 mm deep.

Figure 1 shows tests on model slabs for internal columns. These specimens were 120 mm deep and 1800 mm in diameter. The loading was applied at points equally spaced on a circumference of a circle 1.5 m in diameter, through prestressing strands acting against a reaction block.

Figure 2 shows tests on full scale slabs for internal columns. These specimens were 250 mm deep and 3 m square. The loading arrangement was similar to the model slabs. The specimens were supported on a 300 mm square block in the centre representing a column. Load was applied by means of hydraulic jacks at eight locations on the circumference of a circle of diameter 2.4 m, acting through load cells and prestressing cables. The hydraulic jacks were linked to a common supply so that the force on each cable was the same. The load was generally applied in ten equal increments. At each load stage the crack pattern was recorded, along with the deflections.

Figure 3 shows the test arrangement for the edge column specimens. The test specimen was 4.8 m x 2.4 m and 150 mm deep. The two edge columns were 300 mm square with the outer faces flush with the slab edge. Below the slab the

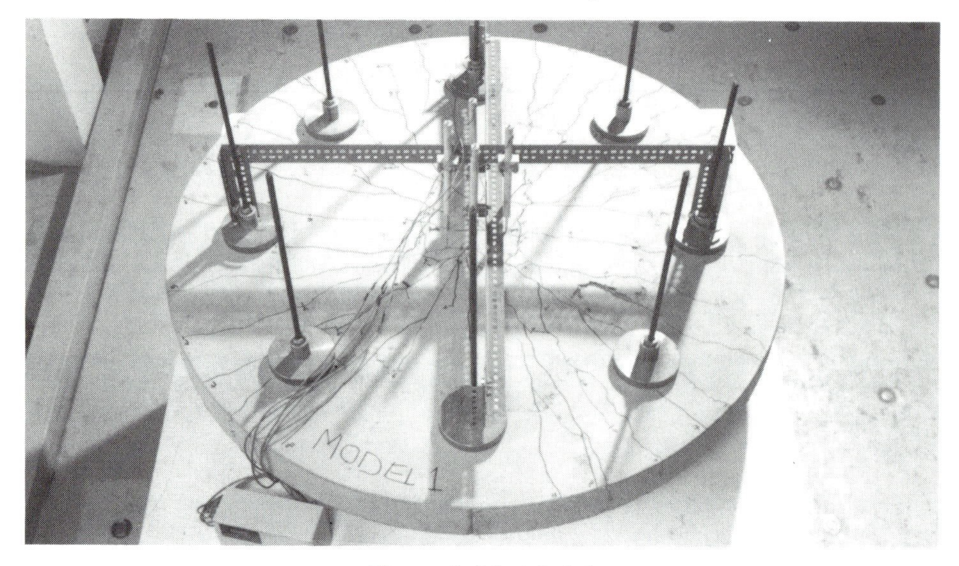

Figure 1. Model slab test

Figure 2. Full-scale slab tests

Figure 3. Edge column slab tests

columns were tied by two 16 mm diameter bars with pin-jointed connections to the columns. The columns were restrained in a similar manner above the slab by a strut with pin-jointed connections. The test arrangement enabled the natural restraining moment to develop as the slab was loaded. The strut and tie forces were monitored by load cells enabling the restraining moment to be calculated. Loads were applied at eight points close to the slab edges through an arrangement of jacks and prestressing strands. The load applied by each jack was measured by a load cell. The test rig enabled a maximum column shear force of 800 kN to be applied and could simulate, at half scale, a variety of column slab connections up to a span of 10 m. Deflections were monitored at each load stage and give an indication of the change in stiffness due to the provision of a shearhead. It was possible to test both slab column connections by placing a support across the width of the slab to isolate the damaged end. Figure 4 illustrates the increase in failure perimeter when a shearhead is provided. A similar observation was made for the internal column slab tests.

SHEARHEAD TYPES AND ADVANTAGES

Some of the shearhead devices tested in the programme are shown in Figures 5 to 8. The test work demonstrated various advantages of providing shearheads such as:

i) they increase the punching shear resistance
ii) they reduce the slab deflection under applied loads

Figure 4. Failure perimeter for edge column connection with shearhead reinforcement

iii) they carry some of the support moments thereby reducing the amount of reinforcement over the supports
iv) they give the opportunity for providing openings close to the columns.

The type of shearhead which was found to maximise these advantages is shown in Figure 9. This is prefabricated from channel sections (or I sections for larger loads) with the 'primary' arms generally passing through the column. The 'secondary' arms consist of a square outer frame of steel sections which can be identical to the primary arms or a reduced section with a lower stiffness. Welding of these units is to BS 5135 and in accordance with recognised QA procedures.

When these shearheads are used, failure occurs outside the shearhead area starting at the outside of the shearhead and forming a typical cone into the slab. The failure cone and hence the failure load depends on the shearhead plan size only and bears no relationship to the size of the column.

Openings can be provided within the shearhead. Where large openings are provided, the design assumes that all the loading is taken by the steel membrane through the secondary and primary arms in turn. Hence, the following aspects of design have to be considered: (a) punching shear resistance of the slab outside the shearhead; (b) design of steel framework to take moments and shear; (c) transfer of load to column; (d) reduction in support steel, if desired. The test work carried out in this project has led to a design method being developed. This has formed the basis for the preparation of tables of standard shearheads with given load capacities to be produced. It is proposed to publish detailed test data and design rules in a Manual.

Figure 5. Shearhead device (I section)

Figure 6. Shearhead device (Channel section)

Figure 7. Shearhead device (Boxed type)

Figure 8. Shearhead device (composite type)

Figure 9. Prototype shearhead for internal column

An innovative feature of these shearheads is that it is not necessary to pass the primary arms through the columns. These arms can be located touching the column perimeter or with the cover region of the column. Shear stress is transferred between the column face and the slab by welding a wedge shaped section onto the web. The structural adequacy of this detail is to be confirmed by further tests.

Patent applications have been filed in Europe and North America. It is expected that shearheads will be available for use by designers later in this year.

REFERENCES

1. Chana P.S. and Clapson J. Innovative shearhoop system for flat slab construction, 'Concrete' (Jan/Feb 1992).
2. ACI 318-89. Building code requirements for reinforced concrete 'ACI' (1989).
3. Corley, G.W. and Hawkins, H.M. Shearhead reinforcement for slabs. 'ACI Journal' pp 811-824. (October 1968)

DISCUSSION PAPER 27

I.M. Watters, The Scottish Office

I had used Hyrib successfully some years ago in a situation where unskilled labour was available. Its use was as Soffit shuttering to a floor slab, and a super-plasticiser had been added to the concrete, making vibration unnecessary and levelling very simple.

I noted that the speaker had said that vibrators must not be used within 2 feet (?) of Hyrib shuttering, and asked how necessary this restriction was. The speaker confirmed that in no circumstances should vibrators be used within this specified distance.

28 COMBINED NUMERICAL–EXPERIMENTAL DEVELOPMENT OF INNOVATIVE STRUCTURAL CONCRETE CONNECTIONS

T. Krauthammer and E. Marx
The Pennsylvania State University, USA

The need for the development of structural concrete connections for severe loading conditions has been achieved by a combined numerical and experimental effort. The first step consisted of the numerical evaluation, by advanced finite element analyses, of candidate connections. The second effort consisted of experimentation, in both the static and dynamic behavioral domain, for the physical verification of the numerical simulations, and for obtaining additional data on the components actual responses. These approach has provided the required structural concrete connections.

INTRODUCTION

The present summary describes several important performance aspects of wall to slab connections using numeric simulation. For simplicity, the model examines a three-dimensional slice of the wall-to-wall connection subjected to the highest loading area. The reinforcement within the connection is then examined in order to establish the strength of the particular reinforcement detail and to determine if failure occurs. Without experimental test results to validate to the numeric simulations, the results should be appraised with respect to other results produced from within the group of simulations. The numeric simulations, together with the laboratory testing data, will produce an insightful examination of reinforced concrete connections subjected to blast loading conditions.

There exists a great deal of uncertainty in the design of structures subjected to dynamic loads. Uncertainties emanate from limited understanding of material behavior under high loading rates as well as uncertainties associated with calculating the load magnitude and duration. The design of connections in the static domain was not well understood until about the mid 1970's. Dynamic effects (primarily that related to seismic loading) were not considered until some time later, however many issues have not yet been resolved. The effects of short duration dynamic loads are at an even earlier state of understanding.

Information on the properties of structural concrete can be obtained from several books on the subject[1,2]. Most of the available data is isolated to the response of concrete in the static, one-dimensional domain. The dynamic behavior and multi-directional loading response have been described in various other publications for both concrete and steel[3-6].

Joints are often the weakest link of a structure. Until recently little attention was given to connections. It appears that after evaluating the working stresses of the adjoining members, most designers assumed that the conditions within the joint were not critical. Perhaps this was due to the fact that most joints were somewhat larger than the members framing into them. The adoption of the limit state design philosophy, a method that utilizes the maximum strength of the member, has exposed the weakness of this practice. Work by Nilsson[7], and Schlaich et. al.[8] has done much to improve upon this area of design. However, the need to increase our understanding of joint behavior remains. By defining certain performance criteria, a logical design method may be defined, as described in[1].

Pantazopoulou and Bonacci[9] stated that joint behavior can not be fully understood without considering joint deformations. This is due to the fact that computations of average principal stresses require knowledge of the pattern and magnitude of deformations. This is an aspect of response usually omitted from design methods because of its complicated computations. Even if the direction and magnitudes of the principal stresses could be calculated, it would be difficult to establish whether these stresses were too high or if they would even develop considering the overall behavior of the structure. They suggested that if the deformations could be monitored, it would be possible to determine the sequence of tensile and compressive reinforcement yielding and concrete crushing regions. Also, if the deformations could be linked to the overall structural response, then the likelihood of these occurrences could be established.

Those previous studies showed that joint deformation is a strong indicator of joint performance. In cases of inadequate joint design, the contribution of joint distortion to the total drift is observed to increase with increasing magnitude of total displacement. In these cases, the joint is often responsible for much of the inelasticity detected in the overall response. By confining the concrete in the core of the connection, the performance and ductility of the joint may be enhanced. Current design philosophy combines the principles of statics with a beam model acting on each face of the joint. The method proves to be reasonably accurate in determining regions of tensile stress when compared with experimental results in linearly-elastic regions and is increasingly less accurate in non-linear regions.

It is desirable for the compression forces generated by the beam models to follow the compression steel reinforcement in such a way as to utilize the entire connection instead of the triangular area formed by the beam and column compression zones. Simply stated, the moment forces from the beam must be turned through the connection to the column and then to the supports. With the addition of a diagonal tensile reinforcement bar, the compression forces develop to utilize the entire joint region. However, diagonal tension cracks parallel to the diagonal bar may form inside the joint if the compression forces are permitted to increase substantially. Radial stirrups must then be provided to prevent opening of the diagonal crack.

In this study, which is a continuation of a previous investigation[10], two design approaches were examined. Design Approach I was compiled from the research of Nilsson[7], Park and Paulay[1], and the current ACI Code[11] code as well as considerations obtained from various other researchers. Their experiments were aimed at determining an understanding of joint behavior and developing a rational design approach. Design Approach II was compiled based on recommendations by the Naval Civil Engineering Laboratory[12], and based on recommendations for the design of blast containment structures in[13]. The approach used here is the same as in Design Approach I except for the amount of steel reinforcement in the corner. The placement of bars has the same geometric orientation as Approach I, however larger quantities of steel may be necessary.

Explosive load functions near the source are not fully understood and measuring the effects is difficult due to the severity of the environment close to the blast. TM 5-1300 (1990) contains such data for calculating blast loading forces as a function of explosive size, shape, weight, and distance to the structure. Typical load functions experience very high magnitudes of load initially and quickly decay.

RESEARCH APPROACH

The analysis approach employed in this study utilizes the finite element method for simulating and studying the structural response for each connection detail. The simulation analyzes the structural connections defined above. From the results of the numeric tests, strength characteristics of the different structural details were quantified and examined to determine the most efficient use of materials. The most efficient detail was selected for further physical study in the laboratory. The finite element code used in this study was DYNA3D[14]. The geometrical approach used to model the wall to slab connection behavior considers a slice of one quadrant of a rectangular structure subjected to an internal blast load, as shown in Figure 1. Although the characteristics resemble a beam to column joint, the actual simulation models the wall to slab connection when plane strain calculations are used. The end of each leg is fixed for all degrees of freedom except translation perpendicular to each leg. These boundary conditions reproduce the symmetry present in the physical structure.

The concrete was modeled with a three dimensional mesh of 8-node solid elements and 6-node prismatic elements. Steel reinforcement was modeled using beam elements connected to the node points of adjacent concrete elements. By using discrete beam elements as reinforcement, stresses in the bars can be measured directly. From this data determinations can be made regarding the efficiency of the reinforcement details under investigation. Although, provisions for bond and slip were not implemented, the use of an average stress-strain relationship with strain hardening represented this effect. The use of accurate material models is the most important factor in obtaining reliable solutions from a finite element analysis. An accurate model for concrete was generated using a

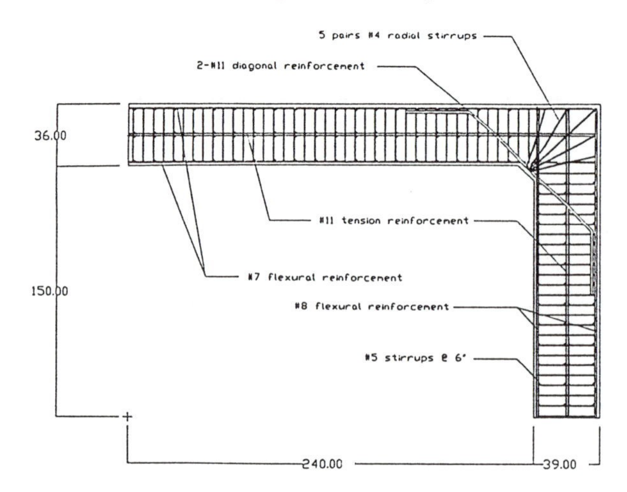

Figure 1 Basic Connection Design Detail

soil/crushable foam plasticity model incorporated in the DYNA3D code. Input variables include: bulk and shear modulus, ultimate uniaxial compressive strength and yield function parameters, and tensile strength cutoff. Stress-strain relations are based on the yield function, although a concrete pressure-volumetric strain curve was required to provide the equation of state (the state at which all equilibrium states exist in a material). The design loads used in this study were specified as an equivalent 300 lbs TNT detonation. The bilinear curve represents the initial blast load accompanied by a gaseous overpressure described previously.

The accuracy of the analysis procedure was checked by modeling two simply supported reinforced concrete beams subjected to impact load at midspan. The models simulated experimental beams C-1 and H-1 of tests completed by Feldman and Siess(15). Beam deflection, steel strain, and load function history plots were recorded in the tests, and the numeric simulations were highly accurate.

RESULTS AND DISCUSSION

Response to Full Blast Loading

Design Approach I: The design case in which one No. 6 bar was provided as diagonal reinforcement was simulated. The loads applied were the two bilinear load curves provided by the specifications. Excessive deformation occurred in the legs and in the connection itself. This indicated that the details was inadequate.

Design Approach II: A numeric simulation was performed on the structure recommended in(12). The diagonal reinforcement provided in this model were two No. 11 bars. The deflections correlate to large steel strains causing structural failure prior to the termination of the loading functions and indicated the excessive deformation in the walls of the test cell.

The numeric simulations predicted that the structure would fail unless modifications are introduced. The most practical approaches to producing an acceptable design would be to either increase the load carrying capacity of the structure or to reduce the load applied. These options are discussed next.

A case providing four No. 11 bars for the diagonal reinforcement was considered. This case was presented to indicate the response of the connection when more diagonal steel was added in an attempt to strengthen the connection. However, it should be noted that the use of four No. 11 bars would not be feasible for construction.

Response to Reduced Blast Loading

The previous data demonstrated that much of the structure's deflection response occurred 10 to 20 ms after the initial blast load had terminated, indicating that the gas overpressure may be responsible for much of the damage. Further investigation was then conducted to evaluate the influence of the gas overpressure duration on the response of the structure. A relationship between the loading function and diagonal reinforcement could be examined to produce a design in which none of the reinforcement would fail in response to a loading function which could be generated. Several configurations of diagonal reinforcement size and gas overpressure duration were simulated. A case in which one No. 11 bar was provided as diagonal reinforcement was subjected to a modified load curve. The vented area was increased by 400% (venting through the roof) such that the total gas overpressure duration was reduced to 44 ms. The results indicated that the structure could resist this applied load. The deflection response for this case was less than in all of the cases were the gas overpressure duration was 165 ms (the original case). A maximum deflection under 17 inches was produced at a peak response time of around 65 ms. These results are acceptable when considered against the design specification simulation in which the deflection were over 30 inches prior to failure of the reinforcement.

Increased Haunch Size

Increasing the size of the haunch would increase the stiffness of the joint while effectively shortening and stiffening the walls. Also, the increased haunch size would help alleviate some of the reinforcement congestion within the connection. This would make it possible to supply two No. 11 bars as diagonal reinforcement without displacing other steel components. The haunch and diagonal reinforcement in the first case was extended to 18 inches (about 9 inches greater than that specified by the original design). The bilinear loading function were applied and two No. 11 bars were supplied as diagonal reinforcement. Around 70 ms, a shearing failure in the side wall began to develop illustrating the weakness within the walls. The diagonal reinforcement did not fail and reached a peak response near 70 ms which is well below the fracture strain. The flexural reinforcement within the walls remained below failure strain up to 90 ms, however, the localized shearing in the plastic hinge region, at the ends of the haunch, suggested the failure of the structure.

The second case extended the haunch further to over 36 inches. Two No. 11 bars were supplied as diagonal reinforcement across the joint and the bilinear load curve was again used. As in the previous simulation, the flexural reinforcement within the wall was highly strained. Although the localized shearing was not present in this case, the forces in the reinforcement was greater, but still below the failure strain. The diagonal reinforcement exhibited almost the same force-time history as for the previous case. This data indicated that extending the haunch may not be enough to save the structure without increasing the amount of reinforcement within the walls. Even though the extended haunch would relieve some of the reinforcement congestion in the joint, the need for additional steel in the wall could negate that benefit.

Static and Dynamic Test Simulation

In preparation of experimental tests, numeric simulations were performed to predict response in the test situations. Both a static and dynamic numeric simulation were performed to help predict structural response. These simulations were also performed to indicate where possible complications may arise in the testing situation.

The results of the static simulation indicated that the diagonal reinforcement is over-sized. A 200 kip capacity hollow-core hydraulic jack was placed 66 inches up the stiff-back to provide the force required to test the response of the connection without causing failure of the legs, as shown in Figure 2. The peak deflection of 0.120 inches occurred at the location of the hydraulic jack. The maximum stress of the diagonal and flexural reinforcement were 14.0 ksi and 54.7 ksi, respectively. This indicated that though the flexural wall reinforcement was near yield, the diagonal reinforcement was stressed to less than 25% of its yield strength.

An experiment setup similar to the static test simulation was used to perform a dynamic test simulation. A 5400 pound pendulum was used to supply the

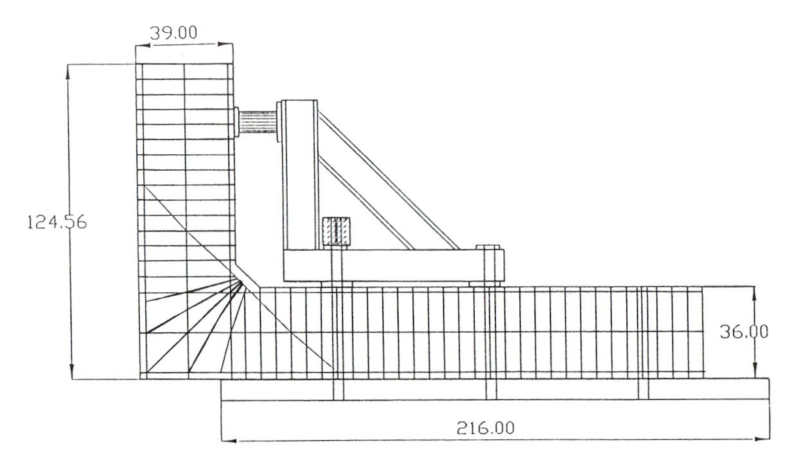

Figure 2 Test Configuration

dynamic load. A 16 inch by 18 inch steel plate acted to distribute the load in order to avoid intense stress at the point of contact. The test model under investigation included one No. 11 bar as diagonal reinforcement. Experimental data gathered from previous tests was used to determine the required drop height of the pendulum to supply the required force and duration to the test simulation. By dropping the pendulum from 13 feet, a 1000 kip load with a duration of 44 ms was produced. The load was applied at the same location as in the static simulation. The joint opening force caused by the pendulum was 17% of that created by the design load. The diagonal reinforcement across the connection yielded very early and failed near 35 ms. The reinforcement within the wall indicated that failure occurred in this section also, however the diagonal reinforcement was the primary cause of failure.

CONCLUSIONS AND RECOMMENDATIONS

The numerical simulations are an attractive option for studying the structural behavior of members subject to dynamic loading, especially those subjected to severe blast loading. These simulations help prepare for physical testing in the laboratory while yielding significant insight into the test situation before a model is constructed. Despite the advantages, however, the results from such numeric simulations must be accurate in order provide useful insight into the problem. An important part of this study was dedicated to establishing these requirements. The method for modeling the concrete, steel, and mesh was validated through the reenactment of two experiments for which experimental data was available. It was observed that the structural response is sensitive the loading function applied to the structure. Therefore, the conclusion derived concerning the legitimacy of the approach method and its practicality are apparent. It is acceptable to perform numeric simulations of real structures provided that the required conditions of the simulation are met and that, as in this investigation, the results are only as accurate as the precision with which the loading function is applied.

A large amount of data was used to evaluate the different structural details under consideration. The evolution of these details illustrated how changes in the design would effect the observed response. The result from the study demonstrate that careful attention to detailing is critical for any structure subjected to such severe loading. Although this conclusion is not new(1)(7), the principal conclusion indicates that the design of connections subject to dynamic loading should not be based on the design recommendations for static design.

By decreasing the duration of the gas overpressure (via additional venting through the roof), the effects of the load are reduced to a level at which the reinforcement is not expected to fracture. This modification would decongest the reinforcement observed in the joint and require the least amount of additional materials. One of the objectives of this study was to help predict the response of the experimental test specimens. Although only two test simulation were examined, other cases could be simulated to better predict the behavior of the test specimen. Such physical experimentation would prove beneficial in calibrating the finite element model in order to produce a more accurate representation of the structure as a whole entity. After the laboratory evaluations, and possible modifications, field tests should be performed for the final validation.

REFERENCES

1. Park R. and Paulay T. *Reinforced Concrete Structures*, John Wiley and Sons (1975).
2. Collins M.P. and Mitchell D. *Prestressed Concrete Structures*, Prentice-Hall (1991).
3. Chen W.F. *Plasticity in Reinforced Concrete*, McGraw-Hill (1982).
4. Soroushian P., Choi K.B. and Alhamad A. 'Dynamic constitutive behavior of concrete', ACI Journal, Vol 83, No. 2, pp. 251-259 (1986).
5. Soroushian P. and Choi K.B. 'Steel mechanical properties at different strain rates', Journal of Structural Engineering, ASCE, Vol. 113, No. 4, pp. 663-672 (1987).
6. Ross C.A., Kuennen S.T. and Strickland W.S. 'High strain rate effects on tensile strength of concrete', Proc. 4th Intnl. Symp. Interaction of Non-Nuclear Munitions with Structures, Vol. 1, pp. 302-308 (1989).
7. Nilsson I.H.E. 'Reinforced concrete corners and joints subjected to bending moment', Document D7, National Swedish Building Research, Stockholm (1973).
8. Schlaich J., Schäfer K. and Jennewein M. 'Toward a consistent design of structural concrete', PCI Journal, pp. 74-150 (May-June 1987).
9. Pantazopoulou S. and Bonacci J. 'Consideration of Questions about Beam-Column Joints', ACI Structural Journal, Vol. 89, No.1, pp. 27-36 (Jan.-Feb. 1992).
10. Krauthammer T. and DeSutter M.A. 'Analysis and design of connections openings and attachments for protective structures', Final Report No. WL-TR-89-44, Weapons Laboratory, Kirtland AFB, New Mexico (October 1989).
11. American Concrete Institute. *Building Code and Commentary*, ACI 318-89, (1989).
12. Naval Civil Engineering Laboratory. 'Basis of design for NAVFAC type I Test Cell', Technical Note N-1752R, Revision 1 (April 1990).
13. Department of the Army. *Structures to Resist the Effects of Accidental Explosions*, TM 5-1300 (November 1990).
14. Whirley R. and Hallquist J.O. *DYNA3D User's Manual*, Lawrence Livermore National Laboratory (1991).
15. Feldman A. and Siess C.P. 'Investigation of resistance and behavior of reinforced concrete members subjected to dynamic loading', Part II, University of Illinois, Civil Engineering Studies, SRS No. 165 (30 September 1958).

29 AN ASSESSMENT OF IMPROVED RESISTANCE TO DYNAMIC LOADS IMPOSED ON ORDINARY DWELLINGS

R. Delpak and D. Poullis
University of Glamorgan, UK

Simple tests were devised to assess the potential resistance of ordinary dwellings in the Mediterranean coastal belt which may be subjected to dynamic loads. Many of the buildings which are occupied, impose cost and logistic constraints on any proposed improvement. The project has been laboratory based and has consisted of: (a) identifying a simple design which typifies both the foundations and the structural layout, (b) choosing a practical scale so that method of construction and material properties could be modelled faithfully and, (c) repairing the ensuing damage after the initial dynamic loads and assessing the improved resistance due to renewed loading. The repairs were found to be simple, inexpensive and effective, improving the structural stiffness by over 20%.

INTRODUCTION AND AIMS

Background

The zone known as the Cyprus Arc is positioned between the large African plate to the south and the smaller Anatolian plate to the north. This earthquake zone is ranked to be the second most active both in frequency of occurrence and in damage caused. Figure 1 is a schematic representation of major recorded earthquake location(s) and intensity on a time-march sequence. According to historic information[(1)], 125 earthquakes are recorded from 180 B.C. to 1900 A.D., 30 of which are regarded as powerful. Table 1 lists the more recent activities. The largest earthquake in Cyprus within the last two centuries, was on 10th September 1953 in Paphos. Villages of Stroumbi, Lapithou, Fasoula, Ascilou and Kitasi (amongst others) were destroyed, causing 63 deaths and 200 injuries[(1)].

The frequency of occurrence of earthquakes in Cyprus is such that it would be imprudent not to implement seismically resistant measures to upgrade traditionally constructed dwellings. These houses are constructed using mudbricks. In particular, within the Walled City of Nicosia, the strengthening and upgrading activity is synonymous with preservation of the historic dwellings with distinctive architectural features. Difficulties are exacerbated since most historic buildings are inhabited making extensive improvements costly.

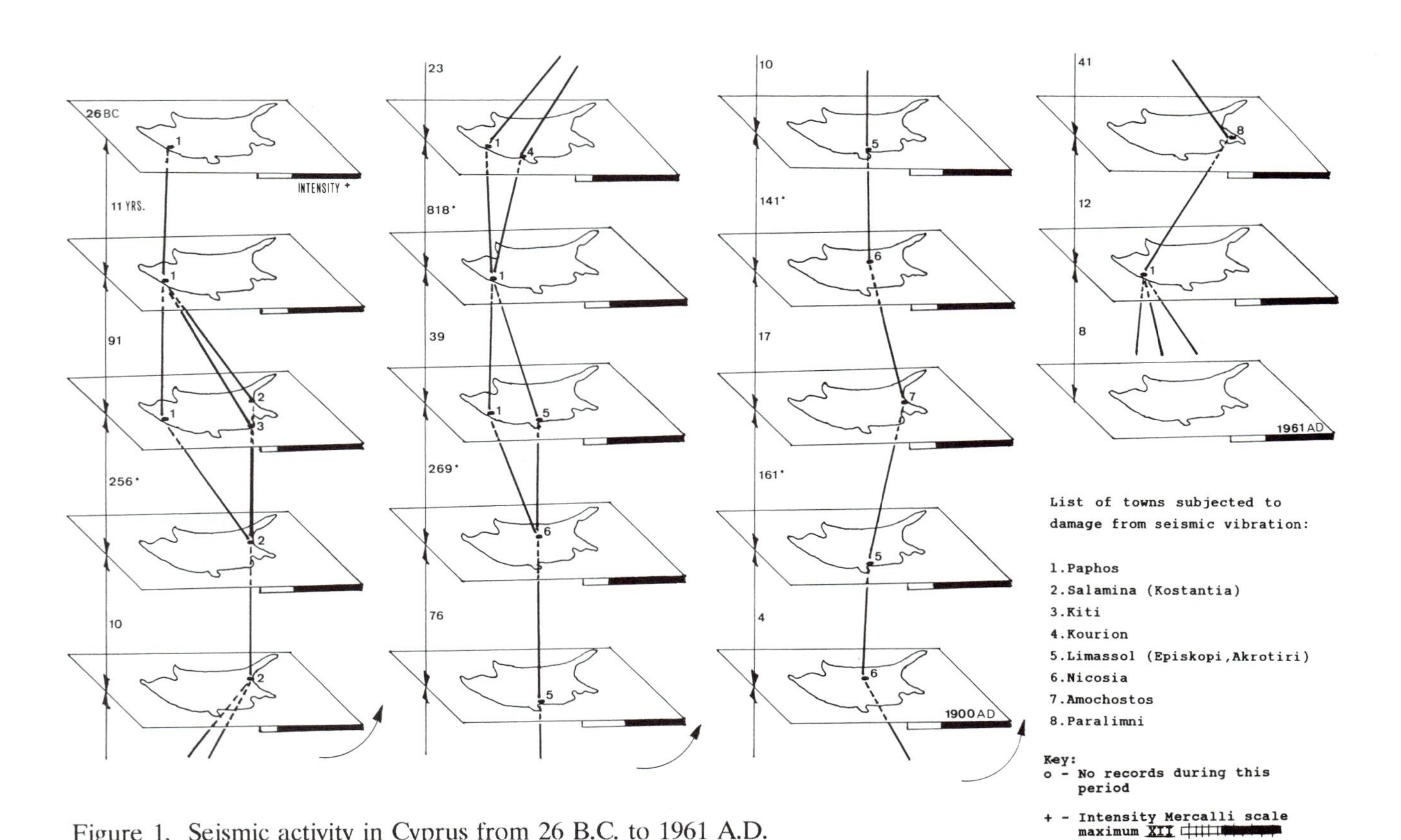

Figure 1. Seismic activity in Cyprus from 26 B.C. to 1961 A.D.

Table 1. Four of the more intense 755 earthquakes at a distance of 50km or fewer from the coasts of Cyprus between 1900 - 1961

DATE	Geographical Coordinates		Depth (Km)	Intensity (Mercalli Scale)	Comments
	Length	Width			
5.1.'00	35.15	33.15	unclear	VII-VIII	Dstrn Nicosia
20.1.'41	34.99	33.58	100	IX	Dstrn Paralimni 24 psns injured & 450 hs's demolished
10.9.'53	34.50	32.21	33	IX	Wide dstrnin Paphos, 95 vllgs destroyed
15.9.'61	34.94	33.78	36	VIII	Powerful e/q in all of Cyprus

Assessment of means and options available

The purpose of the present note is not to propose new methods of antiseismic design. However, it is intended to initiate a feasibility study, in order to assess if traditionally constructed buildings could be upgraded seismically in an inexpensive manner with minimum disturbance to the inhabitants or the structure. It is also hoped that the method devised should be sufficiently simple so that the benefits could be felt not only in inaccessible areas of Cyprus but also in other nearby countries where there are severe resource and cost constraints.

The original construction process was thought to have predated the existing planning permission procedure. Hence finding any possible drawings of the layout was regarded as remote. This imposes an additional constraint that any possible upgrading solution must be applicable systematically to the substantial majority of structures. Since an accurate performance estimate of structural components (eg beams, columns, walls...) was unrealistic, it was thought that the assessment of remedial measures and the associated effectiveness have to be empirical in substance. Nevertheless the senior author is aware of similar research projects also in Europe with greater emphasis on theoretical development. It was thought that model analysis was a simple and relatively inexpensive method of; (a) investigating any structural weaknesses and; (b) possible effectiveness due to any upgrading measures. The project planning was narrowed down in identifying the following:-

(i) Was it possible to define an 'average' house in terms of layout, dimensions, materials and method of construction?

(ii) Could the above details be scaled down to manageable overall dimensions suitable for controlled laboratory testing?

(iii) Could the strength characteristics of the prototype construction materials be modelled satifactorily?
(iv) Would the simplified dynamic loads prove to be unrealistically simple?
(v) Could the laws of similitude remain inviolable through careful choice of test conditions?
(vi) Is it possible to extrapolate the model results, to the prototype in a meaningful way?

The answer to the majority of the above questions is a qualified yes, provided that the expectations are confined to such interpretations which are qualitative in essence.

SIZE AND MATERIAL PROPERTIES DETERMINATION OF AN 'AVERAGE TRADITIONAL' BUILDING AND THE CORRESPONDING MODEL

Inner and other dimensions of a typical prototype

The layout of a typical traditionally constructed dwelling was confined to Cyprus, in particular the Walled City of Nicosia. The assessment and measurements were carried out by a team of diplomates from Higher Technical Institute (HTI) in Nicosia. The assessments were later confirmed by then Head of Department of Construction at HTI. Figure 2 represents a typical 'unit' where the plan view and the elevations are given. Building walls (external and internal) rest directly on strip foundations penetrating generally to about 0.5m into the soil. The foundation depth in sometimes extended to 0.75m but seldom deeper. Excavations show that the foundation width matches the wall thickness closely. The walls are made of sunbaked mudbricks which support timber trusses which in turn are covered by clay tiles.

Engineering properties of the mudbrick

The traditional method of mudbrick production in the region is to dry the briquettes in the direct sunlight, once removed from the moulds. A typical brick size is about 400 x 300 x 6mm which could contain straw, for additional fibre reinforcement. Clay mortars are used for brick laying which make load bearing characteristics fairly uniform. For the present study, a number of samples which were from Cyprus were tested.

Young's modulus (from 3-point loading): Three-point loading tests on prepared specimens were carried out, after which the load/deflection graphs were used to calculate the E-values. The initial part of p/δ curves, displayed remarkable linearity from which the near consistent values of 27.39, 28.33 and 35.50 N/mm^2 were obtained. This enabled E_{av} (brick) = 30.41 N/mm^2 to be estimated. Bearing in mind that E_{av} from bending, contains compressive and tensile components of linearity, an assumption of the relative magnitudes is necessary, in order to estimate the above two components separately. Hence assuming

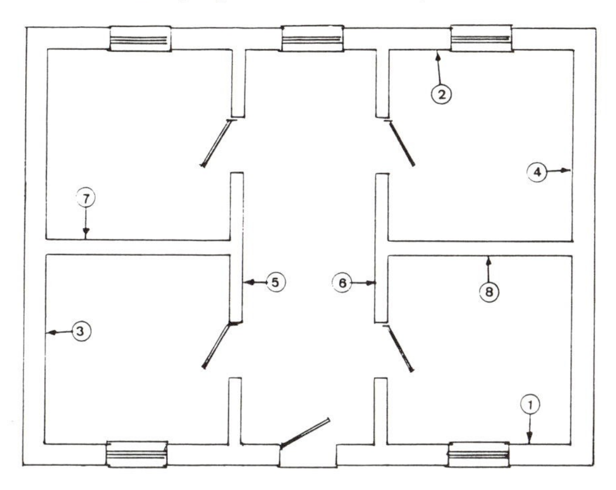

Dimensions of a typical building and model dimensions						
Wall	Dimensions of walls			Scaled down dimensions		
	Length	Height	Thickness	Length	Height	Thickness
1	10.3m	3.0m	0.40m	51.5cmm	15cm	2cm
2	10.3m	3.0m	0.40m	51.5cm	15cm	2cm
3	8.0m	3.0m	0.40m	40cm	15cm	2cm
4	8.0m	3.0m	0.40m	40cm	15cm	2cm
5	8.0m	3.0m	0.25m	40cm	15cm	1.25cm
6	8.0m	3.0m	0.25m	40cm	15cm	1.25cm
7	3.3m	3.0m	0.25m	16.5cm	15cm	1.25cm
8	3.3m	3.0m	0.25m	16.5cm	15cm	1.25cm

Figure 2. Outline dimensions of a typical traditionally constructed dwelling with 1:20 scaled model sizes.

$$E_{comp} / E_{tens} = 20 \text{ and } E_{av} = \sqrt{E_{comp.} \; E_{tens,}}$$

modulus in compression is estimated to be $E_{comp} = 0.136 \times 10^3$ N/mm^2.

Young's modulus (from compression test): There are difficulties in preparing compression samples of near identical cross-section. Control of the cross-section uniformity became a major exercise, since the clay flaked frequently making quality control difficult. Load/compression plots provided near-linear variations which were used to determine E_{comp}. The four values were 1229.2, 1396.8, 1214.7 and 1843.8 which average to $E_{comp} = 1.421 \times 10^3$ N/mm^2.

Unconfined compression test: Four specimen provided the compressive strengths which were 1.301, 1.152, 1.413 and 1.478 N/mm^2, averaging to σ_{comp} = 1.34 N/mm^2. The average cohesive strength was hence estimated to be c = 0.67 N/mm^2.

Young's modulus from other data: There are two sources of published data namely Simons and Menzies(2), and Tassios and Chronopoulles (3) who provide the following formulae respectively, E/c = 1000 - 1500 and E/σ_{comp} = 1200. The E determination from cohesion was based on a revised value from Simons(4) which gave E/c = 1575, hence E''_{comp} = 1.055 x 10^3 N/mm^2. Based on compressive strength, another E-value is extimated to be E'''_{comp} = 1.608 x 10^3 N/mm^2.

Density Determinations: There seems to be a considerable variation in estimating densities. The present density determination is specifically aimed at the Cyprus brick and not the constituent clay, so that ρ_{av} = 1500 kg/m^3. Another estimate (5) based on clay dry densities in Nicosia gives ρ_{av} = 1842 kg/m^3. Yet a third estimate(5), when considering sand stone used in walls, provides ρ_{av} = 2307 kg/m^3. It is difficult to justify the last two densities in the light of mixing and preparation processes involved locally. It is thought that ρ(brick) = 1500 kg/m^3, is a realistic figure.

Summary: Table 2 includes various engineering properties of the mudbrick obtained by averaging where appropriate.

Table 2. Engineering properties of the mudbrick

Density, ρ_{av}	Compressive strength, σ_{comp}	Young's modulus, E_{comp}
1500.kg/m^3,estim	1.34 N/mm^2, (4 samples)	1055.N/mm^2, (4 sources)

Scale parameters in the model

After considering few sizes, it was decided to choose a 1:20 scaled model, while retaining an open mind in implementing possible changes where necessary. The sizes and dimensions are scaled as faithfully as practicable but regrettably there are violations in dimensional analysis, making the model far from perfect.

Model material choice and properties

The substantial amount of clay present in the original structure, necessitated the use of granular material in the model with increased strength properties. It was decided to use clay (pulverized red marl)/lime mixture and was anticipated that reasonable workability and strength would be resulted.

Determination of percentage lime required: The minimum percentage lime needed was determined using a method suggested by Eades and Grim(6). Calculations indicated that a pH of 12.3 corresponded to 7% lime content, see Figure 3. Mixtures in excess of 7% did not change the pH value of the blend. The above limit was regarded as the optimum lime content for clay stabilization. The lime used was purchased locally from commercially available supplies.

Determination of moisture content: Standard Proctor Compaction tests were carried out to determine the moisture content for maximum dry density (BS1377: Part 4: 1990). A dry density of $\rho_{c/l}$ = 1800 kg/m^3 corresponding to 14.5% moisture content was determined see Fig 4. To facilitate workability, a moisture content of 17.5% was adopted which corresponded to $\rho_{c/l}$ = 1700 kg/m^3.

Compressive Strength: The clay/lime specimens were cured at 50°C in humid environment. The 28 day compressive strength was expected to be around 3.6 N/mm^2, Figure 5 after Arabi (7). The specimens failed at 3.37, 3.71, 2.97, 3.58 and 3.69 N/mm^2 with an average of σ_{comp} = 3.46 N/mm^2. The average shear strength is estimated as c' = 3.46/2 =1.73 N/mm^2.

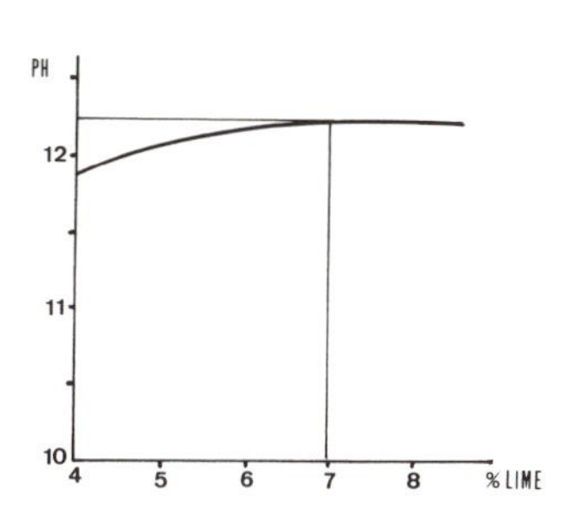

Figure 3. Variation of pH versus % lime added for red marl stabilization optimum moisture content

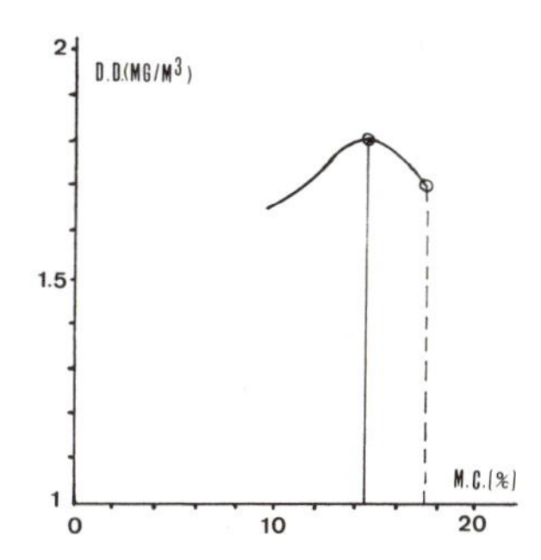

Figure 4. Dry density versus moisture content variation to determine the

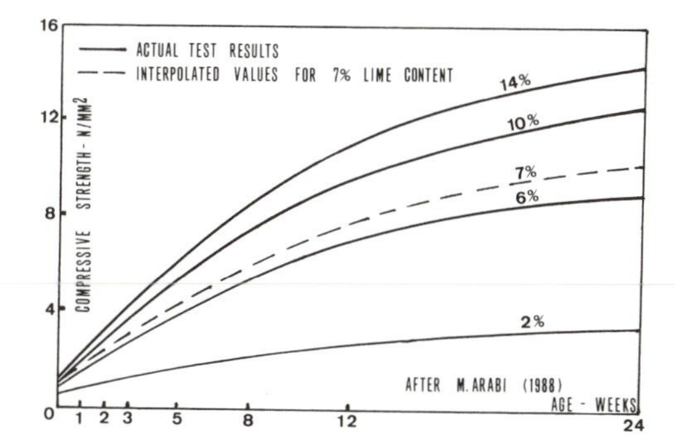

Figure 5. Compressive strength characteristics for cured clay/lime mixtures.

Young's moduli for clay/lime: Again, the above empirical methods were deployed for the clay/lime system. Using Ref.(4), E (c/l) /c = 1575, with c = 1.73 N/mm^2; E (c/l) is calculated to be 2725.5 N/mm^2. E-values from the compression tests are 5011., 5011., 4296., 5011. and 5011. N/mm^2 respectively. Hence E′(c/l) averages to 4868. N/mm^2. Formula E/σ_{comp} = 1200, provides E″(c/l) = 4152. N/mm^2, Ref.(3).

Summary: Despite considerable numerical variations in Young's Moduli, an average was taken. Table 3 contains the engineering properties for the clay/lime mixture.

Table 3. Engineering properties of the model material, i.e. clay/lime mixture.

Density, ρ'_{av}	Compressive strength, σ'_{comp}	Young's Modulus, E′(c/l)
1700 kg/m^3	3.46(4wks), 3.6(6wks)N/mm^2 est	3915. N/mm^2 (3 sources)

By forming the ratios using Tables 3 and 2 , it is seen that the structural model is not mathematically valid[8]. The summary of parametric ratios is given in Table 4.

Table 4. Ratio of the engineering properties of model/prototype.

ρ'/ρ	$\sigma'_{comp}/\sigma_{comp}$	E′(c/l)/E(mudbrick)
1.133	2.6(4wks),2.7(6wks)est.	3.7

Six week test data are also included, since; (a) subsequent to expiry of a four week curing period, the model was assembled in the fifth week, and (b) within nine working days it was loaded to failure.

MODEL CONSTRUCTION

A brief survey of few damaged buildings had shown that the wall corners and connections had mostly retained their L or T- shaped preloading integrity. This had meant that the individual internal or external walls could be cast individually and bonded at appropriate points. After curing the clay/lime slabs, the corners and the extremities were cemented using 'Araldite Rapid' to form the outline of the building shown in Plate 1. The walls were connected to the base using 'Silicon Compound' which once cured, formed a permanent flexible seal with satisfactory adhesive properties.

Actual timber trusses were simulated using Balsa wood. Once the roof was positioned on the structure, a total of 1.0kg of mild steel rods were added to the truss members so that the total roof weight was uniformly distributed. The $(Mass_{roof}/Mass_{building})_{model}$ was 1.0kg/19.25kg = 5.19 %. The following points have been noted in preparation of the model.

(i) Material strength parameters, although different numerically, have near identical characteristic variations.
(ii) The integrity of wall connectivity, either as L or T-joints, have been retained by adequate application of adhesive.
(iii) The flexible interaction between the foundation and the adjoing wall has been retained.
(iv) There is a measure of external wall restraint at the top, through connection to the roof truss system.

There is a given sequence adopted for wall/foundation assembly but the completed model is shown in Plate 2.

ENHANCEMENTS FOR THE 'SEISMICALLY UPGRADED' MODEL

The term 'upgrading' has been adopted to mean repairing and reinforcing the damaged structure by making good the overall shape, so that repaired geometry - as represented by the centre lines - matches that of the original.

Most of the damage was observed to initiate from the door and window corners at about 45° as might be expected. The remedial measures consisted of reinforcing the area likely to be damaged and hence limiting any future excessive strain owing to anticipated dynamic loads.

The influence of an installed reinforcing bar was scaled to a 0.5mm diameter piano wire with average tensile strength of 473.0N. A square mesh formed by soldering cut lengths of the wire was prepared in order to model the square steel meshes which are available commercially. The ratios of the ultimate strength of the reinforcement to the host material are given as follows:
$r_{\sigma m}$ (ratio of model stresses) = 2409./3.46 = 696,
$r_{\sigma p}$ (ratio of prototype stresses) = 250./1.34 = 187,
$r_{\sigma m}/r_{\sigma p}$ = 3.7 (also see the ratios given in Table 4).

The upgradings fell within the following categories:

Plate 1. Assembled slabs of cured clay/lime mixture modelling the footings

Plate 2. The completed model with the weighted roof trusses

Single and double lintel reinforcement
The procedure consisted of etching a channel 3mm wide and 2mm deep located 10mm above the wall opening, so that its length extended beyond the 45° potential cracks in both directions, see Plate 3. The piano wire was cemented in position and excess adhesive was shaved off subsequently. The method would have been repeated 10mm above all the openings. The above repair was carried out in some instances to prevent the potential damage owing to cracks propagating underneath the window corners, again at about 45°.

Strengthening by mesh provision
The square mesh, discussed earlier, was used to reinforce an area with multiple potential faults. The procedure consisted of applying the adhesive to the area in need of reinforcing and positioning the mesh carefully until the adhesive was set. The excess mesh covering the opening would have been cut out initially but the protruding adhesive was removed subsequently. Typical remedial measures are shown in Plate 4.

TEST PROCEDURES AND INSTRUMENTATION

From the outset, the model was intended to be excited horizontally. To minimise any vertical loading, the model was suspended and the connection to the horizontal active shaker was eventually via a rigid base plate. A bent connection between the base plate (containing the foundation) and the shaker element, aligned at the acceleration the models centre of gravity level. Two loading were methods used:

Fatigue loading of the original model
Through a suggestion by Mr W.M. Hague, the model was vibrated with the frequency range set from 8Hz to 200Hz and again to 8Hz at 1.0g acceleration. The above loading sequence was intended to be repeated until failure occurred.

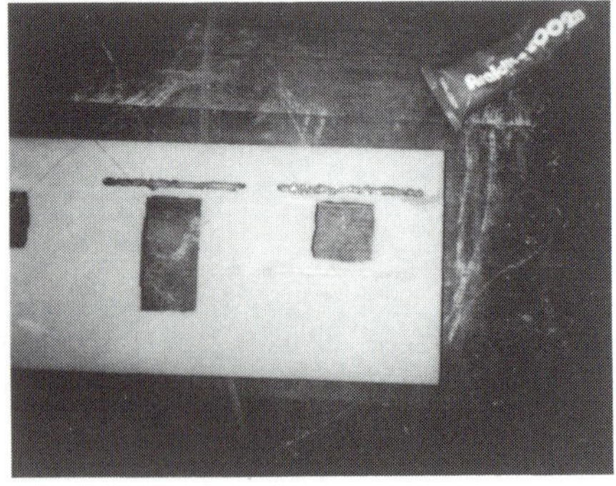

Plate 3. Bonding the reinforcement at the lintel level for openings

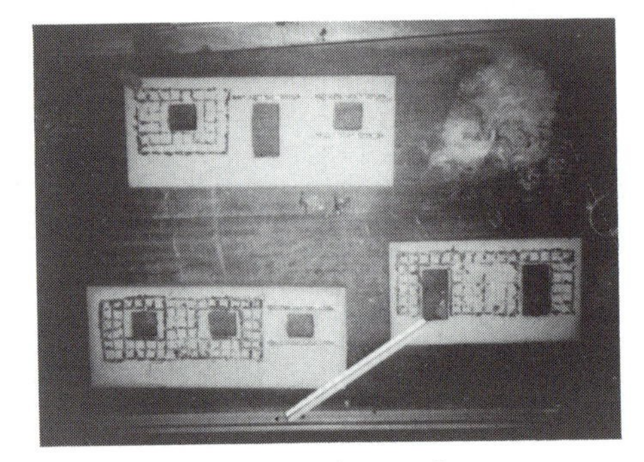

Plate 4. Upgrading by mesh reinforcement

It was observed that neither weaknesses nor cracks appeared on any of the walls. Loading was therefore intensified as follows; (a) 1.55g for 2½hrs., and (b) 1.70g for 4 hrs. In the absence of any slight sign of weakness, the fatigue method was abandoned.

Failure test method of the original model

Using accelerometers, the resonance frequency of the model was judged to be f_n = 23Hz. The model was vibrated at the resonant frequency with various acceleration values. At 1.7g, cracks began to appear and the model failed shortly afterwards.

Failure test method of the upgraded model

By attaching accelerometers the resonance frequency was judged to be f'_n = 26 Hz. The loading was carried out on a trial and error basis as follows;
(a) 1.50g for 2¾hrs., (b) 2.00g for 8hrs., (c) 2.20g for 2 hrs. and finally,
(d) acceleration was increased to 3.00g for 2 hrs, all at 23 Hz.

In the absence of any visible cracks, the model was vibrated at the upgraded frequency of f'_n = 26Hz for g = 3.00 m/s^2. Cracks began to appear and the upgraded structure collapsed subsequently. The model and shaker layout are shown in Plate 5, whereas a typical failed configuration is given in Plate 6. Resonance frequency determinations were confirmed after consultations with Mr W.M. Hague.

RESULTS, OBSERVATIONS AND CONCLUSIONS

The primary weaknesses in the original model were observed to be as follows;
(i) stress concentration cracks emanating from the corners of doors and windows at near 45° angle (sometimes extending to 60°).
(ii) sheer failure on internal walls and on both above and below the openings, followed by immediate collapse of fragments from the unsupported internal door openings,

Plate 5. Shaker and model layout ready for dynamic loading

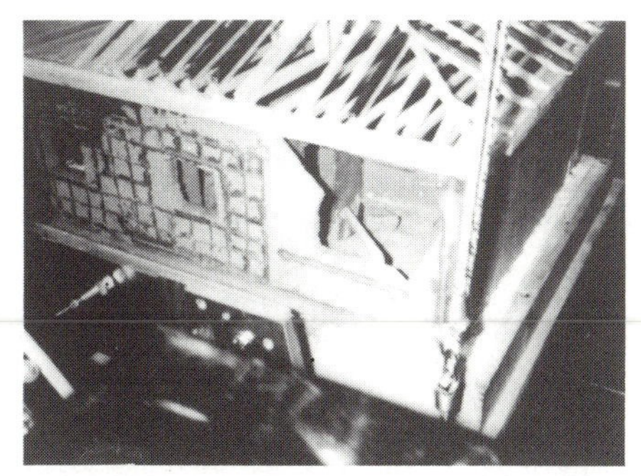

Plate 6. Typical failure illustration, the relative performance of upgraded

(iii) failure due to horizontal shear forces both at the floor and roof levels occurring after collapse in (ii), and

(iv) failure owing to lack of connectivity at wall intersections which is subsequent to completion of shear damage in (iii).

Failure patterns noted in the model, bear a close resemblance to the seismically afflicted dilapidated buildings in Cyprus.

Upon upgrading, the failure tendencies shifted to the secondary lines of weakness identified as:

1) shear failure above internal doors, above and below windows (in some cases at mid-window height) being intiated at the unreinforced side of the wall but displaying a delayed collapse,
2) horizontal shear damage observed at both the floor and roof levels, once shear failure had occurred for walls with opening(s),
3) horizontal shear damage causing disconnection from the supproting edge corners (wall intersections) for walls without openings, and
4) extensive failure and collapse elsewhere.

Lintels which were the mainstay of strengthening of the primary weaknesses, yielded eventually in a similar corner crack mode. Of the three measures outlined, mesh strengthening appeared to be the most effective. However, upon sustained vibration, mesh reinforced areas were also observed to be vulnerable to corner cracks. Wall integrity become more of a problem than anticipated at both phases of model testing where the appearance of diagonal cracks exacerbated the weaknesses.

Upgrading efforts were regarded to be an unmitigated success. To highlight the improvement, the original and upgraded characteristics are given in Table 5. It is seen that inexpensive and low-tech improvements suggested in the present note, have enhanced the structural performance of a model of the traditionally constructed buildings.

Table 5. Characteristics at failure for both models.

	Resonance frequency in Hz (judged)	Collapse acceleration m/s^2
original model	23.0	1.7
upgraded model	26.0	3.0

The increase in structural stiffness is estimated using the general formula ,

$$\omega_n = \alpha\sqrt{k/m}$$

where ω_n is the natural circular frequency and α an empirical shape constant; k and m are the effective stiffness and mass respectively, all for the original model. The frequency of the upgraded model could be written similarly as

$\omega'_n = \alpha' \sqrt{k'/m'}$ k'/m' where $\alpha' = \alpha$ and $m' \simeq m$. Hence, by forming the

ω'_n/ω_n, the following ratio is resulted, $\sqrt{k'/k} = 26/23$. The increase in stiffness can be estimated as $k' \simeq 1.278k$, namely a near 30% improvement. A higher g-value needed to cause collapse of the repaired model is an indication of improved safety features. Clearly the benefits are such that either the structure can resist a higher seismic load or a longer evacuation time for a given seismicity.

ACKNOWLEDGEMENTS

The authors are grateful to Professor P.S. Coupe, Head of Department of Civil Engineering and Building in allowing the use of the resources for the present work. They wish to thank the technical staff both in Civil and Mechanical Engineering Departments for their active support. The guidance of the senior Academic Staff at HTI and the Municipal authorities in Nicosia, Cyprus is acknowledged. Miss M Leonidou's compilation of the relevant information has been an essential part of this note. They are indebted to Miss M. Gapper for implementing various specified constraints in the present typescript most ably. Finally the efforts of Mr W.M. Hague, Senior Lecturer in Mechanical Engineering is gratefully acknowledged, since without his sound technical advice and permission for generous use of his busy Dynamics Laboratories, the present work would have remained inconclusive.

REFERENCES

1. Extracts from: *Cyprus Great Encylopedia*,Vol.12, pp.169-179 (1991) Publs, Philoki Bros, Cyprus.
2. Simons N E and Menzie B K: *A short course in Foundation Engineering*, IPC Science and Technology (1975).
3. Tassios T P and Chronopoulos M P: 'Aeismic dimensioning of interventions (repair/strengthening) of low stength masonary buildings,' Proc. Middle East and Med. Regional Conf. on Earthen and Low-strength Masonary Buildings in Seismic Areas, 31 August - 6 September (1986), Ankara, Turkey.
4. Simons N E: 'Shear strength of stiff clay', Proc. Geotech. Conf., pp159-160 (1967), Oslo.
5. Leonidou M: 'Methods and procedures for repair and strengthening of old traditional buildings in seismic regions, Nicosia, Cyprus', BEng project 1990, the Polytechnic of Wales, Wales, UK.
6. Eades J L and Grim R E: 'A quick test to determine lime requirements for lime stabilization', Highway Res. Rec. No. 139, pp. 61 - 72 (1964), London.
7. Arabi M: *Fabric and strength of clays stabilized with lime*, Vol.II, pp. 61 - 67, PhD thesis (1988), The Polytechnic of Wales, Wales, UK.
8. Davies J D and Preece B W: 'The laws of similitude for structural models', Bull. Mech. Eng. Educ., Vol. 6, pp. 357 - 368, Pergamon Press (1967), UK.

30 INNOVATIVE DEVELOPMENT IN MASONRY

G. Shaw
Curtins Consulting Engineers plc, London, UK

The paper describes the Author's experience in the innovative development of structural masonry through design, research and application over a 30 year period. The cross fertilisation of ideas and developments from other materials, and the unique advantages this revealed for masonry applications are described. The techniques are illustrated with examples of applications. The paper concludes with a description of the most recent developments in prefabrication techniques which place masonry on a new horizon for exciting further developments and applications.

INTRODUCTION

Observations and Related Engineering Principles
The engineering principles which are the basis of the developments discussed in this paper were built upon and extended through analysis, application and design to solve structural problems. The initial observations related to the stability and strength gained from the geometry of layout of bonded masonry units which ran parallel with the author's experience in the analysis and design of prestressed concrete.

During this period the author examined the principles of prestressed concrete and the structural efficiency of geometric sections such as tubular steel, timber stressed-skin plates, BSB and channels, concrete Tee and L beams, and prestressed hollow sections.

It became apparent in this study that sections with a high Z/A ratio have greater efficiency in bending resistance and have an improved radius of gyration and hence resistance to buckling (provided that the section maintains local stability).

In general the Author's early designs of masonry structures were basically gravity structures with compressive loading on the external cladding and internal walls, these designs where backed up by research and testing at Liverpool University on various walls. As the buildings became taller and of lighter construction the problems relating to lateral loading became more critical. The problem for masonry was particularly serious in tall single-storey open-plan structures where compressive loads on the walls was minimal and lateral load due to wind was significant.

When lightweight roofs were adopted the vertical load, at roof level, could reverse to an uplift force during wind loading. The design criteria for these buildings was not the axial compressive strength but the flexural strength of the wall section. This is fundamentally important since the flexural strength of masonry is only approximately 1/30 of its compressive strength. The governing structural design factor at first, became to limit the tensile stress and this was achieved by increasing the section modules of the structural element by using Fin Wall and Diaphragm wall sections, see table 1.

Table 1.

Wall section	Area A m2/m run	Section Modulus Z x10^3 x m^3/m run	2nd Movement of area 1 x 10^3 x m^4/m run	Z/A Ratio
260mm	0.205	3.50	0.179	13.94
665mm Diaphragm section 9	0.251	51.77	17.21	206.25
Fin wall section M @ 3m c/c	0.236	z_1 48.41 z_2 75.33	36.16	205.13 319.19

As is well known the moment of resistance, MR, = f x Z

Thus an increase in Z (the section modulus) results in a decrease in the bending stress f and hence the bending tensile stress.

The normal masonry wall is a 'plate' section which has a low Z/A ratio and is not structural efficient compared with box, Tee and L sections. This therefore influenced the need to investigate and critically analyse section shapes. The use of section geometry seemed an obvious way forward but in a number of situations section sizes became cumbersome. In many cases it was apparent that the compressive stress of the masonry was not being fully developed and the governing factor was its very low bending tensile strength. Masonry being strong in compression and weak in tension is similar to concrete and, whilst construction methods are different, the basic methods of overcoming the weakness are related.

ADJUSTMENTS AND CHANGES TO DEVELOP MASONRY'S STRUCTURAL POTENTIAL

Material Displacement in Prestressed Masonry

Masonry is constructed from small preformed standardised units, ie various types of bricks and blocks, however to make economic geometric forms, it is necessary to arrange these preformed units into larger and more efficient sectional shapes. Masonry does not require shuttering or other complicated methods to form the shapes for vertical elements and therefore rectangular, triangular, box, Tee and L sections can, relatively easily, be constructed. The increase in setting out and plumbing of corners is the main 'extra-over' cost. These cost implications however are insignificant when related to the overall economy of the section.

Masonry Strengthening

In concrete, as is well known, the tensile stresses are either carried by steel reinforcement or eliminated by prestressing - similarly the same process can be applied to masonry. Reinforced masonry has been used quite extensively particularly in the seismic areas of India, Japan, America and elsewhere. In the UK however reinforced masonry has not usually been shown to be cost effective.

In addition to construction and supervision difficulties (with reinforced vertical masonry elements) there is the disadvantage that over 50% of the cross-section is structurally wasted in the tensile zone where it makes little contribution in resisting bending.

Prestressing on the other hand utilises the whole cross-section and, too, can reduce the amount of micro-cracking by maintaining compression in the section.

For sections subjected to compression and single axis bending the use of prestress applied eccentrically to counteract the applied bending moment is generally the most economic. For sections subjected to reversal of bending moments, for example an element resisting wind loading, concentric prestress is often more suitable.

The analysis of the cross-section can be based upon limiting the magnitude of the tensile stresses, due to the applied loading, to acceptable limits. Alternatively tensile stress limits may be exceeded for the ultimate load condition and elements designed on the basis of a cracked section analysis. The cracked section analysis should however only be applied to the ultimate condition and tensile stress should be limited to zero for the serviceability state, thus ensuring the design section does not crack.

Additional analysis is carried out to check the conditions for shear, bond deflection and other design criteria.

Comparison of a traditional cavity wall construction with a diaphragm or fin wall section combined with eccentric prestresses shows that the section can be increased in bending strength by 300 to 600 times. This is a very significant strength increase.

The above basic principles lead the designer away from grouted reinforced masonry sections, such as grouted cavity construction, and towards hollow prestressed masonry sections.

The theoretical basis for the design principles adopted for the case studies can be obtained from the combined stress formula for a section subjected to direct force and bending.

$$\text{i.e } f = W/A \pm M/Z$$

The critical stress condition for masonry where the moment, M, is dominant is W/A - M/Z. Considering this formula there are a number of design aims to consider to improve the condition where tensile stress is critical.

Design aim 1a Increase the load W
Design aim 2a Reduce the applied bending moment M
Design aim 3a Reduce the affect of M by an opposing moment Pe
Design aim 4a Improve the geometry of the section ie Z/A ratio
Design aim 5a Resist the tensile stress with reinforcement
Design aim 6a A combination of 1a, 2a, 3a, 4a and 5a

To indicate the increase in bending resistance of an eccentrically prestressed hollow masonry section from that of a similar cross section of solid plain masonry assume the following:

fbt of masonry = 1. Therefore, since masonry's bending compressive resistance is approximately 30 times that in tension assume fbc = 30.

Assume Z of the plain masonry section to = 1. Therefore since a hollow diaphragm can have a Z of approximately 15 times that of a similar cross-sectioned area solid section then assume Z of hollow section to = 15.

Bending resistance of plain solid wall = f x z = 1 x 1 = 1

Bending resistance of prestressed hollow diaphragm wall = 30 x 15 = 450

This massive increase in bending resistance is achieved with minimal increase in material and relatively insignificant increase in construction costs.

As experience and understanding improved it became apparent that vertical shear stress, principal tensile stress and prestress losses required further research. In addition, the limitation of the BS code approach to slenderness ratio, based upon 'effective thickness', appeared to be both restrictive and illogical and the 'radius of gyration' concept seemed more appropriate for future design.

MASONRY DEVELOPMENT THROUGH PROJECT APPLICATIONS

The Author's development of masonry through applications has been continuous for over 30 years and the following small number of projects selected from that experience are those which tended to punctuate that development.

Freedom Gardens, Ashton-Under-Line

Prior to 1970 the author had gained experience of using plain masonry diaphragm walls and prestressed cavity walls on numerous sports and educational buildings to resist lateral wind loads. In 1970 an opportunity arose to exploit this resistance for much larger lateral loads applied to the wall structure from earth pressure and a combination of design aims 1a, 4a and 5a was adopted in a masonry section. Freedom Gardens was constructed in 1970 and employed curved mass-filled, diaphragm retaining walls as the main feature of a landscape scheme. This was prior to the substantial experience gained on prestressed hollow sections.

Oaktree Lane Community Centre

In 1978 an opportunity arose to use a combination of the prestressing experience with that from the use of geometric forms in the Oaktree Lane Community Centre, see Figure 1.

In this contract the Author incorporated the first use of post-tensioned diaphragm walls, which were adopted as the perimeter walls to the main hall.

Post-tensioning was used to overcome the diagonal tensile stresses produced in the masonry panels during mining subsidence. The prestressing also resisted wind loading but this was secondary to the mining effect. The site was situated in an active coal mining area with both past workings and future extraction to consider - causing further differential settlement. Three seams were due to be worked within the first five year period and the calculated subsidence for the first wave indicated a maximum of 1,080mm and a differential subsidence across the site which would leave the building in a tilted state after the first wave had passed through. The effects of other workings, within the five year plan were less severe and predicted to have a righting effect on this initial tilt.

At Oaktree Lane Community Centre a combination of design aims 1a and 2a was adopted.

Piaus X Church

In 1980 experience gained repairing the failing stack bonded masonry arches to the main frame of a church provided an insight into the possibility of resisting large eccentric thrust lines even in non- conventional bonding, this experience was later to be exploited for prefabricated masonry panels.

Warrington Citadel

In 1981 the Author again used prestressed hollow sections to resist lateral wind loading, in this case for free standing diaphragm walls. At the Warrington Citadel the post-tensioned masonry was restricted to the main hall which was approximately 25m long by 15m wide and 8.5m high, see Figure 2.

Braintree Ambulance Port

In 1982 the use of prestressed hollow sections to resist both normal and accidental load criteria for an ambulance port proved economical see Figure 3.

Figure 1 - Oaktree Lane Community Centre

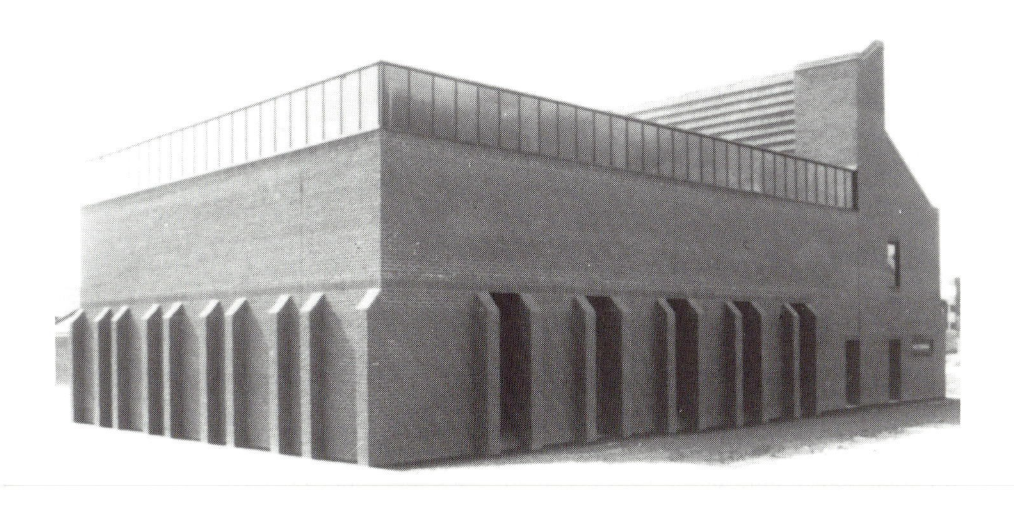

Figure 2 - Warrington Citadel W:G7

The ambulance port forms the main entrance for both ambulances and the general public into the new rehabilitation department. The intention was to provide an open structure on all four sides. The structural design had to solve the problems of overall stability and accidental damage from impact.

Osborne Memorial Halls, Boscombe

In 1984 an opportunity to investigate slender masonry sections using research results and the experience gained to refine the design criteria arose in the design of the Osborne Memorial Halls in Boscombe were tight site dimensions demanded minimum section sizes.

The project adopted design aims 1a, 2a and 4a in the use of a framework of small post-tensioned channel sections (the first use of such sections) to resist lateral wind loading. The introduction of post-tensioning for a small number of sections within the total development made masonry economic and competitive for the total scheme.

The critical structural element where the brickwork channel sections were to support the roof and resist lateral wind pressures.

Because of the restriction on any projection it was essential to use the full potential from the channel cross-section. This was achieved by propping the cantilever channel, against a braced steel roof deck to reduce the wall bending moments, and calculating the required size based upon a cracked section at ultimate load and zero tension at serviceability limit state, see Figures 4 and 5.

A BREAK THROUGH FOR HORIZONTALLY SPANNING MASONRY

Up until 1990 the Author had recognised the simplicity of applying vertical prestress to masonry elements and the economic structures this produced. The difficulties relating to buildability of insitu horizontal members however had made prestressing in this direction economically unviable.

In 1990 an opportunity to combine the experience from Piaus the X Church using stack bonded masonry with the experience gained from other vertically prestressed geometric sections, provided an insight into the use of vertical prestress for prefabricated horizontal members. The job was the proposed cladding panels for the Cowcross estate in London. Architect Alex Lifschutz had detailed his proposals for the prestressed masonry cladding panels and was having difficulty finding an Engineer to make his proposals work structurally.

By introducing shear keys and adjusting the arrangement of the prestress the Author was able to provide a suitable solution and a prototype panel was constructed vertically. The handling and loading tests which followed indicated its suitability see Figure 6.

CONCLUDING REMARKS

The above experience indicates clearly to the Author numerous other possibilities for prestressed masonry, including the use of prefabrication for horizontal

Figure 3 - Braintree Ambulance Port

Figure 4 - Osborne Memorial Halls, Boscombe

Figure 5 - Osborne Memorial Halls, Boscombe

Figure 6 - Horizontally Spanning Masonry

members constructed vertically, and proposals are now progressing adopting this form of construction on future projects.

REFERENCES

1. Curtin, Shaw, Beck and Bray. 'Medical staff residence at Royal Liverpool Hospital'. Ibmac, Rome (1982)
2. Curtin, Shaw, Beck and Bray. 'Loadbearing Brickwork Crosswall Construction' The Brick Development Association Design Guide (1983)
3. Curtin and Shaw. 'The Development and Design of Brick Diaphragm Walls' Ibmac, Washington (1979)
4. Curtin, Shaw, Beck and Bray. 'Fin Wall Construction in tall single storey buildings' Ibmac, Rome (1982)
5. Curtin, Shaw, Beck and Parkinson. 'Masonry Fin Walls'. The Structural Engineer, vol 62. No. 7 (1984)
6. Curtin, Shaw, Beck and Bray. 'Design of Brick Fin Walls in tall single storey buildings'. The Brick Development Association Design Guide (1980)
7. Curtin, Shaw, Beck and Bray. 'Design of Brick Diaphragm Walls'. The Brick Development Association Guide (1982)
8. Shaw and Beck. 'Design of Concrete Masonry Diaphragm Walls'. Concrete Society Technical Report no. 27 (1985)
9. Curtin, Shaw and Beck. 'Design of Reinforced and Prestressed Masonry' Thomas Telford (1988)
10. Curtin, Shaw, Beck and Bray. 'Structural Masonry Designers Manual' 2nd edition. BSP Professional Books (1988)
11. Curtin, Shaw, Beck and Parkinson. 'Structural Masonry Detailing'. Granada Publications (1984)
12. Shaw, Curtin, Priestley and Othick. 'Prestressed Channel Section Masonry Walls'. The Structural Engineer Vol 66 No. 7 (1988)
13. Shaw, Othick and Priestley. 'The Osborne Memorial Halls at Boscombe'. The Brick Development Association Engineers File Note No. 6 (1986)
14. Curtin, Shaw, Beck and Howard. 'Design of Post Tensioned Brickwork'. The Brick Development Association Design Guide (1989)
15. Tsui, Harvey, Mortan and Shaw. 'A Preliminary Investigation of the Vertical Shear Strength of Brick Masonry' Ibmac, Rome (1982)
16. Curtin, Shaw, Beck and Bray. 'Modern Philosophy of Structural Brickwork Design' Ibmac, Rome (1982)
17. Curtin, Shaw, Beck and Bray. 'Designing in Reinforced Brickwork'. The Brick Development Association Design Guide (1983)
18. Curtin and Shaw. 'Designers Experience of Workmanship in Reinforced and Prestressed Masonry'. BMS/BRE Symposium 'Workmanship in Masonry Construction' (1987)
19. Curtin and Shaw. 'Designers Practical Experience of Workmanship' Ibid
20. Shaw. 'Practical Application of Post Tensioned and Reinforced Masonry' ICE Symposium (1986)

21. Shaw. 'Modern use of Reinforced and Post-Tensioned Masonry'. Ibmac, Dublin (1988)
22. Shaw. 'Post Tensioned Brickwork Diaphragm Subject to Severe Mining Settlement' ICE Conference (1982)
23. Curtin, Shaw, Beck and Pope. 'Post Tensioned, Free Cantilever Diaphragm Wall Project' ICE Conference (1982)
24. Beck, Shaw and Curtin. 'The Design and Construction of a 3m High Post Tensioned Concrete Blockwork Diaphragm Earth Retaining Wall in a Residential Landscaping Scheme' ICE Symposium (1986)
25. Curtin, Shaw, Beck and Bray. 'Post Tensioned Brickwork'. Ibmac, Rome (1982)

31 TREATMENT OF FOUNDATION MOVEMENT DAMAGE USING THE HELIBEAM REMEDIAL SYSTEM

J.M. Golding
Curtins Consulting Engineers, UK

Foundation movement of domestic properties is a common problem in the United Kingdom. Remedial solutions, which typically involve underpinning, are costly and disruptive. The Helibeam system is an alternative approach, which uses and enhances the inherent strength of existing masonry walls by the insertion of helical bedjoint reinforcement, to form reinforced and prestressed beams. This paper covers the initial test programme and design principles developed for the system, and describes its installation and applications.

INTRODUCTION

In recent years, a combination of a series of hot dry summers and a more claims-conscious attitude among policyholders, has given rise to a significant increase in the level of subsidence claims (and a corresponding rise in insurance premiums!). The widespread use of traditional (but expensive) underpinning methods of mass concrete or mini-piling is being called into question more and more frequently.

One of the alternatives which are being investigated is the Helibeam system. The system involves the introduction of proprietary helical reinforcement into the bedjoints of existing masonry buildings. The reinforcement and grout act compositely with the existing brickwork or blockwork, both to repair cracks and to form deep reinforced masonry beams. Where appropriate, the reinforcement can be post-tensioned to give additional enhancement to performance.

The beams enable the elevations to span locally over areas of the foundations suffering from subsidence or other ground movement. This can enable the need for (and high cost of) traditional underpinning to be greatly reduced or done away with completely, and helps to avoid differential settlement problems with adjacent structures.

Curtins Consulting Engineers have been appointed by the manufacturers Helifix Ltd, to act as technical advisors for the research, development, and application of the Helibeam system. This paper gives an overview of the work carried out to date.

Causes of local foundation movement
Subsidence of the ground below a foundation can be due to a wide range of causes, eg long-term consolidation of the ground due to the building loads, settlement of the ground under new loading, mining activities, slope movement, removal of ground particles due to inadequate or leaking drains, shrinkage of clays due to seasonal drying or the action of tree roots, etc.

Heave of the ground below a foundation - usually clay soils - can also have a number of causes, eg long-term movement due to the removal of previous overburden loading, expansion due to removal of existing trees, expansion due to seasonal wetting, expansion due to leaking or inadequate drainage, or 'pushing-up' action of tree roots. Heave below a building can result in lateral as well as vertical movement of foundations.

THE HELIBEAM SYSTEM

The Helibeam system is a method of creating structural masonry beams within existing brick or block walls. The proprietary HeliBar reinforcement and HeliBond grout are inserted into panels of existing brickwork or blockwork, transforming them into reinforced or prestressed beams. The reinforcement will be predominately located within bedjoints, but may also include the use of drilled and resin-fixed bars into the depth of the wall, for example to tie together two leaves of a cavity wall. The reinforced bedjoints will typically contain 2 helical bars per bedjoint.

The typical method of installation is as follows (see Figure 1).

1. Rake out mortar joint to the required depth. (This is of the order of 50mm within a half brick or whole brick wall.)
2. Inject grout into back of joint.
3. Insert first HeliBar, and fully grout around the bar. (Note: the helical cross-section enables the grout to flow around and behind the bar.)
4. Insert second HeliBar, and fully grout around the bar.
5. Point up the face of the bedjoint, with a suitable mortar to match existing.

APPLICATIONS

The main uses for the Helibeam system fall into four categories.

- Local subsidence or heave
- Widespread foundation movement
- Crack repair
- Lintel repair

Generally these will need to be used in conjunction with one another to fully deal with the problems of a particular property. In addition there are a variety of 'one-off' engineered solutions where the system is used.

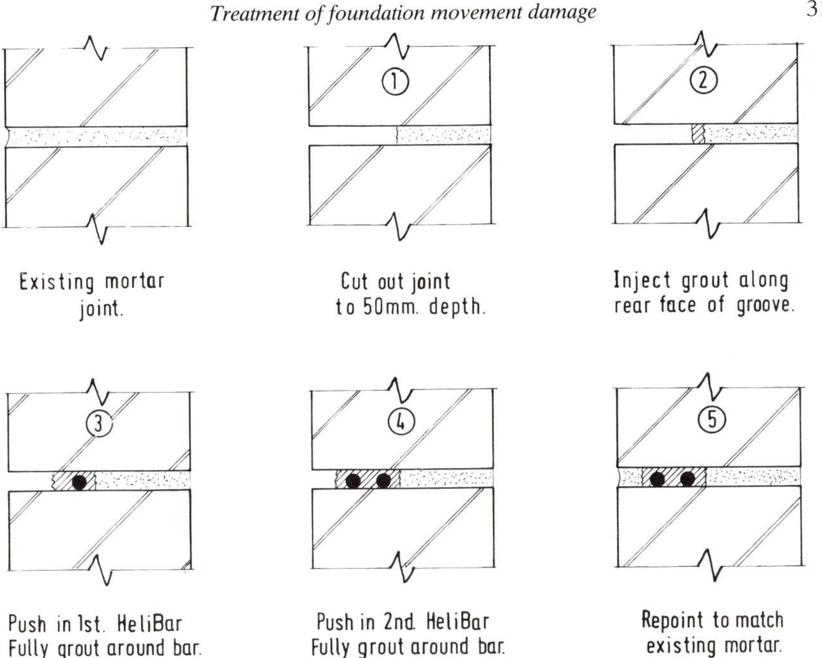

Figure 1. Installation sequence

Local subsidence or heave

If it can be established that the cause of the subsidence is local in nature - this is often as a result of tree root action or leaking drainage - and is not going to affect the remainder of the property, it is possible to repair the affected area solely using the Helibeam system. Helibar reinforcement is introduced into selected bedjoints, generally immediately above foundation level, to form reinforced or prestressed masonry beams, as shown in Figure 2. These beams support the remainder of the elevation, allowing it to span (in the middle of walls) or cantilever (at the corner of walls) over the local areas affected by subsidence or heave. Areas affected by heave may require the introduction of a compressible layer at the underside of the foundations, to accommodate future heave movement.

Besides the cost advantages associated with the Helibeam system, the Helibeam system also overcomes the problems associated with partial underpinning. Partial underpinning - underpinning of only certain lengths of the building's foundations - is sometimes used to deal with this type of localised problem. It can however lead to differential movement and cracking between underpinned areas and adjacent untreated foundations, whether within the same property or adjacent attached properties. The alternative of complete underpinning to avoid this problem results in even greater expense, and may involve underpinning of the adjacent unaffected properties as well.

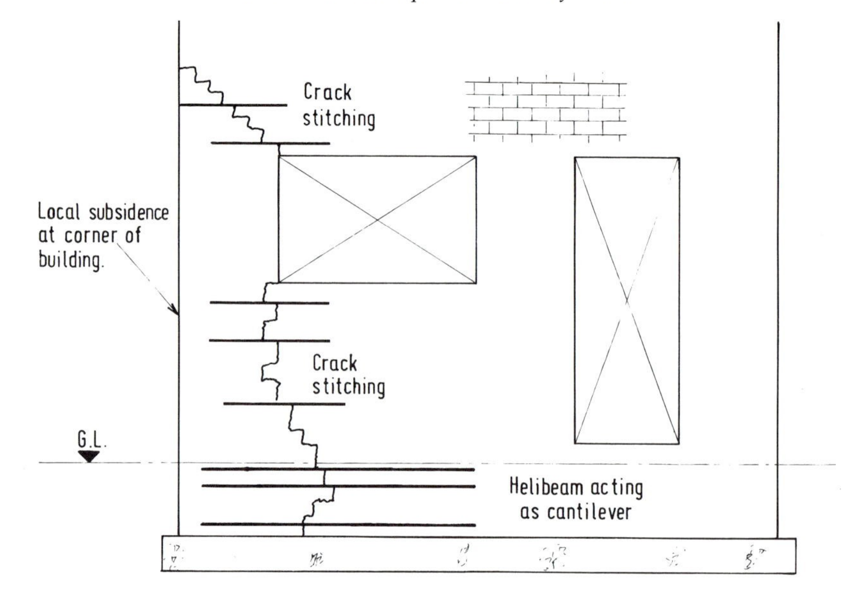

Figure 2. Helibeam solution to local subsidence

Widespread foundation movement
In many situations the ground movement is not confined to a localised area of the property, but affects long lengths of one or more elevations. In such situations the magnitude of the movement, and the beam stiffness needed, are such that the Helibeam system on its own cannot provide the total answer. It will however often be possible for the Helibeam system to be used in conjunction with more traditional methods of underpinning, to produce a cost-effective solution to the foundation movement.

Once the cause of the movement has been identified, and if underpinning is considered necessary to deal with the movement, the Helibeam system can be incorporated to reduce the high costs of either continuous mass concrete underpinning, or of groundbeams used in conjunction with mass concrete pads or mini-piling. The Helibeam alternative involves the use of isolated mass concrete pads or mini-piling as appropriate (mini-piling is more economic for greater depths of underpinning), but with the costly groundbeams being omitted. Instead, the Helibeam system is used to reinforce the walls being underpinned, enabling them to act as beams and to span between the supports provided by the concrete pads or mini-piling. This is illustrated in Figure 3.

Crack repair
Crack repair ('stitching') is only effective in situations where the cause of the movement is no longer active, or where the movement with time is of very small magnitude. Typical situations would be the repair of cracking due to:

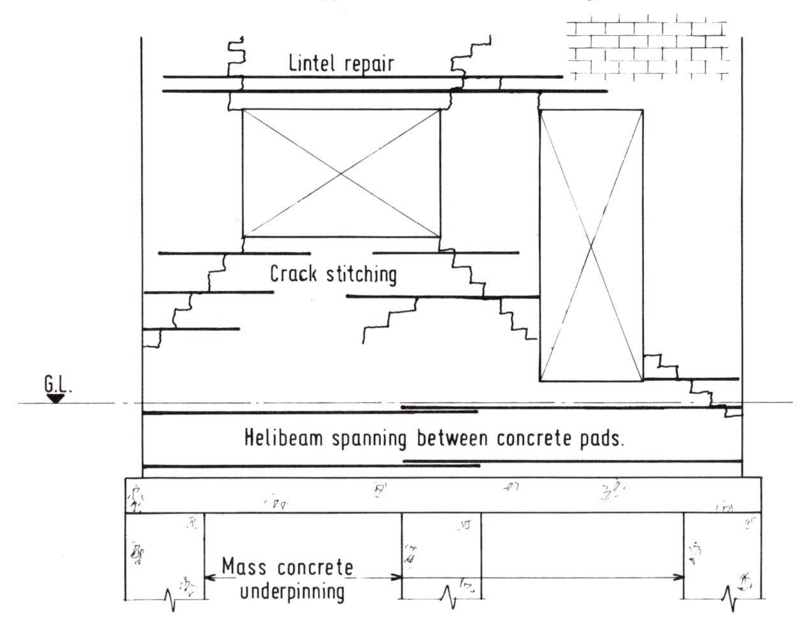

Figure 3. Helibeam in conjunction with concrete pad underpinning

- Subsidence, following measures to arrest the subsidence movement via underpinning, HeliBar reinforcement, removal of offending trees, etc.
- Heave, following measures to arrest the heave movement via underpinning, HeliBar reinforcement, repair of drains etc.
- Expansion/contraction movement of brickwork or blockwork, following the introduction of movement joints or HeliBar reinforcement, etc.

Stitching the cracks restores the strength of the masonry to at least its original pre-cracked level, and acts to prevent those cracks reopening. A typical detail is shown in Figure 4.

Figure 4. Typical crack stitching repair

Lintel repair

Lintel repairs are carried out under similar conditions to crack stitching, ie they are only effective in situations where the cause of the original movement has been dealt with. They apply where the original support over openings was non-existent to start with, has deteriorated with time, or has been damaged by foundation or other movement (see Figure 2).

Panels of masonry over the affected openings are reinforced for their full length with HeliBars top and bottom, as shown in Figure 5. Cracks are additionally stitched with shorter lengths of HeliBar as necessary.

THE TEST PROGRAMME

Four reinforced brick beams were constructed and tested at Middlesex University. The beams were 102.5mm x 290mm deep, ie a half brick wide by 4 courses deep. The beams had a clear span of 3m, and a surcharge load was applied at the bearings to simulate a 'built-in' end condition. The bricks were second-hand London stocks, constructed with a 1:2:9 mortar. 2 no. 6mm diameter HeliBars were then fixed and grouted into each of the top and bottom bedjoints. These bars were located *eccentrically* within the width of the beam, as would be the case when used in practice. In two of the four beams, the bars were prestressed to a load of 4kN per bar prior to grouting. The test arrangement is shown in Figure 6.

Tensile tests were also carried out on 6 no. Helibars, to determine the stress-strain characteristics of the bars.

The principle aims of the test programme were to determine the following

- The performance of the Helibeam system compared with predictions based on normal design methods
- The effect of locating the helical bars eccentrically within the bedjoint (the centroid of the bars was outside the middle third)
- The enhancement contributed by post-tensioning the bars
- The method of failure.

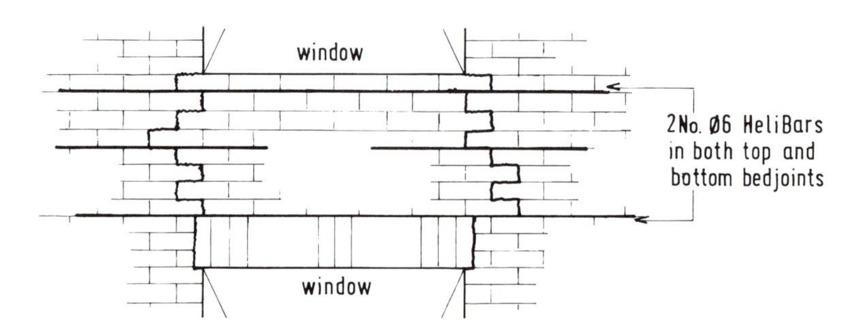

Figure 5. Typical lintel repair

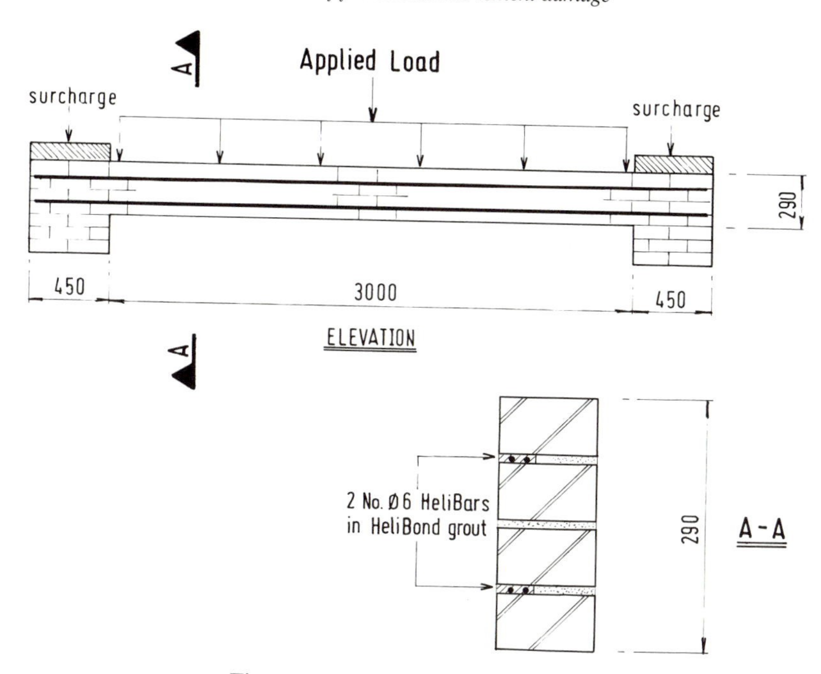

Figure 6. Test beam arrangement

Results

The beams were loaded vertically to failure, using a simulated UDL load. The results are shown in Table 1.

Table 1. Test results for helibeams

Beam No.	Beam type	Onset of cracking			Failure		
		Load (kN)	Vertical Defl.(mm)	Lateral Defl.(mm)	Load (kN)	Vertical Defl.(mm)	Lateral Defl.(mm)
1	Reinforced	11.0	2.3	0.1	21.0	4.9@17kN	0.2@17kN
2	Reinforced	17.0	2.8	0.6	23.5	2.8@17kN	0.6@17kN
3	Prestressed	16.0	4.0	1.0	23.5	5.1@17kN	1.1@17kN
4	Prestressed	13.0	2.4	0.8	27.9	9.5@22kN	1.7@22kN

Notes:

i) The lateral deflection values for beams 3 & 4 do not include the initial deflections due to the prestressing, which were 0.22mm and 0.18mm respectively.

ii) Beam 3 experienced a failure of a prestressing anchorage during the stressing. This was made good and the beam re-stressed.

Typically tensile cracking initiated in the bottom perpend joint at midspan. Small shear cracks developed in the mortar joints near the supports; these latter cracks slowly propagated, and sudden failure occurred with horizontal delamination of the beam along the midheight (unreinforced) bedjoint over the middle two-thirds of the beam, joining up with the stepped diagonal shear cracks towards the supports. The failure mode was judged to be shear failure.

Anticipated failure loads were also calculated using the design equations in BS5628: Part 2, with all partial safety factors set to unity. These values are shown in Table 2. The calculated values indicate that the (shear) strength of the test beams was approximately twice that predicted by theory.

Table 2. Calculated results for helibeams

Serviceability vertical deflection at 10kN	1.9mm (test results averaged 1.65mm)
Failure load based on limiting bending capacity	16.5kN
Failure load based on limiting shear capacity	11.2kN

The tabulated loads indicate the prestressing appeared to enhance the strength of the beams at failure. In addition, although the 'onset of cracking' loads were of a similar order for both the reinforced and prestressed beams, the development of the cracking with increased loading was noticeably reduced for the prestressed beams.

The low value of the lateral deflections indicates that the eccentricity of the HeliBars within the mortar joints did not have a significant effect on the behaviour of the beams.

The tensile test results are shown in Figure 7. The ultimate tensile strength of 995N/mm^2 show that the HeliBars, which are made from ordinary Grade 304 stainless steel, are significantly enhanced by the work-hardening process which produces the helical profile. The initial stiffness of 115N/mm^2 is however considerably less than the 200kN/mm^2 of the parent material. This is explained by the helical shape of the bar, whose effective stiffness is a combination of the material stiffness and the geometric (ie 'spring') stiffness of the bar.

DESIGN PRINCIPLES

The starting point for the design of the Helibeam system is the British Standard for reinforced and prestressed masonry, BS5628: Part 2 (1985).

The bending capacity of the Helibeams is calculated using the masonry rectangular stress-strain block in the British Standard, but making due allowance for the actual stress-strain behaviour of the HeliBars, as given in Figure 7.

The cross-sectional area of the HeliBars is too small to contribute to the shear capacity of the beams via dowel action, and the shear capacities are therefore based on the unreinforced capacity. The Helibeams are typically relatively deep in section, and it is usually possible to enhance the shear capacity near the supports based on the shear span to effective depth ratio (a_v/d), as allowed in BS5628: Part 2.

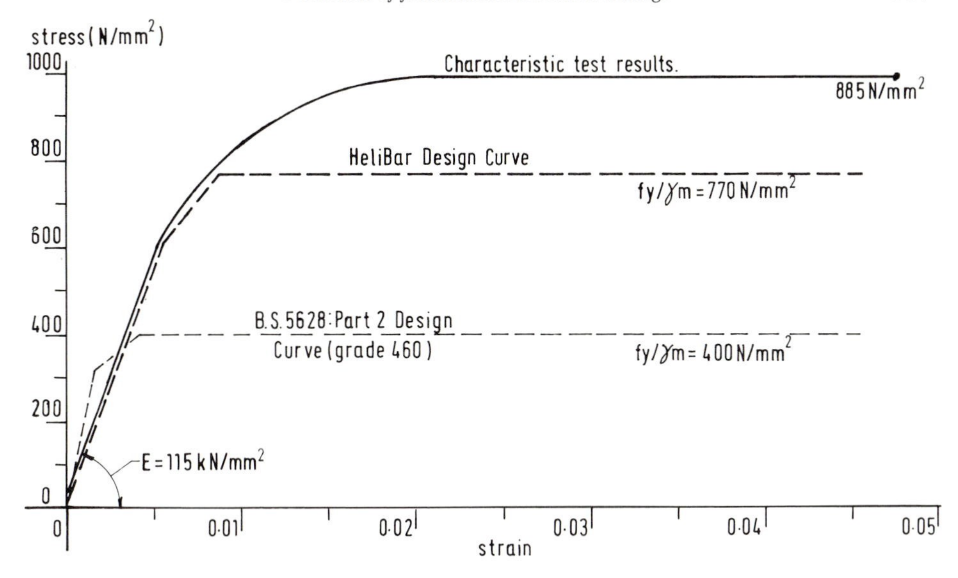

Figure 7. Stress-strain characteristics of helibar

It is generally the case that the shear capacity, rather than the bending capacity, is the limiting condition for Helibeams. Since, according to BS5628: Part 2 this is independent of the strength of the HeliBars, further work needs to be done, to find ways to increase the shear strength of the Helibeam system.

Deflection is not generally a problem for Helibeams, due to the low span to effective depth ratios (L/d) which are typical. In all cases however, the design does not exceed the allowable L/d ratios in BS5628: Part 2.

CONCLUSIONS

The Helibeam system makes good use of the inherent strength of brickwork and blockwork, and the strength and ductility of reinforced masonry, to provide a cost-effective variation on traditional underpinning practices.

Laboratory tests on Helibeams have confirmed the strength enhancement due to the HeliBar reinforcement.

Further experimental work needs to be done to optimise the shear strength, so that the capacity of the Helibeam system is not limited by the low shear capacity of the unreinforced masonry.

32 PROTECTIVE MATERIALS, PURPOSE AND FAILURE

D.R. Plum
University of Newcastle upon Tyne, UK

The problems encountered by concrete structures in recent years have resulted in a renewed interest in the behaviour of both base and protective materials. New materials have been developed to give an improved base concrete, and in addition some materials have been developed for use in various surface protections. The purposes which these materials are intended to serve vary greatly, and may result in a wide variation of required material properties. The paper suggests ways of defining the necessary properties, depending on the proposed application. Suggested theoretical approaches are examined and evaluated. The paper also comments on test results in relation to some of the effects of environmental conditions.

INTRODUCTION

The advent of new materials throughout technology has resulted in a range of special materials suitable for the construction industry. In many cases these materials may be tailored to the needs of a particular project. In other cases general formulations having wide application are available. These special materials for use in the construction industry are principally cementitious/ pozzolanic, or pure polymer/polymer modifier. They include materials as diverse in structure and properties as silica fume, epoxy resin, and styrene butadiene rubber.

The function of these materials may be to improve a principal construction material, such as concrete, or to repair an existing damaged material, or to provide a protective envelope to a lower grade material. In all cases it is necessary for the designer to understand the function of the material with regard to its own properties as well as their inter-relation with the substrate properties. The need for repair and/or protection must be part of an overall maintenance strategy if the materials are to function to their greatest advantage.

The performance of special materials must be proved before use in a particular application, and quality control is required during any period of use. Testing for performance is therefore of great importance, and careful correlation is required between any laboratory tests and site conditions. The structural designer must therefore not only understand the function of each material in use, but also how their performance may change with site conditions.

CAUSES OF STRUCTURAL DETERIORATION

Structures do not last forever, although some have performed better than others. Castles, palaces, cathedrals and temples, have been built on behalf of clients for many centuries. The owners of these structures, however, had different expectations of each building's durability. In fact it is in the matter of expectation (of the client) that we should note a common deficiency in Structural Engineering, and a prime source of structural deterioration. In many branches if engineering an operating and maintenance schedule is issued on completion of each project. Inspection of all plant and buildings is listed, with all necessary cleaning even to external maintanance. In marked contrast, Structural Consultants often fall into the trap of describing a structure as maintenance free.

The discipline of producing a maintenance schedule has two effects. Firstly, it reminds the client that all things need maintenance, and makes clear to him what his duties are, thereby shifting some responsibility away from the designer. Secondly, it concentrates the Consultant's eforts to produce structures that are maintainable, and details that are accessible. Deterioration, therefore, begins in concepts of maintenance (or lack of it), and in honesty in facing the need for it. There is a mismatch between expectation and fulfilment.

We see, therefore, that design concepts which pay inadequate attention to future maintenance and accessibility lie at the root of some current problems of structural deterioration. For example the repair of many highway viaducts in the UK show up some of these deficiencies, such as leaking expansion joints, no drain provision, and no attempt to protect the concrete. Of course the complication of ill-informed joints, sub-standard concrete and low cover to reinforcement does not help the situation, and the contractor is usually blamed. The client too does himself a disservice if he insists always on the lowest possible price, thereby ensuring the highest possible probability of a corner-cutting contract.

Structural deterioration worsens as a result of changing use, misuse, and the steady onset of natural decay. When industrial structures are built clients try to ensure that aggressive chemicals are completely sealed within vats and closed pipelines. The occasional joint leak, pipe fracture or vat maintenance, however, ensures a heavy dosage of such chemicals (e.g. chlorides) at irregular intervals. Hosing down the floor ensures a splash zone 1m up all walls and columns, and corrosion of reinforcement and loss of cover often follow within a short time. Natural weathering, carbonation and mild corrosion can be equally unsightly. Maintenance and protective measures eventually adopted in remedial work could have taken the form of prevention rather than cure. With hindsight it would be better to assess durability and adopt a policy of maintenance by design, rather than by reaction to the inevitable decay.

Market forces, in the guise of the cheapest contract price, do not always produce quality, but often the reverse. Our problem is that what appears today to produce a cost benefit, is in 20 years time an in-built defect. Examples of this in recent years have been system building, the use of high alumina cement and the use of woodwool formers. UK government interference in the market to favour some new thing, such as system building, does not help. The problem is that valuation of the present is difficult, and valuation of the future nearly impossible. Unfortunately the real cost of any new wonder material can only be established after many years experience. Long-term evaluation of any new product comes at the rate of one year per year.

REPAIR FUNCTIONS

In many cases the first objective of a repair programme must be to restore some of the omissions which led to the present situation. A maintenance strategy needs to be developed, access to vital parts of the structure arranged, and protective envelopes considered. Then removal of defective or impregnated materials will be needed, followed by restoration of the structural members to their former function.

The restoration work requires a clear concept of function, in particular of the division between the structural and non-structural or cosmetic applications. In the structural application, concrete must be replaced in order to carry compressive stress and the principal question of the designer is, 'Does the repair carry all the load formerly carried by the removed concrete ?' Of course many subsidiary questions may arise concerning the loading condition during the repair operation, and whether reinforcement or concrete cross-section can be increased. If no load may be removed during repair then no stress is carried by the repair, except due to secondary effects, or when some load is later removed. Full load removal during repair allows full stressing of the repair. Partial load relief is of course common practice, in which imposed or live loads only are removed during the repair operation. Whether or not the repair carries all the compressive load desired can be assessed by a theoretical study[1,2], which defines the Force ratio (F_r/F) relative to the Area ratio (A_r/A). In general terms the dominant factor in this assessment may be shown to be the creep of the material, with load relief being less significant (Figure 1). This leads to the concept of repair efficiency which measures the degree to which the original load carried by the defective concrete is replaced by the repair material. A 100% repair efficiency means that all the load appropriate to the repair cross-section is actually carried by it, and this is clearly desirable from the design point of view. If repair efficiency reduces to say 50%, then half of the load which should be carried by the repair is in fact being shed to the core concrete, and this becomes overstressed.

By contrast the cosmetic application assumes that stress carrying by the repair material is not required. In the event of some applied expansion, due to increases in temperature or moisture content or both, constraint from edges or overbreak may give rise to stresses in the repair. Again a theoretical approach[1,2] is possible giving a means of relating an applied expansion to the bond strength of the repair to the substrate or core concrete. Creep, bond strength and expansive strain are all significant (Figure 2), but for the cosmetic application creep reduces the required bond strength, i.e. it allows stress relaxation to occur. A measure of the effectiveness of a cosmetic repair is the failure factor which compares the

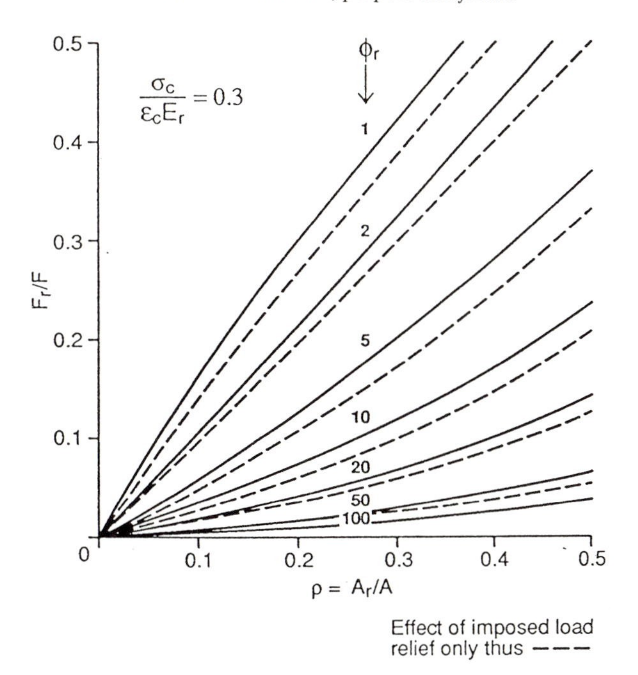

Figure 1. Force ratio

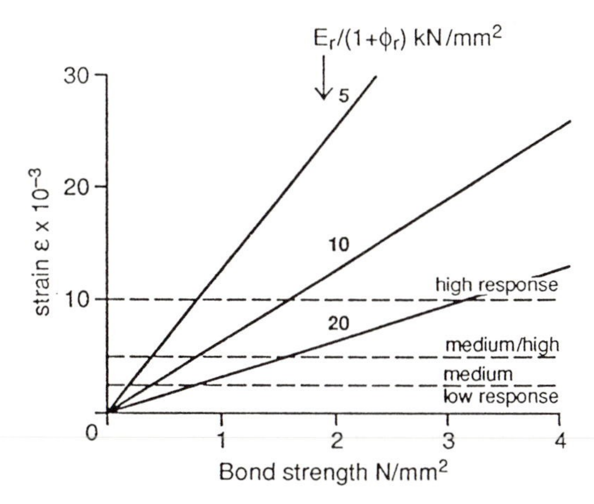

Figure 2. Tolerable expansion

tolerable expansion, obtained from theory, with the imposed expansion, estimated to occur in practice. Ideally a ratio in excess of three might be desirable, bearing in mind the approximate nature of the theory and the variableness of bond strength. Imposed expansions will vary with the environmental conditions and the type of material. Two common sources of expansion are increase in temperature and saturation. Different repair materials will respond differently to these stimuli, and different levels of response are shown in Figure 2. In general low response materials rarely cause difficulties, while those with a medium/high response commonly fail. The medium response zone forms a transition. Most materials available commercially today exhibit a low response.

A special, and rare, application exists somewhere between the structural and the cosmetic. This case is the unusual patch repair in a compression zone which lies in the cover concrete. In most cases of structural repair treatment of the reinforcement is needed which demands removal of concrete behind the bars. As a result any repair is substantially locked into the structure mechanically and interface bond strength is of lower priority. Repairs in the concrete cover clearly do not have this advantage and depend on interface bond strength. As a result the cosmetic approach is more in keeping with their behaviour, hence this in-between case.

Assuming then that we can assess the behaviour of a repair whether structural or cosmetic, what properties might we look for in a repair or protective material? Current needs indicate that improved durability of cover concrete is much sought after. This is achieved very commonly by use of polymer materials in pure form (epoxy resin, polyester etc.) or as concrete modifiers (SBR's, acrylics, etc.) These are resistant to many chemicals and also form barriers to water movement. An alternative material is a microsilica concrete, but the polymers generally have the edge due to improved bond strength with a concrete substrate. Alternatively protective films may be obtained by use of polymer coatings and other surface treatments, with the many new materials now available.

Further improved properties which may be sought include abrasion resistance, skid resistance, water shedding and anti static. Abrasion resistance usually involves the use of hard aggregates from granite and bauxite to slag and ceramics. The softer matrix of cement or polymer is generally the weaker link so that optimum aggregate packing is important. Clearly the overall objective of repair is to lengthen useful life. In some cases improved material properties are essential, in others improved properties are a desirable extra.

MATERIAL PERFORMANCE

Repairs may be carried out using a number of different methods, principally injection, casting and trowelling, each of which requires the use of different variations in the materials. The many materials currently available form three main groups, the cementitious (including the cement replacements), the pure polymer, and the polymer modified cementitious. The behaviour of the first group is generally well understood by Structural Engineers, even though some of the new materials available have some special properties. The danger is that because we understand cement based materials we assume that all aggregate plus binder materials behave the same. Bitumen binders should serve as a warning to us, and perhaps we would do well to mentally group polymers with bitumens. It is clear however that polymer and polymer modified materials really belong in a class of their own.

The pure polymer materials based on epoxy resin, polyester or polyurethane plus a hardener, a graded aggregate, and some 'extenders', are valuable in repair work. They have high chemical resistance and can be made to form an effective moisture barrier. Their bond strength to a concrete substrate is usually so good that tensile failure of the concrete precedes bond failure. Their weaknesses lie in their dependence on specialist workmanship, and the variability of their properties with environmental conditions. Concrete has proved itself as reasonably tolerant of both workmanship and conditions, and in addition our experience has taught us to avoid situations in which it might be more sensitive. The polymers are not tolerant of either workmanship or conditions, and our experience is meagre.

Test results for epoxy resin materials show a high compressive strength, achieved at early age, and a tensile strength much higher than that of concrete. The tensile strength will commonly be 20% to 40% of the compressive strength for commercially available mixes. This is of course ideal for repairs in tensile zones or shear zones. The problem illustrated by test results is that the properties are dependent on environmental conditions[3]. Curing under slightly different conditions of temperature and humidity has some effect on strength. Generally this effect is not too serious. Elastic modulus shows a similar effect, but creep shows itself to be considerably affected by these small temperature/humidity changes. This is particularly apparent in flexural creep.

In a similar way test results for polymer modified materials show a marginal increase in compressive strength compared with unmodified concrete, and a greater increase in tensile strengths. These effects also are inhibited by changes in the environment. In particular increased humidity reduces the possible gains to near zero. But as for the epoxy resin materials the greatest variations occur in the creep property. Under dry conditions values for creep coefficient similar to concrete are recorded. But increasing humidity pushes the material into an entirely new mode of behaviour (Figure 3). This diagram represents flexural creep behaviour, and a similar pattern is given by compressive creep, but with lower values.

Clearly the environmental conditions of a test affect the results, in some cases markedly. Of course the conditions actually obtaining on site may be quite random. Both temperature and humidity may fluctuate, and conditions during the cure period may differ from those in service. Tests conducted so far have been concerned with behaviour at early age, during cure and the early stages of loading. The data presented so far raises the matter of laboratory test conditions vis a vis site conditions. Unfortunately all Standard specifications, e.g. BS 6319[4], are rather lacking in this respect, and appear to carefully avoid the matter of humidity, which turns out to be crucial. Thus rather naive use of 'ambient'

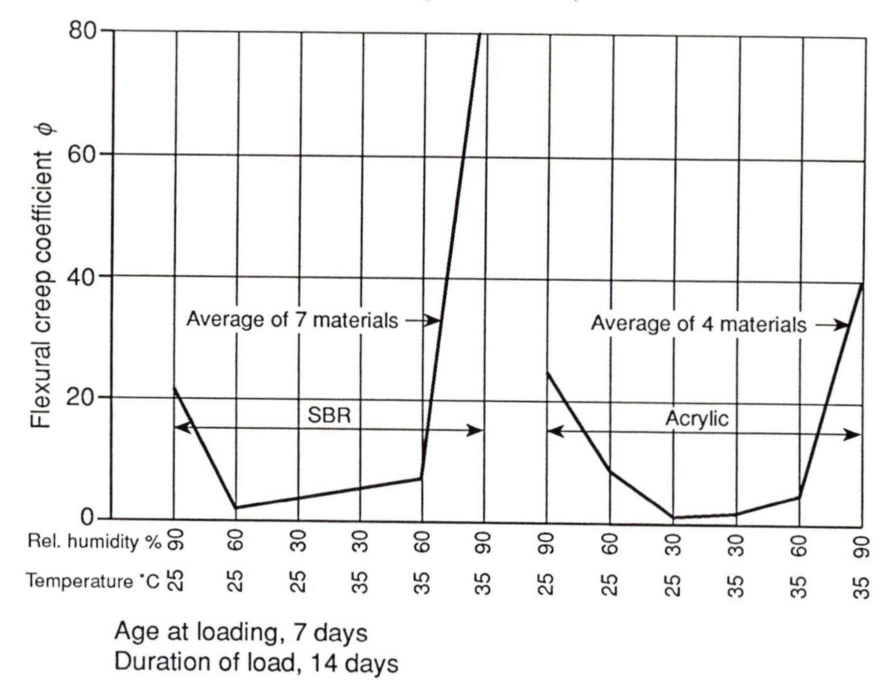

Figure 3. Creep of polymer modified concretes

laboratory conditions for testing may explain why most of the polymer materials perform well in these conditions, but far less well in what might be called 'site conditions'.

PERFORMANCE RATINGS

As described earlier the repair efficiency is a way of measuring the performance of a material in the structural application. Figure 4 uses repair efficiency to show how some of these polymer materials perform. As the properties are changed by the environment, the repair efficiency changes with them. So for epoxy resin materials (Figure 4) the efficiency is excellent (100%) at ambient, but less good (below 50%) at 35°C 90% r.h. In a similar way the polymer modified materials show declining efficiency at higher temperature/humidity. But individual performances vary and some resist the effects of raised humidity very well. Hence before using a polymer or modifier in a structural application, it is advisable to obtain test results over the full range of site conditions of temperature and humidity. Especially is it advisable to obtain creep data.

The cosmetic application may be tested by a failure factor. From methods described earlier the results for epoxy resin materials may be found (Table 1). In

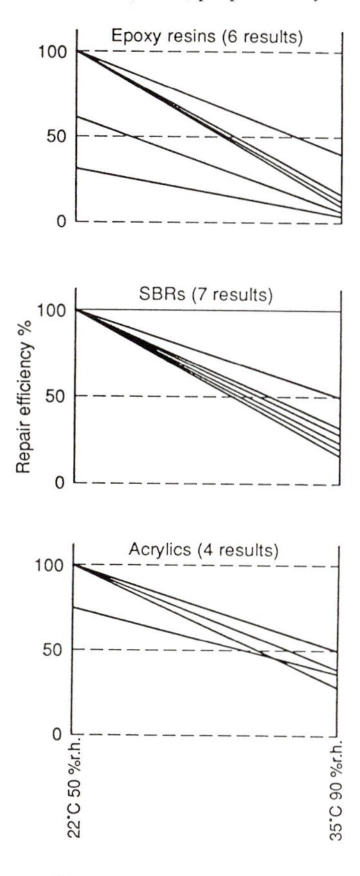

Figure 4 Repair efficiency

general raised temperature/humidity increases creep which is beneficial. The conditions considered therefore are ambient, but with variations in bond strength. Despite the fact that good bond strength is usually associated with these materials, some are workmanship sensitive. The bond strength is particularly sensitive to proper preparation of the surface, and correct use of the primer. Others are much less sensitive and appear to overcome poor preparation and use of primer with little difficulty. Some are formulated for laboratory use, while others readily adapt to real site conditions. In general however a very adequate failure factor is obtained. The lowest value recorded was 2, and this must be considered somewhat marginal, bearing in mind the inexact nature of this science.

Table 1. Failure factor

Polymer material		E4
Saturation expansion		0.90×10^{-3}
Repair function	good bond	27×10^{-3}
	poor bond	6.5×10^{-3}
Failure factor	good bond	30
	poor bond	7.3

CONCLUSIONS

It has been demonstrated that an assessment of likely success is possible and a comparison between materials on offer can be carried out. It is, however, necessary to carefully define the operating range of temperature and humidity, and ensure any test data covers the site range.

REFERENCES

1. Plum D.R. 'The behaviour of polymer materials in concrete repair and factors influencing selection'. The Structural Engineer, Vol.68, No 17 (September 1990)
2. Plum D.R. 'Repair materials and repaired structures in a varying environment'. Proceedings of the International Seminar The Life of Structures, Brighton (April 1989)
3. Plum D.R. 'Environmental effects on polymer modified materials'. Proceedings of the 1st International Conference on Highrise Buildings, Vol II, Nanjing (March 1989)
4. BS 6319:1983. 'Testing of resin compositions for use in construction' (1983).

CHAIRMAN'S REMARKS

Dr Klaus Brandes, BAM, Berlin, Germany

Within Session 5 - New Concrete Techniques and Masonry Structures - the whole range of problems related to the subject 'Building the Future' have been touched.

- New techniques as regarding formwork for concrete constructions for more economic methods on site
- New construction as eg for shearheads and for connections of reinforced concrete members
- Extension of the use of well-tried brickwork to advanced application by prestressing it
- Qualitative investigation of the earthquake resistance of improved strengthening of ordinary dwellings in the Mediterranean area by small scale testing
- Advanced rehabilitation of existing masonry buildings which suffer damage from foundation movement.

In all cases, experimental investigations and modelling and numerical analysis are the ambivalent tools.

As engineers know, the real problems arise when going into the details of construction and performance. Thus, again and again, new solutions have to be created for old and new problems, for existing structures and for those which have to be designed.

The protection of the buildings we design, erect and maintain, has been the subject of a survey, mentioning the expectations of the client, the effort of consultants to produce maintainable structures, because 'deterioration begins in the concept of maintenance (or lack of it)'...

Thus, in this session, as in most of the sessions, are reflected all the facets of the engineer's endeavours to meet his responsibility.

Thanks to the authors for emphasizing this.

PART SIX

STEEL/CONCRETE COMPOSITE STRUCTURES

33 EXPERIMENTAL STUDY OF FLEXURAL CHARACTERISTICS OF STEEL–CFRC COMPOSITE PLATE

H. Sakai
M. Nakamura
T. Hoshijima
Mitsubisi Kasei Corporation,
Y. Mitsui
K Murakami
H. Era
Kumamoto University, Japan

This study aims at developing a lightweight and high-strength concrete-based curtain wall. This paper outlines the test results of the composite panel of thin CFRC reinforced with high performance pitch-based carbon fiber and lightweight H-section steel with welded headed studs. Judging from the results, the composite panel consisting of CFRC and H-section steel can be designed lighter than ordinary precast curtain wall by the efficient use of flexural strength of CFRC.

INTRODUCTION

Because carbon fiber reinforced cement composite (CFRC) is lighter in weight and has a higher tensile strength than ordinary concrete, there are reported many cases of CFRC application as a material for curtain walls, taking advantage of its characteristic higher flexural strength that enables production of thin wall.

However, the CFRC curtain wall having a thinner wall thickness is vulnerable to flexure when a wind load is applied, and there are cases that the deflection exceeds its allowable limit, though it depends upon the type of finishing material used. Accordingly, this study examined the performance of a composite panel reinforced with lightweight H-section steel connected with headed stud connectors. Since in designing curtain walls for high-rise buildings, a negative wind load is made the design load, this experiment was conducted to examine the dynamic behaviour of the composite panel under a negative bending load.

EXPERIMENTAL

Specimens

Panel specimens were plate and made in 120cm x 240cm of two thickness of 40mm and 70mm. The shape and dimensions of composite specimens are shown in Figure 1. A list of panel specimens and composite panel specimens used for

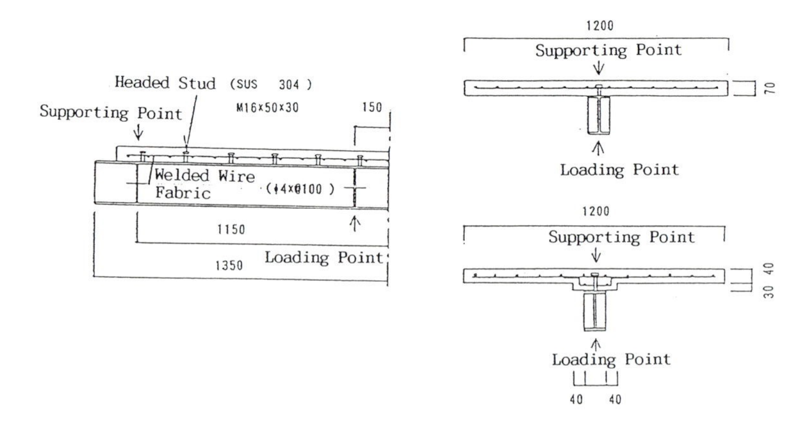

Figure 1. Shape and dimension of composite panel

Table 1. Panel and composite panel

	Panel thick (mm)	Type	Reinforcing material*	Number of studs	Degree of composition
H 40	70	CFRC	-	-	-
H 40	40	CFRC	-	-	-
K70-200	70	CFRC	H-200x100	12	Perfect
K40-200	40	CFRC	H-200x100	12	Perfect
F70-200	70	CFRC	H-200x100	6	Imperfect
F40-200	40	CFRC	H-200x100	6	Imperfect
K70-150	70	CFRC	H-150x 75	10	Perfect
K40-150	40	CFRC	H-150x 75	10	Perfect
F70-150	70	CFRC	H-150x 75	6	Imperfect
F40-150	40	CFRC	H-150x 75	6	Imperfect
PC70-200	70	CFRC	H-200x100	12	Perfect
PC70-150	70	CFRC	H-200x100	10	Perfect

* Lightweight H-section steel
H-200 x 100 x 3.2 x 4.5 H-150 x 75 x 3.2 x 4.5

this experiment is shown in Table 1. As for the codes of the specimens. H and PC stand for panel and concrete, and F and K complete composite and incomplete composite respectively.

Mix proportion and placing and curing methods of CFRC and concrete

The materials and their mix proportion of CFRC are shown in Table 1 and 2. A 300-litre capacity ordinary mortar mixer was used for mixing CFRC. The mix proportion of ordinary concrete is shown in Table 4. An ordinary mortar mixer was also used for mixing concrete. The property of the green CFRC was that the flow value was 146mm/142mm (JIS R5201) and the air content was 4.9% (JIS A1128). The dispersion of CF in each batch was observer visually, which showed that fibers were dispersed fairly evenly without forming fiber balls in all batches. The value of slump of ordinary concrete was 18+2.5cm, ie the target slump. Each specimen was removed from the formwork 24 hours after placing and afterward, left indoors and cured at room temperature between 25°C and 30°C. All specimens tested were aged four weeks or longer.

Table 2. Material of CFRC

Cement	Low shrinkable cement
Aggregate	Silicate sand No. 5:Shirase = 1:1 (weight ratio)
Admixture	Dispersant: Methyl Cellulose water reducing agent
Fibre	High-performance pitch-based carbon fiber Size: $17\mu m\phi$ x $18mm\ell$ T.S: $180kgf/mm^2$ T.M: $18tf/mm^2$

Table 3. Mix proportion of CFRC

Volume fraction of CF	1.8
W/C (%)	45
S/C (%)	25
Water reducing Agent/C%	3.0
Dispersant/C (%)	0.75

Table 4. Mix proportion of concrete

W/C (%)	50
Water content per unit volume of concrete (kg)	177
Cement (kg)	354
Fine aggregate (kg)	870
Coarse aggregate (kg)	966

Specimens for material test and measuring method

A list of the specimens of CFRC and concrete for material test is shown in Table 5. The JIS No 1. specimens of H-section steel were cut out from its flange and web. The mesh cut to a length of approximately 50cm was used as a specimen for the material test. The sketchy method of applying load onto a panel specimen or a composite panel specimen is shown in Figure 2. Negative bending loading was made at four points for the bending test (bending span l = 230cm). A displacement gauge (precision 200μ/mm) was used for the displacement measurement. A strain gauge was used for the strain measurement (gauge length 60mm for the panel section and 6mm for the steel section).

Table 5. CFRC and concrete specimen for material test

Type	Specimen	Dimensions(cm)	Quantity
CFRC	Specimen for compression test	10ø x 2 0 ℓ	15
	Specimen for cleavage test	10ø x 2 0 ℓ	6
	Specimen for bending test	10 x 10 x 40	27
Concrete	Specimen for compression test	10ø x 2 0 ℓ	3
	Specimen for cleavage test	10ø x 2 0 ℓ	6
	Specimen for bending test	10 x 10 x 40	3

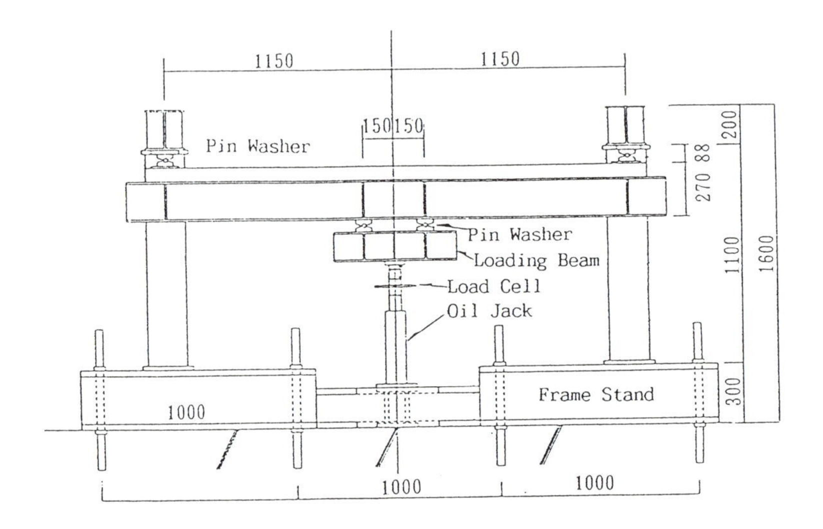

Figure 2. Outline of composite panel test

TEST RESULTS AND DISCUSSION

Material test and bending test

The results of the material test of H-section steel and reinforcing mesh almost satisfy respective specified values. The results of the material test of CFRC and concrete are shown in Table 6 and 7 respectively. The concrete of Table 7 shows a considerable lower compressive strength compared with that of ordinary concrete. This is attributable to the fact that the mixing was made with the proportion of low shrinkable cement as if it were ordinary portland cement. The amount of low shrinkable cement should have been increased and also steam curing be employed.

Results of the bending test are listed in Table 8. Representative examples of load-deformation (deflection at the center of span) relationships of the specimens are shown in Figure 3. While CFRC composite panel specimens maintained a linear load-deformation relationship before their initial cracks reached a certain length, concrete composite panels were deformed abruptly after initial cracks occurred. The maximum load (Pmax) of the CFRC composite panel specimen is approximately 2.2 to 3.2 times its cracking load (Pcr), showing it has a considerable margin before its bearing force is reached after it has developed cracks.

Evaluation of bending rigidity

Table 8 shows calculated values of geometrical moment or inertia at the sections of CFRC composite panels. (In and Ie for perfect and imperfect composite panels respectively.) The geometrical moment of inertia for imperfect composite panel (Ie) was calculated based on the under-mentioned equation given in the society

Table 6. Results of material test of CFRC

		Mean value	Standard deviation
Specific gravity		1.46	-
Young's modulus ($x10^5$ kg/cm^2)	Tension	2.26	0.15
	Com-pression	0.99	0.072
Compressive strength (kg/cm^2)		253.3	14.18
Cleavage strength (kg/cm^2)		49.8	10.66
Bending strength (kg/cm^2)		116.4	23.4

Table 7. Results of material test of concrete

	Mean value
Specific gravity	2.27
Young's modulus (x 10^5 kg/cm)	1.67
Compressive strength (kg/cm^2)	140.7
Cleavage strength (kg/cm^2)	13.6
Bending strength (kg/cm^2)	29.3

Table 8. Test results

Type	No.	Pcr (t)	Pmax (t)	σ_b (kg/cm^2)	σ_b Panel kg/cm^2)	In (cm^4)	Ie (cm^4)	Iexp (cm^4)	cI+sI (cm^4)
CFRC Panel	H70-1	0.53	0.916	-	34.6	-	-	-	-
	H70-2	0.53	0.97	-	34.6	-	-	-	-
	H40-1	0.143	0.238	-	35.6	-	-	-	-
	H40-2	0.134	0.236	-	34.2	-	-	-	-
CFRC Composite Panel	K70-200	4.58	10.83	25.4	-	3625	-	3364	1221
	K40-200	4.0	9.9	24.3	-	3210	-	3140	1082
	F70-200	4.1	10.36	30.0	-	-	2616	2944	1221
	F40-200	3.78	10.55	29.4	-	-	2577	3196	1082
	K70-150	2.55	6.94	25.8	-	1686	-	1650	603
	K40-150	2.24	6.25	23.3	-	1685	-	1799	464
	F70-150	3.0	6.7	35.5	-	-	1319	1682	603
	F40-150	2.2	7.1	25.8	-	-	1318	1639	464
Concrete Composite Panel	PC70-200	2.6	9.25	21.0		3732	-	3001	1393
	PC70-150	1.75	5.9	24.1		1979	-	1585	775

Pcr, Pmax: Cracking load and maximum load
σ_b : Flexural strength at initial cracking of composite panel
σ_b Panel : Flexural strength at initial cracking of CFRC panel
Iexp : Geometrical moment of inertia of composite panel at its section inversely obtained from the initial slope of load-deflection relationship
cI : Geometrical moment of inertia panel
sI : Geometrical moment of inertia of steelwork

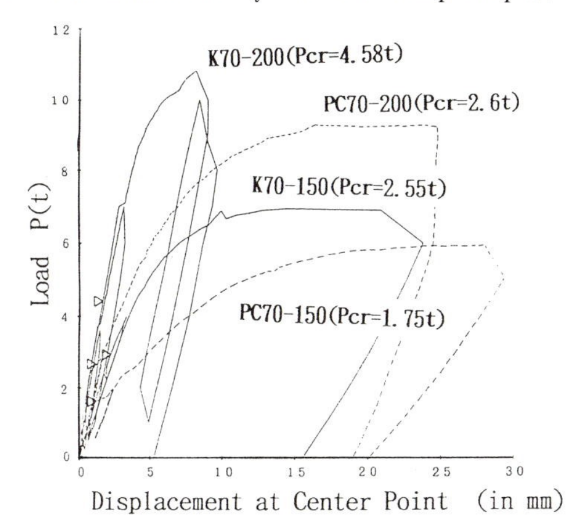

Figure 3. Load deflection curve between CFRC and concrete

Guidance[1]. For both calculations, modular ratio of CFRC (n) was made 20, substituting the equivalent section of steelwork.

$$Ie = sI + \sqrt{np/nf} \cdot (In - sI)$$

Where, np: Number of studs for imperfect composite panel
nf: Number of studs for perfect composite panel
sI: Geometrical moment of inertia at H-section steel

The table also shows the geometrical moment of inertia (Iexp) of composite panel at its section inversely worked out from the initial slope of the load-deformation curve. Comparing In, Ie and Iexp, it is seen that In and Iexp nearly correspond with each other, indicating that all of CFRC composite panels. In the table, the relative ratios of the sum of the geometrical moments of inertia of both CFRC and H-section steel CI + sI to Iexp for each panel specimen fall in the range of 2.41 to 3.88 indicating that the effect of composition upon the section is outstanding. Furthermore, for concrete composite panels, calculated values of geometrical moment of inertia including the tensile stress of concrete are shown. Modular ratio of concrete (n) was made 10.

Evaluation of flexural strength at the occurrence of cracks.

The flexural strengths of CFRC and composite panel specimens at the time cracking occurs are shown in Table 8. For composite panel specimens, modulus of section for the perfect composite panel was used in the calculation. Meantime, it has been known that the flexural strength of CFRC members are affected by scale effects such as the size of its section and the length of bending span[2]. With this in mind, the method of evaluating flexural strength at the time cracking occurs is reviewed, also comparing the results of the bending test of CFRC panels which were reported before [3][4]. Figure 4 shows experimental data plotted in logarithm with its y-axis standing for flexural stress at the time cracking occurs and x-axis standing for:

panel axis x panel thickness x length of bending span

(called the 'volume of specimen' for the sake of convenience). For composite panels, a height including the height of H-section steel was made the panel thickness. Although data fluctuate widely, there exists nearly a linear relationship between the both, ie it is clear that flexural strength decreases as the volume of specimen increases. Because high-stressed volume is in proportion to the above mentioned volume of specimen[5], the linear equation $\sigma_b = 1618.1 \times (1 \bullet D \bullet t)^{-0.313}$ given in Figure 4 represents the scale effects of panel upon the flexural strength of CFRC. It is assumed that the lower flexural strength of the composite panel is attributable to the increase in the volume of the part of CFRC subjected to a tensile strength due to the composing effect at its section.

CONCLUSIONS

CFRC composite panel specimens were placed under a negative bending load (CFRC side pulled) and their dynamic behaviour was examined. The results of these tests are as follows:

1) With CFRC composite panels, load-deformation relationship maintains linearity after initial cracks occur and as far as they are short, and residual

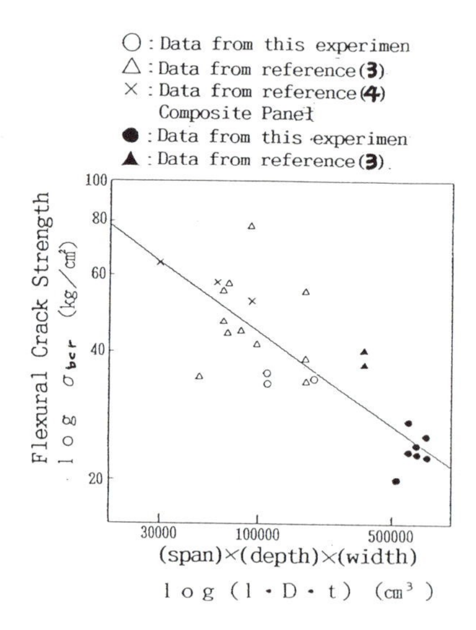

Figure 4. Scale effect of CFRC

deformation after loading has been released is also small. With concrete composite panels, deformation takes place abruptly immediately after initial cracks occur.

2) There was an agreeable correspondence between the flexural rigidity (geometrical moment of inertia) obtained from the initial slope of the load-deformation curve of CFRC composite panels and the calculated value obtained assuming that they have perfect composite sections. Therefore, composite panels fitted with the number of stud connectors within the range of this experiment can be considered perfect composite panels.
3) The flexural rigidity of CFRC composite panels is 2.41 to 3.88 times higher than the sum of individual flexural rigidities of CFRC panels and H-section steel, demonstrating that the effect of composition is outstanding.
4) The flexural strength of CFRC composite panels at the time cracking occurs can be evaluated by a flexural strength taking account of the dimensional effects of CFRC panels. For this evaluation, the total height including the height of H-section steel is made the panel thickness, and the modulus of section for the perfect composite panel is usable.

ACKNOWLEDGEMENT

The authors would like to express their thanks to the technical staff of Kumamoto University and Mitsubisi Kasei Corporation.

REFERENCES

1. AIJ, 'Design recommendations for composite constructions' (Revised 1985).
2. Akihama S., Kobayashi M., Suenaga H., and Suzuki K. 'Experimental study of carbon fiber reinforced cement composite (Part 4). Scale effect of CFRC test specimen on flexural properties'. Proceedings of the Annual Report of Kashima Corporation, vol 32 (1984)
3. Sakai H., Mitsui Y., Murakami K., Urano T., Takahashi K. and Nakamura M. 'Experimental studies on flexural characteristics of steel-CFRC composite panels (Parts 1 and 2).' Proceedings of the Annual Meeting of Architectural Institute of Japan (Touhoku) (September 1991)
4. Masuda K., Sugamata S., Morohashi S., Usui K., Yanase T., Sakai H., Takahashi K., Kuribaysshi H. and Nakamura M. 'Experimental studies on flexural strength of CFRC'. Proceedings of the Annual meeting of Architectural Institute of Japan (Touhoku) (September 1991)
5. Urano T., Murakami K., Mitsui Y., Shigaki T. 'Experimental study on toughness design of steel fiber reinforced concrete members'. Proceedings of the Annual Concrete Research and Technology, Vol.14 No.1 JCI. (1992).

34 DEEP SECTION COMPOSITE FLOORING PROFILES

L.H. Humphries and C.K. Jolly
The University of Southampton, UK

Current composite floor deck profiles permit a maximum unpropped span of about 3.6 m. This paper shows the design of an experimental deep section composite floor deck to span 6.0 m unpropped during the construction stage. Its profile is based on a popular existing re-entrant profile. Shear and bending tests on samples of this deep profile show a satisfactory performance in bending and deflection during the wet stage of floor construction. The paper proposes improvements to the shear stiffness of such a deep section profile.

INTRODUCTION

Composite steel sheet flooring systems are widely used, both in the USA and Europe. These floors have many advantages for fast-track projects, providing a rapidly constructed working platform for following trades.

Standard profiles used in the UK fall broadly into two categories, namely trapezoidal and re-entrant (or dove-tailed). These profile shapes are illustrated in Figure 1. Profile depths currently used are between 50 mm and 100 mm. Profile depth is governed by minimum constraints given in BS 5950:Part 4,(1) and by concrete thickness requirements for fire protection. Spans are limited to between 3 m and 4 m if unpropped, or 6 m and 8 m when propped.

There is constant pressure from design engineers to increase the permissible span for this type of flooring, thereby reducing the number of intermediate beams required. There are three methods by which this required increase in span may be achieved:

a) The use of more extensive propping under the existing profiles,
b) Increased steel thickness (or gauge),
c) Increased profile section depth.

Figure 1. Common Profile Types. (A = Trapezoidal, B = Re-entrant)

The first method is unpopular in this country, although it is efficient and widely used on the continent. The second is structurally inefficient, and therefore costly, using a greater weight of steel per unit area which is approximately proportional to the increase in strength. The third is efficient since the strength increases with the square of the profile depth, and maintains all the advantages of unpropped construction.

This research, therefore, pursued this third option, by investigating potential deep section shapes. Profiles chosen were those which could be formed by reshaping a common existing profile. This clearly minimises the cost of additional rollers if the profile is factory produced. Alternatively, on-site equipment could be devised to modify the existing profile immediately prior to installation. This latter method permits more economic transportation to site since the existing re-entrant profile can be nested compactly.

PROFILE CHOICE & MANUFACTURE

The profile types investigated are illustrated in Figure 2. The most common re-entrant profile was used as a basis, as it had significant advantages over existing trapezoidal profiles when re-formed into a deeper profile. It provides better adhesion between the sheet and the concrete, and gives better shear-bond characteristics. The underside of the sheets include slots into which wedges may be fixed, providing an excellent means of supporting service ducts. A trapezoidal profile, however, would give significant problems with vertical separation of the concrete from the sheet in bending.

Different values for the angle theta, Θ, in Figure 2 were considered. Sections were chosen with an overall width of some multiple of 75 mm in keeping with other standard widths. The use of shear studs also constrained the range of angles, as the maximum spacing from BS 5950:Part 3[(2)] is 600mm. The resulting admissible profiles were three of the 'triangular' type, and one of the 'trapezoidal' type, with angle theta of approximately 39, 60, 76, and 84 degrees respectively. Overall profile depth (excluding concrete) ranged from 140 mm to 354 mm, substantially deeper than the current standards.

Figure 2. Deep Profiles Geometries. (A = Triangular, B = Trapezoidal)

The relative second moments of area of the profiles are compared in Table 1. It is clear that the 'deep trapezoidal' profile provides the best span increase. Therefore this profile was selected for production despite the fact that, being made from two standard sheets, it would present greater problems in on-site fixing than the single sheet options. Preliminary calculations from BS 5950:Part 4[1] indicated that a 6.0 m span was feasible, and so this was prepared for testing.

The profile was produced from two existing profiled sheets. It should be noted that introducing further bends into a 6.0 m long cold-formed sheet of 1.25 mm thickness required more than a little effort! Profiles were bent with a series of passes of a small jig, connected to an hydraulic jack. The resultant shape was a close approximation to the desired profile geometry.

Table 1. Relative Second Moments Of Area For Profiles Investigated.

Profile Type	Cover Width (mm)	Height (mm)	Area (mm^2)	Second Moment Of Area (mm^4)
Triangular	525	140	1 332	2 113 849
Triangular	450	174	1 307	3 433 345
Triangular	375	190	1 299	4 194 800
Trapezoidal	675	354	2 437	40 765 156

TEST PROGRAMME

A profile's maximum span limitation in composite floor construction is most often constrained by the moment capacity of the deck when supporting the wet concrete during the construction stage. Therefore it was this stage which had to be simulated to test the new profile. To model the uniformly distributed load of the wet concrete, steel ball bearings were used to load the sheets. Since the density of ball bearings is around 2.5 times that of lightweight concrete, polystyrene blocks were placed in the profile troughs at regular intervals to take up part of the volume, as shown in Figure 3. This allowed the ball bearings to reach a much greater depth than they would otherwise have done, for the same load. Hence a more accurate representation of the pressure distribution on the webs was provided. The ball bearings were contained using plywood shuttering.

The internal stiffening provided beneath the deck was kept to an absolute minimum, see Figure 4. The reason for this is to enable the stiffening to be provided at low cost and without obstructing the space between profiles above the supporting beams. In this way useful service openings can be maintained. The profile was minimally braced on its under-side. The sheets were side-stitched together at the top using self-drilling, self-tapping screws, and tied to the correct

Figure 3. Partially Loaded Bending Test.

Figure 4. View Of Bracing, Ties, And Shutter.

soffit width using steel straps, which also simulate the effect of connection to the beam flanges.

The deflections of the profile were measured using six dial gauges, located at midspan and quarter points of the profile, on each side. They monitored the bottom flange, the remainder of the profile being enclosed by the shuttering.

A 2.1 m sample of the profile was also tested to determine the shear capacity of the profile. A central point load was applied, with deflections measured at midspan, and at the profile ends on the top flanges. Load was spread locally over the profile using an arrangement of timber beams and struts, over a length of 500 mm. As load was applied the shape of the profile at the ends was also recorded.

Additional tests from samples of the sheets were carried out, to determine the actual thickness, t, of the sheets and the actual yield strength, p_y.

RESULTS

The 6.0 m test to determine the moment capacity of the sheet was more successful than expected. The sheet was loaded with a uniformly distributed load until a reasonable depth of concrete had been simulated. Thereafter, the sheet was loaded with a hydraulic jack, giving a central point load to find the excess bending capacity.

Failure was by compressive buckling of the top flanges around the point of application of the load, as shown in Figure 5. However on removal of the polystyrene and shuttering, bearing failure at the ends of the profiles was also

Figure 5. Exposed, Buckled Top Flanges Of The Failed Bending Sample

observed. At one end of the profile, there was a large crease in the lower part of the web, which extended approximately 500 mm along the span. The re-entrant at the bottom of the web had also deformed at that point. Distortion of a lesser degree had also occurred at the other end. Unfortunately it was not known at which load the creases had occurred since these were completely obscured during the test.

The 6.0 m sample reached 83% of its predicted moment capacity from BS 5950:Part 4[(1)], with an ultimate failure moment of 49 kNm. This discrepancy could be explained by any one or a combination of the following. Firstly, if it is assumed that compressive buckling occurred after the bearing failure in the webs, some twisting in the profile would have reduced its ability to carry load. Secondly, the re-entrants at the corners of the top flange rotated downwards under load, becoming virtually aligned with the top flange. This reduction in profile height would have caused a reduction in second moment of area, and depth to the neutral axis, both affecting the moment capacity. Thirdly, it is clear that the empirical design equations in BS 5950:Part 4[(1)] were derived for the present profile geometries, and have not been proved applicable to such a deep section.

Failure of the 2.1 m sample was at a load of 28 kN. Buckling of the deep web had started, when the studding which formed part of the bracing made contact with the supporting beam. The remaining web, and the studding then formed a sway mechanism.

There are four distinct sources of deflection in these tests; bending deflection, shear deflection, profile geometry changes, and non-linear effects (which include changing section properties due to buckling distortion and yield).

At early stages of loading the lower stresses in the compression zone result in greater effective compression flange widths and hence increased second moments of area. At zero load the second moment of area is that of the gross cross-section, and it reduces to the value for the net cross-section calculated using the effective compression flange widths at ultimate load. Consequently, there is a gradual reduction of bending stiffness as the load is increased. The effect of the anticipated non-linearity of bending deflection due to this change is negligible. The bending deflections can be predicted using the section properties calculated at the theoretical ultimate moment. The maximum elastic bending deflections predicted are 20.40 mm for the 6.0 m test and 0.54 mm for the 2.1 m test.

Shear deflections based on the assumption that the webs were rigid plates proved to be extremely small, at 0.85 mm and 0.24 mm for the two tests, respectively, based on a shear modulus of 80770 N/mm^2.

Analysis of the 2.1 m test load versus total deflection minus bending deflection curve showed a linear response after the bearing settlement occurred during the first 3 kN of the total 28 kN applied, as shown in Figure 6. Using this curve's slope to calculate an effective value for the shear modulus produces a value of 1274 N/mm^2. This shear modulus value is considerably lower than the value for the webs as a single flat plate, due to the distortion facilitated by the re-entrant portions at the top, middle and bottom of the webs. This reduced shear modulus predicts a shear deflection of 15.03 mm in the 2.1 m sample over the full load range to failure.

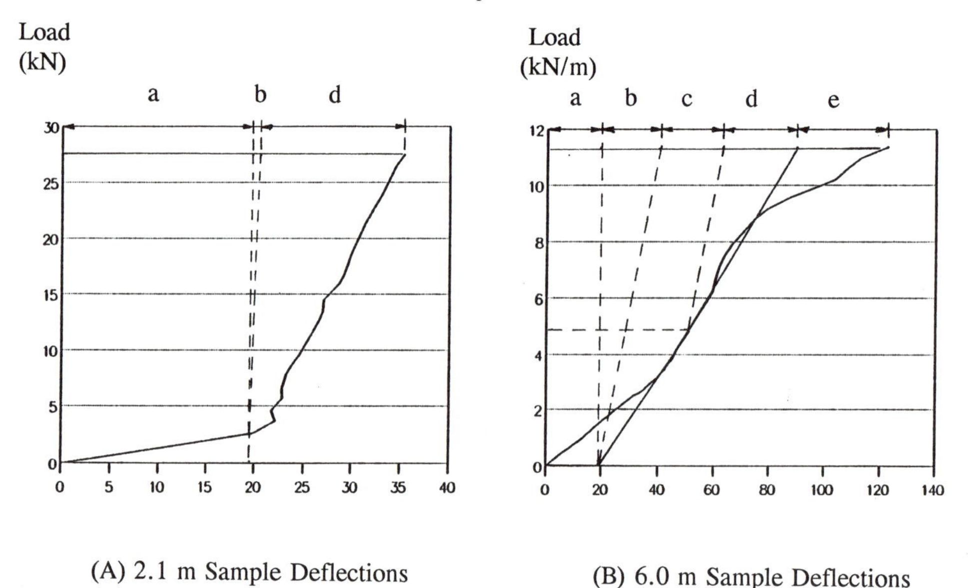

(A) 2.1 m Sample Deflections (B) 6.0 m Sample Deflections

Figure 6. Load - Midspan Deflection Graphs For (A) The 2.1 m Sample and (B) The 6.0 m Sample. (Key to deflections: a = web distortion, b = elastic bending, c = distributed load shear, d = point load shear, e = yield)

For the 6.0 m sample, there are two conditions to consider. Up to a uniformly distributed load of 4.46 kN/m the shear deflection is predicted by the low modulus to be 21.22 mm. Higher loads are applied as a central point load, and a further 27.24 mm shear deflection is predicted before failure.

During the application of the first 3 kN to the 2.1 m span sample, there is a deflection of 19.85 mm which can be attributed to profile geometry charges under load. This is believed to be due to the load concentration on the lower flange causing the re-entrant sections in the web to pull open and thereby increase the midspan section depth.

For the 6.0 m span sample, the transfer from uniformly distributed load to point load was taken as the pivot for the shear deflection predictions. This left a residual deflection of 19.51 mm, which is remarkably similar to the geometry change observed in the early part of the shorter span test. It is therefore attributed to the same cause.

The 2.1 m span test failed suddenly before significant non-linearity had developed. Consequently, all the deflections for this test are explained by predictions of the bending, shear and bearing deflections.

In the 6.0 m span test, there is a ductile failure. Above the 8.5 kN/m load intensity, there are deflections which cannot be attributed to elastic bending, shear or bearing. A further 34.44 mm is caused by the remaining non-linear effects at the measured failure load intensity of 11.36 kN/m.

The moment capacity of the profile tested would enable a construction depth of up to 455 mm. This figure allows for a construction load of 1.5 kN/m^2 and the requisite partial load factors. The comparable figures for shear and deflection are 330 mm and 440 mm.

Minimum construction depth for this profile to satisfy fire regulations would be 400 mm. The ribbed floor geometry is extremely strong against imposed loads, and it is the construction stage loading which will limit capacity (assuming no intermediate propping). It is apparent that the profile tested is perfectly satisfactory for 6 m spans in all respects other than shear. Much of the shear deflection is due to the deformation potential at the web re-entrants. Modifications to the profile shape could easily be made to improve the shear performance to a satisfactory level. The re-entrants in the webs are in any case of considerably less use than those in the flanges for supporting services, though changing their geometry would mean that existing profiles could not be used to form the modified deep profile.

The cost of a very deep profile floor such as that described in this paper is very similar to the cost of a conventional composite floor on a steel frame. Intermediate transverse beams are eliminated, and longitudinal beams reduced. Only a small part of the saving is required to cover the increased cost of decking. The increased concrete volume and associated shrinkage/fire reinforcement (usually close to the minimum percentage) make the overall cost per square metre come within 5% of the current construction cost. Such a small difference will vary according to the individual project geometry.

Where this form of deep profile offers some advantage is in the reduction of floor depth in one direction balanced by an increased depth in the orthogonal direction. This may appear at first to be a disadvantage. However, the openings above the primary beams formed by the deep trough profile are large enough to permit larger pipework and air-conditioning ducts to pass through the building without a need to provide a service zone beneath the beam. An overall small saving of construction depth is therefore possible. In many instances, even a small saving in construction depth is more important than cost per square metre of construction.

CONCLUSIONS

This paper shows that it is possible to design a deep profiled composite floor deck which will span 6.0 m unpropped. The sample profile showing this used an existing shallow re-entrant profile as its basis. Tests on prototypes show that the designed profile was satisfactory for the bending and deflection criteria in the wet stage of construction. It is also satisfactory for all composite stage design criteria.

The profile tested was not strong enough in shear. The problems observed were due to the re-entrant shapes creating excess flexibility in the webs. These problems can be eliminated easily if the deep profile is designed independently of the shallow profile geometry. Alternatively, the re-entrants on the corners could be spot welded across the neck of each corner re-entrant to transfer load directly from the top flange into the web and onto the supporting beams. Spot welding of the re-entrants in this fashion would make the on-site production of a deep profile easier.

Maintenance of the deep profile geometry where it rests on the supporting beams was found to be important to the profile strength. Design of a suitable former to maintain profile geometry at these locations, without blocking the passage of services, requires further attention.

The moment capacity, from existing theoretical prediction, was 59.3 kNm. When tested in the laboratory, the actual moment capacity was 49.0 kNm. Thus the sheet achieved 83% of the theoretical value, which within experimental error, represents a good correlation for a manually produced, prototype profile.

REFERENCES

1. BS 5950:1982 'Structural Use of Steelwork In Building, Part 4, Code of Practice For Design of Floors With Profiled Steel Sheeting', British Standards Institution (1982).
2. BS 5950:1990 'Structural Use of Steelwork In Building, Part 3:Section 3.1, Code of Practice For Design of Simple And Continuous Composite Beams', British Standards Institution (1990).

35 CONCRETE-FILLED STEEL TUBE COLUMNS

M.J. Barker
Mott MacDonald Structural and Industrial Ltd (Connell Wagner Pty Ltd, Australia)

The concrete filled steel tube column, together with it's method of design and construction is considered a viable alternative to the traditional steel or reinforced concrete column. The tube column enables the economies of a reinforced concrete column to be enjoyed whilst providing all the advantages of the steel column's constructability.
This paper describes the emergence of the tube column and it's adoption on the 46 level Casselden Place project in Melbourne, Australia.

INTRODUCTION

Over the past decade, Connell Wagner, Mott MacDonald's Associated Australian Company has been involved in the design and supervision of numerous multi storey buildings. The ongoing exposure to this type of project has enabled a continuous evaluation process into the suitability and relative economies of various structural forms. A particular area of study concerned columns.

The choice between steel and concrete for multi level buildings has never been obvious. The final decision has generally been driven by local conditions, the capabilities of particular contractors and the relative economies of the materials at the time of tendering the project. The traditional steel column has always been recognised as being more expensive than it's reinforced concrete counterpart, however the steel column has considerable constructability advantages, which would be utilised if the relative costs could be equated.

The desire to adopt steel as the structural medium is driven by a number of factors. A major factor is the growing trend to prefabrication of the building's components off site. A further significant factor is the ability to simultaneously work on a number of different levels without the obstruction of formwork supports and backpropping associated with concrete construction. In the case of Casselden Place, further bias was given to steel as at the time of tender, formwork prices were very high.

The factors above were very much in the minds of the Contractors, Baulderstone Hornibrook, when they submitted an alternative design for the all concrete original scheme proposed for this 46 level project. The alternative proposed by the contractor retained the concrete core and substituted the floors to a steel framed composite deck system, and the columns to bare steel tubes filled with reinforced concrete. This philosophy enabled the buildings to be essentially formwork free, with the exception of the core, above the ground floor level.

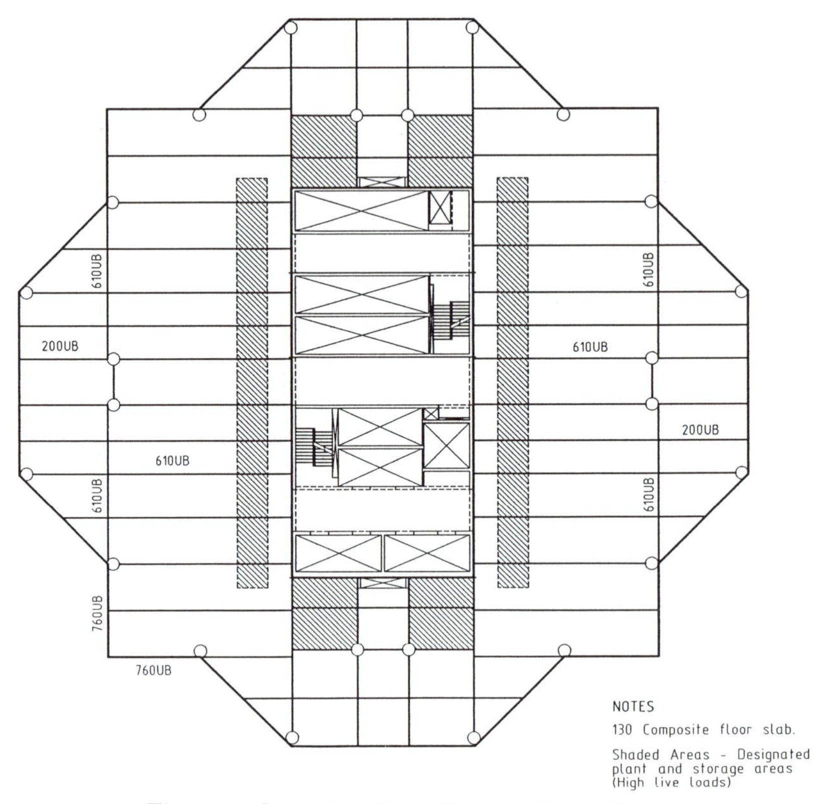

Figure 1. Casselden Place Typical Floor Plan

The alternative was accepted and work on the validation was carried out. Studies were undertaken on the integration of the column into a project of this size. It soon became apparent that if the reinforcement could be omitted, then further dramatic constructability gains could be made. This option was then pursued. The problem however, was fire protection. In a fire, the bare reinforced tube does not provide the required period of protection. (In this case a 2 hour requirement). For a bare tube, fire strength is achieved by the consideration of an 'inner' reinforced concrete column, or by fire protecting the steel tube itself.

The full advantages of the tube are not realised when reinforcement is introduced. The fabrication and placing of the cage slows the construction process and involves more site based operations. More critical however is the requirement of the beam connections to penetrate into the inner "core" of the reduced section column. Refer to Figure 2a.

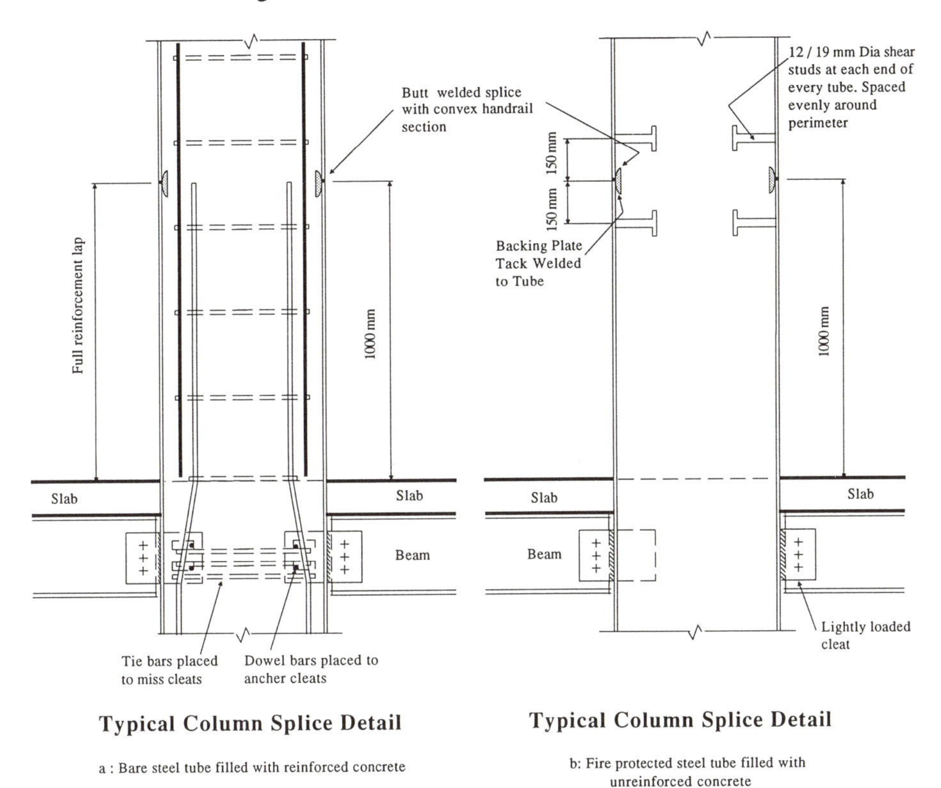

Figure 2. Column Splice Details

RELATIVE COSTS

Prior to the concept of the unreinforced tube being accepted as a viable alternative form of construction, a detailed cost analysis was carried out on comparative column forms.

Figure 3 presents a study of the relative costs of six columns, supporting ten and thirty storeys. The costs are calculated on a unit length basis. The study showed that the reinforced concrete column was the cheapest when examined from this viewpoint, and has been used as the base from which the remaining columns are compared. The detail of the relative costs will vary between different projects as the particular factors individual to the project are taken into account. However on most projects it is considered unlikely that the relationship between the various forms will vary dramatically from that indicated in the study.

The competitiveness of the unreinforced steel tube is more marked in the larger section compared with the reinforced column supporting ten storeys.

However, when the relative economy is coupled with the constructability advantages, the benefits are considerable. The study also does not take into account the considerable dead weight savings of a building constructed with a steel framed floor system over that framed in concrete. The resulting reduction in dead weight has beneficial effects on the foundations, core and of course the columns themselves.

Number of Storeys	Type 1 Conventional R C Column	Type 2 Concrete Column with Steel Erection Column	Type 3 Concrete Encased Steel Strut	Type 4 Un-fireproofed Tube filled with Reinf. Concrete	Type 5 Un-Reinforced Steel Tube filled with Concrete	Type 6 Full Steel Column
	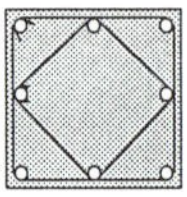 R 10-200 mm links	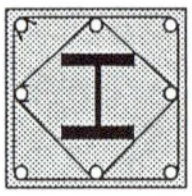 R 10-200 mm links		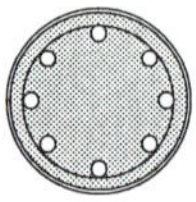 R 10-200 mm links Mild Steel Tube	 Mild Steel Tube with Fire Spray	 Fire Spray to Steel section
10 Levels	450 x 450 8 T 32	450 x 450 8 T 32 203 x 203 UC 46	410 x 410 305 x 305 UC118	500 x 6.4 CHS 6 T 20	500 x 6.4 CHS	305 x 305 UC240
Relative Cost	1.0	1.22	1.53	1.14	1.10	2.27
30 Levels	750 x 750 20 T 36	750 x 750 12 T 36 254x 254 UC 89	570 x 570 Plate Girder 400 x 50 Flanges 360 x 25 Web	800 x 10 CHS 6 T 32	800 x 10 CHS	Plate Girder 500 x 60 Flanges 460 x 40 Web
Relative Cost	1.0	1.13	1.85	1.11	1.02	2.61

Note

Column loaded by 8.4 x 8.4 m bay, steel framed. Concrete Fcu 60 mPa Steel Grade 50

Figure 3. Column Comparison Study

The form of the column does indeed make it very cost effective. The design, fabrication and erection processes of this column are all very efficient.

For design

The tube is fabricated from Mild Steel plate. Plate rolled into a tube is a very effective way of utilising the strength of the steel.

The major portion of the axial load is resisted by the high strength concrete. This is the most cost efficient way of resisting compressive forces.

Confinement of the concrete enhances the capacity of the concrete to resist load. This additional benefit is only possible with this form of column.

For fabrication

Plate rolled into tube with simple welded cleat connections is an extremely efficient way of purchasing the strength of steel. In the case of Casselden Place, the tube with welded beam cleats worked out to be 10% cheaper than an equivalent UB / UC section.

The connections to the beams are very simple. Plate web cleats were used for all except moment resisting joints on Casselden Place. (Figure 2b). Fabrication costs for this type of connection are very competitive.

For erection:
The absence of formwork and reinforcement cage reduces the on site operations with a consequential reduction of site labour.
The on site operations are very simple and quick to complete, and only consist of placing the element, welding to it's lower partner and filling the tube.

BUILDABILITY

The main advantage of the unreinforced concrete tube is it's buildability. The main reasons are:

1. With the lack of formwork, and the ease of placement, the column can be erected quickly. As with most multi storey buildings, the cranage on Casselden Place was a critical resource. The use of the unreinforced column minimised cranage use, freeing it for other operations. With the choice of structural form, there was in fact no formwork used on the project outside of the core.
2. Construction can proceed on multiple levels in the building, prior even to the column itself being concreted. The concreting of the column is then not a critical activity.
3. The ability to be able to pour up to four floors in one operation reduces the column concreting activities to 25% of those in a comparable reinforced concrete building.
4. The lack of a reinforcement cage simplifies the on site construction operations. With this element omitted, the major source of work in the construction of a concrete column has been removed. There is a further benefit, in that the required quality and consistency of the placed concrete is more easily achieved.

The main disadvantage of the unreinforced tube is the requirement for the full strength site butt weld connecting the column sections. This operation however is generally not on the critical path of floor cycle construction.The regular repetitive form of the weld lends itself to automation, although on Casselden Place, this was felt to be unnecessary. A typical splice detail is shown in Figure 2b.

The adoption of the convex handrail section for the backing strip to the site joint fulfils three functions:

1. It assists in the alignment of the upper two storey tube section to the lower, during erection.
2. The handrail provides the backing strip for the site full strength butt weld.
3. The section aids, together with a series of shear studs placed to each side of the joint, the transfer of load from the steel tube wall to the concrete heart of the column.

The tubes on Casselden place were fabricated in two storey lengths (7.5 m). The tubes were produced by rolling mild steel plate into a tube form and longitudinally seam welding the edges using a tandem wire submerged electronic arc process. The fabricator for the project was able to form columns from 488 mm dia. up to 2,500 mm dia. in continuous lengths up to 12 m. The plate thickness able to be handled ranged from 6 to 28 mm. A variety of steel yield strengths could be utilised.

STRENGTH

The strength requirements for both the unreinforced tube and the bare steel, reinforced tube are discussed below. The fire strength requirements for the reinforced tube are also examined.

Table 1. Capacity Reduction Factors

	AS 4100	AS3600	Connell Wagner	BS 8110	BS 5950	Eurocode No. 4
RC in Compression	-	0.6	0.85	-	-	-
RC in Bending	-	0.8	0.6	-	-	-
Steel in Bending	0.9	-	0.85	-	1.0	1.0
Steel in Compression	0.9	-	0.85	-	1.0	1.0
Concrete in Compression	-	-	-	0.67	-	0.67
Reinforcement in Compression	-	-	-	0.87	-	0.87
Dead Load Factor	1.25	1.25	1.25	1.40	1.40	1.35
Live Load Factor	1.50	1.500.6	1.500.85	1.60	1.60	1.50

Unreinforced tube

For the Casselden Place project, a method of analysis and design needed to be formulated, as the Australian Codes of practice(2) and (3) could not be directly applied. The method adopted was based on Eurocode EC4.(4).

The ultimate strength of the section was determined by the use of a rigorous strain compatibility analysis. A program GEXSAN(5) (Kemp 1989), was utilised for this task. The analysis method consists of imposing on the section a large range of axial strains and curvatures, from which a comparable range of N and

M values were produced. The results were then plotted for the range of cases examined and the design interaction curve formulated from the surface of the plots. The curve then provides the relative values of M* and N* (Ref.[2] AS 3600: Where M* is the design bending moment and N* is the axial force, both derived from the appropriately factored service loads).

For the forces in the concrete, the Desayi-Krishnan[6] stress strain curve is used. The forces in the tube and if appropriate, reinforcing steel are evaluated by using a bi linear elasto-plastic stress strain function, E being set at 200 GPa. The forces in the concrete and steel are then reduced by their respective capacity reduction factors, by adjusting the relative stress strain functions.The method of analysis was investigated and verified by the University of Sydney, Centre for Advanced Structural Engineering, through a series of scale model tests.[9].

The partial capacity reduction factors as required in the Australian codes, for concrete and steel are shown in Table 1 and compared with the requirements in Eurocode EC4,[4] BS8110[7] and BS 5950[8].

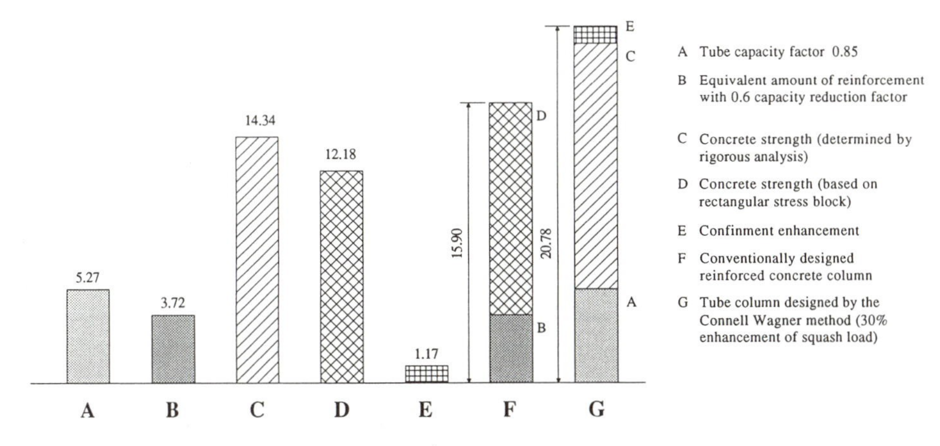

Figure 4. Contributing Factors for Strength

The use of the stress strain curve for the calculation of the strength of the concrete greatly enhances the strength of the column, when compared to that calculated from the rectangular stress block method. This approach holds for axially loaded or near axially loaded columns.The confining effect of the concrete in the tube also helps to enhance the strength of the column. This coupled with the stress strain analysis approach as discussed above, greatly increases the strength of this column over that of an equivalent reinforced concrete column. Figure 4 provides an indication of the increase in strength available from this type of column, and the make up of the strength, from the various contributing sources.

Certain other factors must be evaluated in the determination of the strength of the section. These include the local stability of the tube when empty under construction load, the load transfer between the steel tube wall and the concrete heart, the local concentrations of stress at the connections and the code minimum eccentricity requirements for the determination of minimum design bending moments.

Reinforced tube

The bare steel tube need to be checked for two design cases. The usual strength / serviceability case and the fire strength case. For the fire design case, AS 3600[2] required the column, at the end of the required exposure period of the fire (2 hours) to be capable of sustaining a load not less than 1.1DL + 0.4 LL.

The column was checked in the BHP Laboratories in Melbourne using the TASEF-2 heat flow finite element program. This analysis primarily established the temperature gradient across the section during the fire exposure period. The results indicated that, at the end of the 2 hour period, the steel tube was ineffective, the reinforcement remained only 87% effective and the outer 36 mm of concrete should be disregarded when considering the load carrying capacity of the section.

Table 2.

	Normal strength	'Fire' strength
Tube	5.27	----
Reinforcement	1.15	1.00
Concrete	14.34	11.87
Confinement	1.17	----
Totals	21.93	12.87

If for example, the 800 x 10 tube filled with 50 N/mm^2 concrete is again considered, this time including 1% bar reinforcement, the resulting squash load capacities are presented in Table 2. It can be seen that the fire strength is 58% of normal strength.When this is examined by comparing the proportions of dead to live loads, it indicates that unless the dead load is less than approximately 1.3 times the live, the fire design will be critical. For buildings of the type of Casselden Place the proportions are in the region of 1.6:dead load to reduced live load. When considering smaller diameter columns, the loss of the external 36 mm of concrete has greater effect, making the fire design even more significant.

On the Casselden Place project, it was found that the costs of fire protection and cladding the columns was more than compensated for by the omission of the reinforcement, the simplification of all the beam joints and the subsequent ease of placing the concrete.

CONCLUSIONS

This form of column will, no doubt, be successfully used on many future high, medium and low rise projects. Further potential areas of exploiting this form of column include:

Jumpstart. The use of this type of column is ideal for 'jumping' construction to typical floors straight from foundation level, allowing construction to proceed simultaneously on two fronts.

High Strength Concrete. A concrete heart of very high strength concrete is ideal for this column form. The concept of the tube should gain ecomony as the strength of the concrete increases.

Seismic resisting construction. Conventional reinforced concrete requires complex detailing of reinforcement in the need to provide sufficient ductility to certain locations of structures. The tube provides great ductility, with much greater simplicity.

The tube column has been shown to be a viable alternative form of column, which exploits the constructability advantages of the all steel column, whilst also enjoying the economies of the reinforced concrete column. The column form has been proven to be a success on Casselden Place, and has been subsequently used successfully on the comparatively sized Market City project in Sydney.

REFERENCES

1. Webb J. and Peyton J.J. 'Composite concrete filled steel tube columns' 1990.
2. AS 3600-1988 S.A.A. Concrete Structures Code S.A.A. 1988.
3. AS 4100-1990 S.A.A. Steel Structures Code S.A.A. 1990.
4. Eurocode EC4 Common Unified Rules for Composite Steel and Concrete Structures. EUR 9886 EN Commission of the European Communities 1985.
5. Kemp M.V. (1989) 'General analysis of the strength of cross sections of composite materials'. University of Sydney M Eng.Sc. Thesis 1989.
6. Desayi and Krishnan 'Equation for the stress strain curve of concrete'. Journal ACI Proc. Vol 61 No.3 March 1964.
7. BS 8110-1985 Structural Use of Concrete. British Standards Institution.
8. BS 5950-1985 Structural Use of Steelwork in Building. British Standards Institution.
9. Report S837 'Alternative composite column design' Nov. 1990 Centre for Advanced Structural Engineering, University of Sydney.

36 EXTERNALLY REINFORCED CONCRETE

P.G. Lowe
The University of Auckland, Auckland, N.Z.

Dedicated to the memory of John Fleetwood Baker (1901-1985)

The name Externally Reinforced Concrete (E.R.C.) has been given to a steel-concrete composite in which the reinforcement is provided as thin (steel) casings inside which the concrete is cast. Beams as well as columns and other elements have been made in this manner. If suitable jointing systems can be devised then a new building system could be developed. The prospect is that such a system could be very cost competitive.

INTRODUCTION

This conference is focusing on the future. To better appreciate what this may hold, at least some appreciation of the past, the achievements and the failures, is desirable.

Building technology today embraces two dominant technologies - structural steelwork and reinforced concrete. Both have relevant histories stretching back around 100 years. Each has developed largely independently of the other. In the 1990's they could both be described as **mature** technologies.

The present day cost of established technology can probably not be much changed, except in respects such as savings from multiple repetition or standardisation of components.

If **major** cost savings are to be achieved, assuming an adequate technical standard is reached, then it is likely that some **new** technology will be needed. Structural Steel and Reinforced Concrete Technologies have co-existed and developed side by side throughout the twentieth century. Could it be that in the twenty-first century this independence could or should be replaced by a single **merged** technology which combines the best features of both existing technologies? This is certainly a very ambitious, perhaps even an impossible, scenario.

Introducing change in an Industry such as the Building Industry, in any country and at any time, is likely to be very slow and accompanied by much discussion in technical circles. Studies of the changes brought about in the Industry, for example earlier this century in the U.K., are available[1] and are necessary reading for any proper assessment of the future prospects.

Even after 100 years some technical features of present day technology are incredibly crude. For example, the still widespread practice of achieving force transfer from bar to bar in reinforced concrete by bar lapping is a practice which should not be tolerated on technical grounds although lack of cost effective alternatives may prevent a change on grounds of economics.

In this connection there is some evidence which points to a deterioration of technical standards after the developmental stages of the technology. Thus a recent (1990) demolition of an early (c 1900) reinforced concrete building in Auckland brought to light extensive use of (blacksmiths) welded butt joints in the main (37 mm) reinforcement, thus avoiding lapping of bars even at this early stage for the technology.

There have been concerted attempts from time to time to advance Building Technology. One attempt was by the Steel Structures Research Committee. It was active from 1929 till completing it's Final Report in 1936(2) and first brought the then young research worker J.F.Baker (later Lord Baker of Windrush) to prominence. This paper is dedicated to his memory as a personal tribute to his achievements.

If Building Technology in a general sense is to stay abreast with modern life then it must adapt to the computer age, from the fabrication shop through to the finished product.

Arguably neither present dominant technology has done this. Structural Steel with stock rather than custom sized sections and reinforced concrete with labour intensive bar fixing are just two of many respects in which these technologies are behind their times.

There is a worldwide trend towards codification, with the Building Codes and Standards prominent at National and International level. Many Building controls are prescriptive, rather than performance based, and this poses particular difficulty for Innovative Building Schemes. From this circumstance alone, what chances are there for innovative building technology to succeed?

It is now more than fifty years since Whitney(3) made his proposals about the plasticity of concrete, and about an equal length of time since plasticity effects in steel and ductile structures response were first seriously promoted(4). It is time to consider the shape of things for the next fifty years!

SOME REQUIREMENTS OF AN IDEAL BUILDING TECHNOLOGY

In the barest of essentials the requirements could be said to be, first the technology must be affordable while ensuring adequate technical performance. The raw materials should be obtained in an environmentally sustainable manner and any building which has no further usefulness should be recyclable in some sense. Of these qualities affordability is the most easily quantifiable.

Prefabrication of components is a key feature of the present dominant technologies, and structural steel offers greater scope to prefabricate components. Precasting of concrete components, while very attractive from many viewpoints, still presents technical problems of jointing long after general acceptance of the concept.

In the twenty-first century presumably systems will emerge which better satisfy the criteria than can at present be achieved.

THE SCENARIO FOR EXTERNALLY REINFORCED CONCRETE

In general terms what Externally Reinforced Concrete consists of is thin walled, hollow, box-like members, beams and columns particularly, made from steel strip, jointed together to produce frameworks and filled with concrete.

The novelty of the proposed structural material is that members in **primary flexure** are envisaged, as well as column-like elements, and hence there is scope to develop a **complete building system.**

Compared with traditional structural steelwork, E.R.C. uses **custom-made** hollow sections formed from thin (2-3 mm) steel - essentially sheet metal, rather than much heavier, hot rolled or fabricated three or four plate sections.

Compared with traditional reinforced concrete E.R.C. requires **no temporary formwork** and **no manual bar steel fixing**. Indeed E.R.C. case manufacture can be achieved quite easily and cost effectively on modern numerically controlled sheet metal equipment. Because the casings are prefabricated and are light weight, there is scope to assemble them very quickly.

Concrete-filled Rolled Hollow Sections have been studied for many years, and have found application in some structures as columns. More recently much larger concrete filled steel columns have been incorporated into multi-storey structures in the USA and Australia(5). These are usually of circular section with much higher Diameter/ Casing Thickness ratios than have been used with RHS-filled columns.

We are not aware of earlier proposals of the type described here as E.R.C., where **all** the structural members are concrete filled, hollow steel cases.

E.R.C. BEAM BEHAVIOUR

Crucial to any building material or system is satisfactory performance in bending.

In Figure 1 and Plate 1 are shown experimental data for an E.R.C. beam. In this example the hollow steel case, 100 x 200 mm in section and 2400 mm long had a one piece top flange and webs, 1 mm thick. The bottom flange, 1.6 mm thick, was MIG welded along the two lower edges. The concrete filling was added through holes in the top flange. No special treatment was used on the inside of the casing. Transverse bolts, which can be seen in Plate 1, provided a measure of shear resistance and helped keep the casing in shape during filling.

We shall not elaborate here on the mechanics of stress and force transfer in bending other than to note the results. The steel casings are not bonded to the concrete in the usual sense but true composite action does result.

The test results relate to a test arrangement with a lever arm of 1 m. As can be seen the behaviour was of a ductile and strong beam which shows none of the

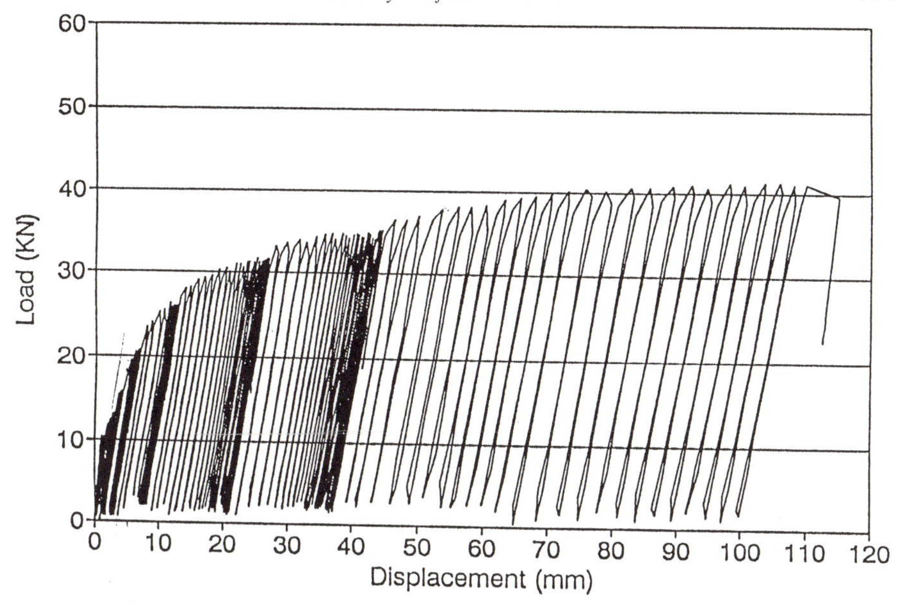

Figure 1. Experimental Data relating to E.R.C. Beam.
After Choong, K.C. M.E. Project, University of Auckland 1990 Fig. 3.5(b)

Plate 1 E.R.C. Beam under test.

deterioration of strength associated with concrete crushing in compression. The concrete does crush eventually but this has essentially no effect on the current beam strength. The capacity of the casing in the compression zone to contain the concrete in the crushed state ensures that this crushed material remains load bearing. This behaviour is very favourable and contributes greatly to the capacity of such members to absorb energy, as for example, in an earthquake.

Members in bending which are deeper in relation to the span have been tested, using span/depth ratios which require special treatment for shear in traditional reinforced concrete, but no equivalent to a shear failure has yet been observed.

JOINTING OF E.R.C. MEMBERS

An extensive series of beam-and-column specimens has been fabricated and tested. The primary type of connection used was beam flange plates passing through the column.

A moment capacity range from about 20 kNm to 400 kNm with members from 150 mm to 500 mm deep has provided data on moment resisting joint behaviour which has been applied to the most ambitious project attempted thus far - the design, fabrication, erection and testing of a near full size two bay x one bay x two storey frame.

AN EXPERIMENTAL COMPLETE FRAMEWORK

In Plate 2 is shown the near full-size E.R.C. two storey framework which has been built and tested.

The overall dimensions were: Height 4 m, Bay Length 4 m, columns square 250 x 250 mm x 2.5 mm, beams rectangular 150 x 200 mm x 1.6 mm. In all there were six columns and fourteen beams. All components were factory made, and 'site' bolted. On one lower level bay a steel trough deck slab was cast between the parallel two-bay frames. This was concreted to 60 mm above the decking. See Plate 2.

The primary objectives were to study the design, erection and behaviour of the frame under simulated (quasi static) lateral earthquake-type loading. This system of loads was applied at beam level from one end by four jacks. The jack loads were cycled with increasing force amplitude in both a pull and a push direction. Eventually approximately twenty complete cycles of such load were completed.

One parameter of interest is the lateral deflection, expressed as a ratio of the first yield value. This quantity, sometimes called the Ductility, is a measure of safety in extreme events such as earthquakes. A value of five or more without loss of strength is probably adequate, and the frame achieved this on the way to a final state of hinge collapse.

Most of the desired features, such as beam hinges forming approximately 600 mm from the column faces, leaving the beam-column junction essentially rigid, were realised. If the hinges can be positioned in this manner there is

Plate 2 Experimental E.R.C Two Storey Framework at time of Test.

considerable benefit in terms of load carrying capacity compared with the hinges nearer the column face.

The beam moment of resistance was calculated at 50 kNm and for the column 90 kNm. From a hinge collapse calculation the loads for collapse (held in the ratio 1 upper to 0.75 lower) gave 180 kN and 135 kN. These values were quite close to the experimental observations. Finally column head deflections of the order of 250 mm were produced as the final hinges formed in the frameworks.

CONSTRUCTION PROCEDURES AND COST IMPLICATIONS

The proposed E.R.C. technology provides the scope to automate fabrication and extend prefabrication. Thus the data for use in fabrication of reinforcement casings can conveniently be supplied direct to the fabricator on disc. Numerically controlled sheet metal equipment can then shape the steel sheet or coil in its flat state. Most casings are made from two components, each shaped by two (parallel) folds from the flat item. The final rectangular shape is achieved by two longitudinal MIG welds.

Typical casings, for example the 4 m long columns for the frame described in the previous section, weigh less than 1 kN. Such lightness is an important feature of most E.R.C. steel components.

Site assembly could proceed by assembling components into final position, or might involve ground level completed subassemblies being lifted into final position. Either way, the average weight/lift compared with many present-day site lifts of prefabricated items is likely to be much reduced. As a guide, the empty casings assembled will probably represent about at most ten percent of the weight of the final filled frame. The options available for assembly, the tolerances and the speed of assembly possible, are at least as favourable as for modern prefabricated conventional steelwork, but with the added freedoms that working with components about one fifth current weight would allow. Once assembled the cases are filled by pumping from one or a few points, perhaps even from ground level.

It is quite simple to arrange that the empty framework is self-supporting while being filled, thus requiring no temporary propping.

One costing exercise which compared a reinforced concrete and an E.R.C. alternative design showed a 16% saving on structure for E.R.C. for the structural cost.[6]

An important consideration is **FIRE** Design of E.R.C.. This is currently being investigated. The susceptibility is likely to be less than for traditional unprotected structural steel but more than for traditional reinforced concrete. The concrete filling is a significant heat sink. There are several possible scenarios to achieve desired ratings. The fall-back position is to treat E.R.C. as traditional steelwork, though this will incur maximum cost to achieve desired rating.

CONCLUSIONS

The viewpoint adopted in this paper is that in the medium term future, if the Building Industry is to achieve some or all of: greater cost efficiency, incorporation of a greater degree of automated manufacture for building components, and/or some significant rationalisation from two competing, separate building technologies to a single hybrid technology, then major changes will be needed to the current technologies.

One scenario for a hybrid technology, here termed Externally Reinforced Concrete, has been outlined, and other scenarios are no doubt possible.

Exciting prospects become apparent once radical alternatives are considered. Of all the prospects two in particular stand out. First, possible major cost reduction. Secondly, the scope to streamline the manufacture, assembly and appearance of the built product, while at the same time achieving quite adequate strength and stiffness. An analogy might be drawn with the automobile industry, which changed from a separate chassied to a monocoque vehicle construction about forty years ago. Thinking of the traditional steel frame or reinforcement cage as the 'chassis', a possible way forward for building technology could be to adopt a 'stressed skin', monocoque structure of the type envisaged with E.R.C.!

ACKNOWLEDGEMENTS

Several graduate students, colleagues and friends have contributed to the E.R.C. Research Programme at Auckland. Reference[6] sets out these contributions. Noted there also are the Industry partners and sponsors who have contributed funds and advice to the programme.

Building represents one of the oldest of human activities, and there have been many major contributors to our present state of knowledge, and level of achievement. This paper is dedicated to the memory of John Fleetwood Baker, one of the leading contributors to our subject this century.

REFERENCES

1. Bowley M. *The British Building Industry*, Cambridge (1964).
2. Steel Structures Research Committee, First, Second and Final Reports H.M.S.O. (1931-36), containing a series of papers by J.F.Baker who at the time was Technical Officer of the Committee.
3. Whitney, C.S. 'Plastic theory of reinforced concrete design' Trans. A.S.C.E. 107 (1942) P.251-282. Discussion P.283-326.
4. Hill R. *The Mathematical Theory of Plasticity*, Oxford (1950) - see also Baker J.F. et al. The Steel Skeleton, Cambridge, Vol.1 (1954) Vol.2 (1956).
5. Webb J. and Peyton J.J. 'Composite concrete filled steel tube columns' I.E. Aust., Conf., Adelaide (1990) P.181-185.
6. Lowe P.G. Externally reinforced concrete - a new steel-concrete composite. Conference Paper I.P.E.N.Z. (1992) P.461-471, also, slightly extended, to appear in Transactions, Institution of Professional Engineering, New Zealand.

37 DOUBLE SKIN COMPOSITE CONSTRUCTION

R. Narayanan
I.L. Lee
The Steel Construction Institute, UK
H.D. Wright
University of Strathclyde, UK

An innovative form of steel-concrete-steel sandwich construction developed in recent years is discussed in the paper. This consists of two relatively thin steel plates and a plain concrete in-fill. Stud shear connectors welded to both the plates ensure composite action. An extensive series of model tests validate the concept and demonstrate that this form of construction is technically feasible and is likely to be an economically viable alternative to the conventional composite designs. This paper reviews the investigations funded by SERC into the potential for this new form of construction. Tentative design recommendations based on these studies are presented in this paper.

INTRODUCTION

During the initial design stages for the Conway river crossing in North Wales, two firms of consulting engineers, Sir Alexander Gibb and Partners and Tomlinson Partnership, considered an alternative form of construction for immersed tube tunnels, which has become known as Double Skin Composite Construction. This form of construction, which consists of a layer of unreinforced concrete, sandwiched between two layers of relatively thin steel plates and connected to the concrete infill by welded stud shear connectors, has the potential to combine the advantages of both steel shell and reinforced concrete construction[1-3]. Uses for this form of construction include submerged tube tunnels, building cores, nuclear shelters and marine quay walls.

As a consequence of the initial feasibility studies, a theoretical and experimental research programme, sponsored by the Science and Engineering Research Council, was initiated in the School of Engineering at the University of Wales College of Cardiff. This research programme, has resulted in a number of papers, dealing with various aspects of the structural performance of double skin composite elements[4-8].

REVIEW OF PUBLISHED WORK

Experiments on the 53 model scale tests and eleven large scales tests conducted at the University of Wales College of Cardiff during 1987-90 have been reported in references[4 to 8]. These experiments were designed to provide an understanding of the structural behaviour of double skin composite elements and consisted of tests on beams, columns and beam-columns. A wide range of parametric variations in steel plates, studs and concretes were employed. Simple design methods have been developed and are based on the observed behaviour of test specimens. Calculated predictions using the proposed design methods are found to compare satisfactorily to the experimentally observed values. The comparisons demonstrate that the proposed methods are suitable for most simple elements.

The following aspects of design have been reported on:

Ultimate strength of sections in bending
Ultimate strength of sections in shear
Ultimate strength of sections in axial compression
Ultimate strength of sections in combined bending and axial compression
Stiffness of double skin elements in bending
Detailing of double skin elements.

BASIS OF DESIGN

The starting point for the design methods is the similarity between the double skin composite system and doubly reinforced concrete construction, and a recognition of some important differences. The hypothesis is detailed below (see Figure 1).

The steel skins may yield in tension and, if buckling is prevented, yield in compression. In this respect, steel skins are similar to the compressive and tensile reinforcing bars in a doubly reinforced concrete section.

Concrete in the compression zone of a double skin composite section may crush. If sufficient compression steel is provided in such a way that the tension steel yields before the concrete crushes, a ductile 'under-reinforced' type of behaviour would result.

Failure caused by longitudinal shear forces between the steel plates and concrete core is another likely mode of failure. In double skin composite

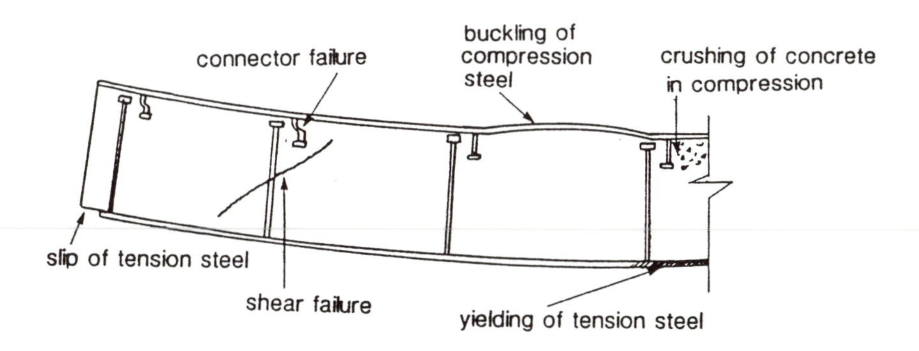

Figure 1. Basic Failure Mechanisms

elements, it is the shear connectors that resist these forces. (In reinforced concrete elements, this type of failure is prevented by well-anchored deformed or ribbed reinforcing bars.)

There should be a sufficient number of studs welded to both the plates. Long stud shear connectors welded to the steel skin in tension with their heads anchored in the concrete near to the compression skin are used to prevent vertical failure of the concrete close to the support or any point load. They act in a manner similar to the shear links in a reinforced concrete beam. For efficient performance, the studs anchored to the compression plate should overlap with those anchored to the tension plate.

When a double skin composite element is subjected to compression, the steel skins can buckle away from the concrete. Failure triggered in this manner is dependent on the steel plate thickness, the pull-out resistance of the stud connectors and their spacing in the transverse and longitudinal directions.

As there are some similarities between DSC elements and doubly reinforced concrete elements, BS 8110 can be considered as a good document for the basic design of DSC elements(9). However, DSC elements do have certain aspects in their construction that are more in common with components covered by other codes of practice which include BS 5400:Part 5 and BS 5950:Part 3.1(10,11).

The design methods suggested as a result of the studies so far indicate the need to select appropriate parts of these Codes, in order to reflect the observed behaviour on test specimens.

DESIGN CRITERIA

The following design criteria and design rules are recommended:

Partial Safety Factors for Loads

These should be taken as those quoted in BS 8110: Part 1.

i.e. Imposed load = 1.6
Dead load = 1.4

Partial Safety Factors for Strength of Materials

Partial safety factors for steel = 1.1 (as provided in BS 5400: Part 5)
Partial safety factors for concrete = 1.5 (as provided in BS 8110: Part 1)
Partial safety factors for connectors = 1.1.

Ultimate Moment Capacity

The ultimate moment capacity of the section is calculated by assuming fully plastic stress blocks for the steel and a rectangular stress block of depth 0.9 times the depth to plastic neutral axis for the concrete (see Figure 2). It is assumed that the concrete below the plastic neutral axis has cracked and does not contribute to the strength of the section.

The position of the plastic neutral axis can be determined from statics in the usual way and the ultimate moment can then be calculated.

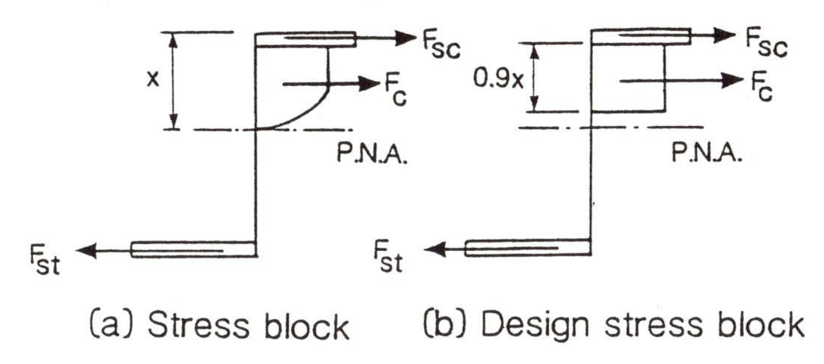

Figure 2. Design Stress Block Diagram

Steel Plate Buckling
The ratio of the centre to centre distance between stud shear connectors to plate thickness must be less than 40.

Vertical Shear Failure
The rules given in BS 8110: Part 1 for vertical shear resistance should apply. For elements subjected predominantly to flexural loads the value of the design concrete shear stress v_c, obtained from Table 3.9 of BS 8110: Part 1, should be reduced by 20%.

Stud Pull-out
The pull-out resistance of the studs is an important factor in Double Skin Composite elements subjected to compression. The stud shear connectors have to be sufficiently long and the concrete sufficiently strong. The suggested detailing rules to prevent studs pull-out are as follows:

(i) All stud shear connectors subject to pull-out forces must be at least 10 times as long as their shank diameter.
(ii) All studs should have a head diameter at least 1.5 times their shank diameter.
(iii) All studs should have a head depth at least 0.4 times their shank diameter.
(iv) Studs welded to plates in tension must extend into the compression zone of the concrete or to the opposite steel face.

Squash Load Capacity for Columns
So long as the stud spacing and stud length criteria referred above are met, the squash load of a short or braced column may be calculated from Equation (38) of BS 8110: Part 1.

Shear Connector Capacity
The number of uniformly spaced connectors between the position of maximum

moment and the support or the point of contra-flexure may be calculated by assuming that the shear connector strength is 0.8 times the characteristic connector strength for connectors in the compression region of the beam and 0.5 times the characteristic strength for connectors in the tension region of the beam.

Partial or Incomplete Shear Connection

The number of shear connectors provided between the critical section and the support or the point of contra-flexure may be reduced as long as the force generated in the steel plate is assumed equal to the reduced connector force. The number of connectors provided in this way should be at least 40% of that required to transmit the full force in the steel plate. For any reduction in shear connection it should be established that the connectors have sufficient ductility to accommodate the slip that would occur between the steel and concrete without any loss of load capacity.

Stiffness

The stiffness of DSC elements subjected to lateral loading may be calculated by assuming that both the steel and concrete behave in an elastic manner, although concrete below the elastic neutral axis should be assumed to be cracked and not included in the calculations. This will apply to sections where the degree of shear connection between the concrete core and steel plate is at least 60%.

Ultimate Strength of Sections Subjected to Axial Load and Bending

The design interaction charts similar to those given in BS 8110: Part 3[(12)] may be constructed and used.

Detailing

The following tentative detailing rules are proposed:

(i) Stud diameter
 The stud diameter should not be greater than:
 (a) 2.5 times the steel plate thickness for studs welded to the compression face;
 (b) 2.0 times the steel plate thickness when welded to the tension face.

(ii) Minimum concrete cover to the side of the studs should be greatest of:
 (a) 50 mm;
 (b) the aggregate size; and
 (c) the stud size.

(iii) Maximum connector spacing must be less than:
 (a) four times the stud height;
 (b) three times the thickness of the slab; and
 (c) 0.6 m.

(iv) Minimum connector spacing must be greater than:
  (a) 1.5 times the stud diameter plus the maximum aggregate size; and
  (b) the minimum safe welding space.

(v) For long studs acting as links the longitudinal spacing must be less than 0.75 times the element thickness.

CONCLUDING REMARKS

This paper has summarised the design recommendations based on studies on steel-concrete-steel sandwich elements carried out at University of Wales College of Cardiff. An extensive study funded by the Commission of the European Communities is currently in progress at the Steel Construction Institute, with the continuing participation of the University of Wales College of Cardiff.

REFERENCES

1. Narayanan R., Wright H.D., Evans H.R. and Francis R.W. 'Double Skin Composite Construction for Submerged Tube Tunnels', Steel Construction Today (1987), 185-189.
2. Narayanan R., Wright H.D., Evans H.R. and Francis R.W. 'Load Tests on Double Skin Composite Girders', *Proceedings of the International Conference on Composite Construction*, Hennikar, New Hampshire, USA, (1988).
3. Tomlinson M., Chapman M., Wright H.D., Tomlinson R. and Jefferson A. 'Shell Composite Construction for Shallow Draft Immersed Tube Tunnels', *International Conference on Immersed Tube Tunnel Techniques organised by the Institution of Civil Engineers*, Manchester, (April 1989).
4. Oduyemi T.O.S. and Wright H.D. 'An Experimental Investigation into the Double Skin Sandwich Beams', Journal of Construction Steel Research, London, Vol 14 (1989), pages 197-220.
5. Wright H.D., Oduyemi T.O.S. and Narayanan R. 'Double Skin Composite Compression Elements', *Proceedings of the International Conference on Steel and Aluminium Structures*, Singapore, (May 1991).
6. Wright H.D., Oduyemi T.O.S. and Narayanan R. 'Full Scale Tests on Double Skin Composite Elements', *Proceedings of the Third International Conference on Steel-Concrete Composite Structures*, Fukuoka, Japan, (September 1991).
7. Wright H.D., Oduyemi T.O.S. and Evans H.R. 'The Design of Double Skin Composite Elements', Journal of Constructional Steel Research 19 (1991) p.111-132.
8. Wright H.D., Oduyemi T.O.S. and Evans H.R. 'The Experimental Behaviour of Double Skin Composite Elements', Journal of Constructional Steel Research 19 (1991) p.97-110.
9. BS 8110:Part 1, 'Structural Use of Concrete, Code of Practice for Design and Construction', British Standards Institution (1985).

10. BS 5400: Part 5, 'Steel Concrete and Composites Bridges, Code of Practice for Design of Composite Bridges', British Standards Institution (1979).
11. BS 5950:Part 3:Section 3.1, 'The Structural Use of Steelwork in Building', Code of Practice for Design of Simple and Continuous Composite Beams, British Standards Institution (1990).
12. BS 8110:Part 3, 'Structural Use of Concrete, Design Charts for Singly Reinforced Beams Doubly Reinforced Beams and Rectangular Columns', British Standards Institution (1985).

38 AN EXPERIMENTAL STUDY ON THE LOAD CARRYING CAPACITY OF COMPOSITE SLABS WITH A REINFORCING LATTICE USED AS CONNECTORS

T.S. Arda and C. Yorgun
Istanbul Technical University, Turkey

The paper describes an experimental study in the positive bending moment zone, to investigate the elastic and plastic behaviour of composite slabs where the adherence between the concrete and the steel sheeting is obtained by welding a reinforcing lattice on the profiled steel sheeting. The compatibility between theoretical and experimental results for the ultimate and serviceability limit states has been evaluated and some recommendations are proposed.

INTRODUCTION

During recent years, there has been an increased use of composite slabs made with profiled steel sheeting in floor construction.

As well known, when the steel deck can provide composite interlocking with the concrete, it is capable of performing the dual role of functioning, as a form during the construction stage, and as positive reinforcement for the slab under service conditions. Thus, the only additional steel necessary in the slab is that required for shrinkage and in the case of continuous construction, to resist negative bending.

The mechanical interlocking which permit the steel deck to serve as tensile reinforcement and act compositely with the concrete, is realized either by the shear transferring devices, the geometry of the steel sheet profile or with the combination of both.

In most cases, the composite action between the steel sheeting and concrete is totally provided by the mechanical shear transferring devices[1]. These are:

- Indentations or/and embossments rolled into the deck
- Stud connectors welded on supports
- Reinforcing steel lattice welded on steel sheeting.

In Turkish market, the steel deck surfaces are commonly smooth, without any embossment and indentations. For this reason, the interlocking between steel and concrete is sometimes provided by welding a reinforcing lattice on the profiled sheet. Since the quality of connection provided by this welded steel lattice was not exactly known, it was a necessity to perform tests on such composite slabs. This welded lattice is assumed to assure a perfect adherence between the two materials and, on the other hand, also to function as a load distributing reinforcement in transverse direction (Figure 1).

TEST SPECIMENS AND LOADING SYSTEM

This paper describes an experimental study in which a preliminary series of 6 tests and a main series of 9 tests are carried out, in the positive bending moment zone, to investigate the elastic and plastic behaviour of such composite slabs where the adherence between the concrete and the steel sheeting is obtained by welding a reinforcing lattice on the profiled steel sheeting(2).

The profiled steel sheetings used in test specimens are galvanized and display three different thicknesses (t=0.75 mm, t=1.00 mm and t=1.20 mm), same widths and depths of 860 mm and 27.5 mm, respectively (Table 1.).

The test specimens with a span of 300 cm, were simply supported at both ends and the test frame was designed to apply two concentrated loads at one thirds of their span lengths (Figure 2). The load P applied to the test specimens was static in character and was increased in steps starting from zero up to the level where the failure mechanism was observed. At each increment of the loading, the deflections at the mid-span (L/2) and at one thirds of the span (L/3) along with the strains at various locations of the steel sheeting and the concrete, were both measured.

The experiments were conducted at the Istanbul Technical University's Structural Laboratory.

TEST RESULTS

In evaluating the results of the preliminary series of tests, a method of calculation has been developed for the selection of a suitable steel reinforcing lattice welded to the folded steel sheeting to provide shear connection. The reinforcing lattice

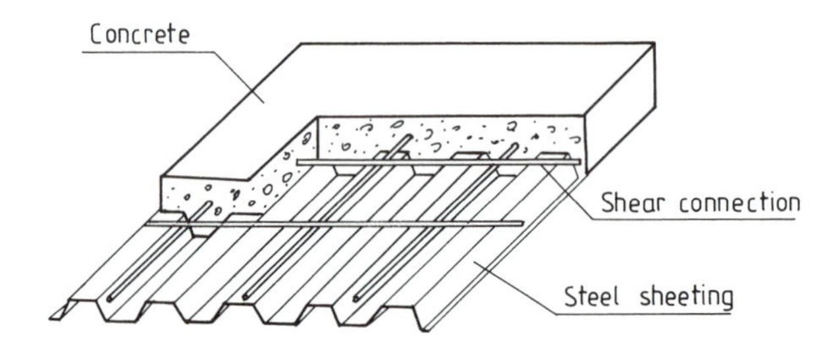

Figure 1. Typical Steel Sheeting-Concrete Slab Floor System

Table 1. Summary of the main series' specimens properties

Slab Element	Shear Connection Lattice	Depth of Slab Element (Measured)	Compressive Strength of the Concrete	Yield Strength of the Steel Sheeting
	mm	cm	daN/cm^2	daN/cm^2
3.A-0.75 3.B-0.75 3.C-0.75	Ø 4.5/150 (150.250.4,5.4,5)	10.5 10 10	300	2850
4.A-1.00 4.B-1.00 4.C-1.00	Ø 4.5/150 (150.250.4,5.4,5)	11 9.5 10	450	2400
5.A-1.20 5.B-1.20 5.C-1.20	Ø 5/150 (150.250.5,0.5,0)	10 10.5 10	350	2850

Figure 2. Overall View of Test Arrangement

designed using this method, shear-bond failure has never been observed in the main series' test specimens before the flexural failure.

The main purpose of the present experimental study is to investigate ultimate limit states of the composite slabs in positive bending moment zone.

The load carrying capacity of the tested slabs was first analysed employing elastic and plastic design methods to explore their suitability in interpreting experimental results. In the elastic design method, the values of stresses were determined assuming triangular stress distribution in the cross section and the calculations were carried out as those of a rectangular concrete section with an effective slab depth of d_s.

The experimental values of bending stresses for the concrete and the steel sheet, σ_{bd}, σ_{sd} were compared with those determined theoretically, σ_{bt}, σ_{st}. It is observed that for the concrete, the σ_{bd}/σ_{bt} ratios for all of the tested slabs are always less than unity, only in a few cases, equal to 1. This means that the calculations made with the elastic method will give results that are on the safe side for the concrete. Contrarily, in the profiled steel sheeting, the experimental stress values measured in test specimens exceed theoretical stress values calculated with the elastic method when they reach at 40 percent of the yield stress. Since the allowable stress in the elastic method is equal to 0.60 σ_{ys}, it can be clearly seen that the elastic method is not in agreement with the experimental results.

On the other hand, when the variation of the strain values is examined, for each load increment along the depth of the steel sheeting, an approach to rectangular strain-distribution corresponding to the yielding of steel, can be easily seen.

This feature in strain distribution can also be observed in stress distribution specifically and at ultimate load level, stress distribution along the depth of the steel sheeting, is clearly a rectangular block diagram (Figure 3).

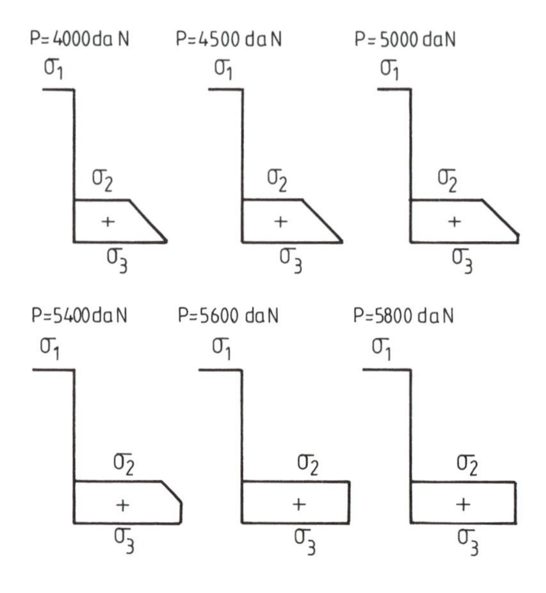

Figure 3. Stress Variation for the Test Specimens

Thus, it can be concluded that the plastic design method was more suitable than the elastic one.

If a rectangular stress-distribution is accepted (Figure 4) as in EC4 (3) and ASCE[(4)], the location of the neutral axis can be found as,

$$y = \frac{\alpha_a\ \sigma_{ys}\ A_s}{\alpha_b\ \sigma_{br}\ b} \le \begin{cases} d_0 \\ d_s/2 \end{cases}$$

And then, the ultimate moment of the slab in positive bending moment zone is

$$M_u = \alpha_a\ \sigma_{ys}\ A_s \left(d_s - \frac{y}{2} \right)$$

Where, A_s is the cross-sectional area of the steel deck, σ_{ys} is the yield strength of the steel, σ_{br} is the compressive strength of the concrete, d_s is the effective slab depth which is defined as the (distance from the extreme concrete compression fiber to the centroidal axis of the steel deck's full cross section), d_o is the depth of the concrete above the top corrugation of the steel deck and, b is the width of the slab.

The theoretical ultimate values are fully conform with the experimental ultimate values if factors of α_b=0.80 and α_a=1.00 are utilized (Table 2).

On the other hand, comparisons are made concerning the serviceability limit states. In transforming the concrete cross-section to an equivalent steel cross-section, the width of the concrete part is divided by 2n, in which $n=E_s/E_b$ is the ratio of the modulus of elasticity of steel to concrete.

To calculate the moment of inertia different assumptions were considered:

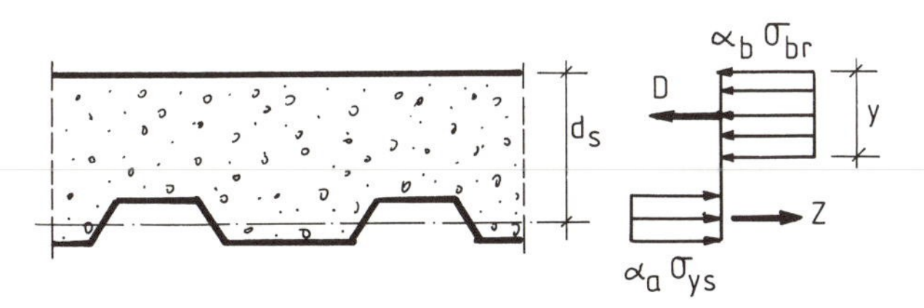

Figure 4. Stress Distribution in the Cross-Section of a Composite Slab

Table 2. Theoretical and experimental ultimate load values

Slab Element	P_{ud}	P_{ut}	P_{ud}/P_{ut}
	daN	daN	---
2A-0.75	3500	3130	1.12
2B-1.00	4100	3566	1.15
2C-1.20	5800	5354	1.08
3A-0.75	3600	3300	1.09
3B-0.75	3500	3100	1.13
3C-0.75	Adherence	3100	---
4A-1.00	4800	4092	1.17
4B-1.00	3500	3380	1.04
4C-1.00	3800	3618	1.05
5A-1.20	5800	5354	1.08
5B-1.20	6400	5710	1.12
5C-1.20	5800	5354	1.08

i) First, moments of inertia being obtained separately for the cracked and uncracked cross-sections, (see EC4 and ASCE) [3,4], their average was used.
ii) Secondly, only the concrete slab located upon the steel sheeting profile was taken into account, without giving any consideration to the location of the neutral axis.

From the practical point of view, deflections are important only in between the two following application limits:

1) The design load P^{i}_{ut} obtained by dividing the theoretical P_{ut} with safety factor 1.7 as required in TS.4561[5].
2) Deflection limit L/300, which is equivalent to P_{us} loads, as required in TS.648[6].

As in these limits, the two methods described above give similar results, the second one which is simpler is concluded to be more appropriate for practical purposes (Table 3).

Table 3. Measured and computed deflections at P^i_{ut}

Slab Element	P_{us}	P^i_{ut}	f_e	f_t	f_e/f_t
	daN	daN	mm	mm	---
2A-0.75	1950	1841	9.20	10.44	0.881
2B-1.00	2250	2098	8.30	10.47	0.793
2C-1.20	2300	3149	14.90	14.19	1.050
3A-0.75	2100	1941	8.95	9.77	0.916
3B-0.75	1750	1824	10.70	10.66	1.004
3C-0.75	1850	1824	9.60	10.66	0.901
4A-1.00	3100	2407	5.65	8.63	0.655
4B-1.00	2340	1988	7.65	11.06	0.692
4C-1.00	2550	2128	7.00	10.17	0.688
5A-1.20	2300	3149	15.10	14.19	1.064
5B-1.20	2460	3359	14.50	13.14	1.104
5C-1.20	2300	3149	15.60	14.19	1.100

CONCLUSIONS

The results of the experimental research permitted to conclude that the plastic design method was more suitable than the elastic one.

When the plastic design method is used, the theoretical ultimate values fully conform with the experimental ultimate values if factors values α_b=0.80 and α_a=1.00 are utilized.

The accordance between the experimental and the theoretical deflections is less remarkable compared to the results reached in ultimate loads values. However, the best correlation is obtained with two methods of calculations in which either the average moment of inertia of cracked and uncracked sections, or only the moment of inertia of the concrete located upon the steel deck are used. The second approach which is simpler, is judged to be more appropriate for practical purposes.

The general evaluation of the test results shows that, for the cases where, t/d ratio (thickness of the steel sheeting/nominal depth of composite slab) is smaller than, or equal to 1 percent, the ultimate limit state is more significant than the serviceability limit state.

REFERENCES

1. Janss J. 'Etude D' une Liaison Acier-Béton Pour Planchers Mixtes Avec Coffrage Collaborant', Centre de Recherches, CRIF, Bruxelles, (1982).
2. Yorgun C. 'The Behaviour and the Load Carrying Capacity of the Composite Slabs with the Reinforcing Lattice Used as Connectors' (in Turkish), phD Thesis, Istanbul Technical University, (1992).
3. Eurocode No 4. 'Régles Unifiées Communes pour Les Conctructions Mixtes Acier-Béton', Commission des Communautés Européennes Rapport EUR-9886, (1985).
4. ASCE. 'Specifications for the Design and Conctruction of Composite Slabs', (1984).
5. TS 4561. 'Turkish Standard', (1985).
6. TS 648. 'Turkish Standard', (1980).

PART SEVEN

STEEL STRUCTURES

39 EXPERIMENTAL STUDY ON BUTT-WELDED STEEL BEAMS WITH HALF NPI CHORDS AND RECTANGULAR OPENINGS

T.S. Arda and G. Bayramoğlu
Istanbul Technical University, Istanbul, Turkey

This paper describes an experimental study on steel beams with rectangular openings. In the paper, the theoretical ultimate loads and deflections of the tested beams have been calculated by using different approaches and compared with the experimental results. The compatibility between the theoretical and experimental results for the ultimate limit states and serviceability limit states has been evaluated.

INTRODUCTION

In Turkey, it is not easy to find great sizes of rolled profiles. They are not produced by Turkish steel-works and their importation takes time. On the other hand, they are generally very heavy. That is why fabricated beams are often used by Turkish constructor.

Due to the advantages they present, structural elements including openings in their webs are used in various applications abundantly. Despite the fact that they require an extra amount of workmanship, openings are almost a requirement in modern buildings where different installations have an ever-increasing importance. Additionally, due to the production techniques, the increasing distance between the flanges results in an increase of moment of inertia; which in turn causes an increased strength and a lesser amount of deformation when compared with a solid web structural element of equal cross-sectional area.

The openings inside the structural elements can be in various shapes such as hexagon, octagon, rectangle, circle or ellipse. This paper describes an experimental study on steel beams with rectangular openings(1). The aim of this study was to determine the ultimate and serviceability limit states of these structural elements which were used in constructions, but not tested for the time being.

TEST PROCEDURE

NPI 200 rolled profiles were cut longitudinally by oxygen torch into two equal parts. The beam specimens were then constructed by butt welding 200 mm high intermediate plates to these half NPI shaped chords in a way to form rectangular holes. A total of twelve beams were tested in three groups with 40 cm, 50 cm and 60 cm opening step lengths (Figure 1), half of them having a span of 4 m and the other half 5 m. They were all of the same depth. The experiments were conducted at the Istanbul Technical University's Structural Laboratory.

The steel material used in the experimented beams is nominally St. 37. However, the tension tests performed on it have given an actual experimental value of 2300 daN/cm² for the yield stress. This value was used for the calculations.

The tested beams were simply supported at both ends and were subjected to a concentrated load applied at the midpoint of the span. Lateral buckling was prevented by a trussed equipment mounted on the top chord of the beam by unfitted bolts.

In order to ascertain the structural stability of the experimented beams, vertical constructive stiffeners were placed at the two supports, at the central point where the load was applied and at the two arbitrary locations between those points closer to the supports.

LOADING AND MEASUREMENTS

The load P applied to the beams was static in character, and was increased in steps starting from zero up to the level where the failure mechanism was

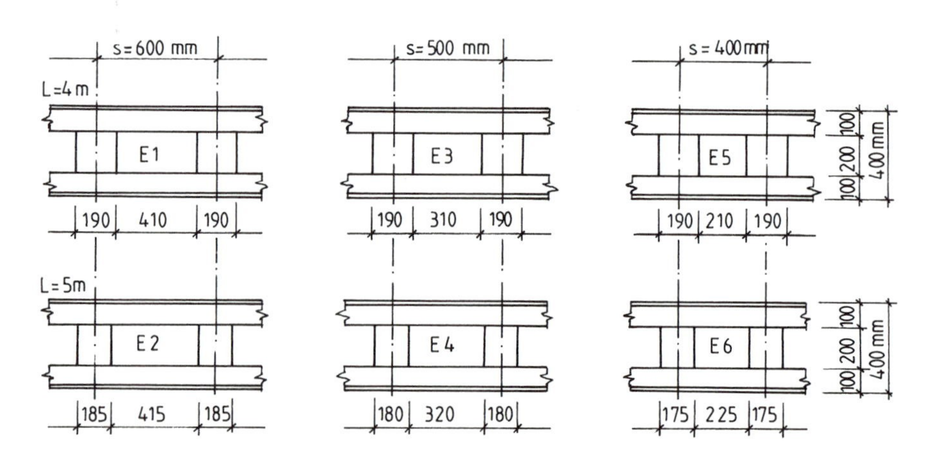

Figure 1 Types of tested beams

observed. At each step of the loading, both the vertical deflections at the mid-span by mechanical comparators with a sensitivity of 1/100 mm and the strains at various locations of the beam by strain-gauges were measured.

EVALUATION OF THE RESULTS

Ultimate Limit States: It is clear that a structural element can exhibit different ultimate limit states under various combinations of internal forces. Different ultimate loads can be calculated relating with the combinations of internal forces and their interaction.

The theoretical ultimate loads of the tested beams have been calculated by using different approaches and have been compared with the experimental ultimate loads.

i) The Mechanism Method and the Shear Panel Mechanism: From the examination of tested beams, it is observed that the shear panel mechanism is encountered in the two openings situated on both sides of the beam mid-span, by the formation of eight plastic hinges (Figure 2). Using the mechanism method, it is possible to calculate the ultimate load P_{1T} (see Table 1, column 3).

ii) The Panel Mechanism Including the $M_u + N_u$ Interaction in the Chord Cross Sections: A more accurate way of determining the beam ultimate load by this method is to use M_u which includes the axial force and shearing force effects in the formation of plastic hinges in the beam chords instead of M_p. Using the panel mechanism method, it is possible to calculate the ultimate load P_{2T} (see Table 1, column 4).

iii) The Panel Mechanism Including $M'_u + N'_u + V_c$ Interaction in the Chord Cross Sections: In addition to the axial force and bending moment effects on the beam chords, there also exists the shearing force (Figure 3). According to Turkish Standards[2] the shearing force V_f was taken into account by a reduction in the chord web thickness. Using this method, it is possible to calculate the ultimate load P_{3T} (see Table 1, column 5).

iv) The Simplified Method, Using Directly the Force-Couple Acting on the Beam Chords: In this method, the shearing force and the bending moment actions

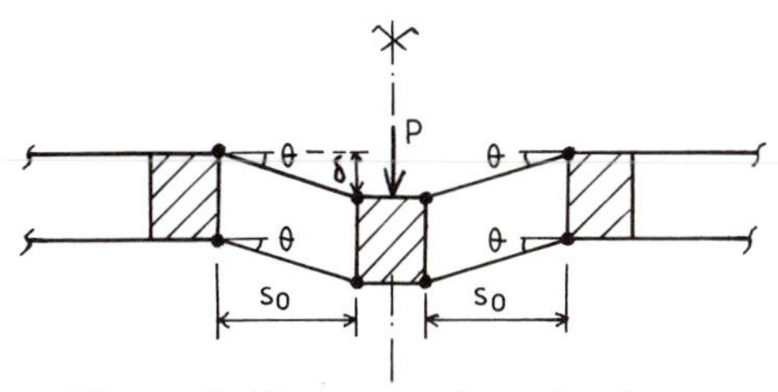

Figure 2 Shear panel mechanism

Table 1. Comparison between experimental and theoretical ultimate capacities

Name of the Beam Specimen	Experimental Ultimate Loads	Theoretical Ultimate Loads (P_t)					Experimental Ultimate Load / Theoretical Ultimate Load				P_e/P_t
	P_e	P_{1T}	P_{2T}	P_{3T}	P_{1M}	P_{2M}	P_e/P_{1T}	P_e/P_{2T}	P_e/P_{3T}	P_e/P_{1M}	P_e/P_{2M}
	x10³ daN	x10³ daN									
E11	11.5	15.58	9.03	8.58	13.45	14.04	0.74	1.27	1.34	0.86	0.82
E12	13			8.47			0.83	1.44	1.53	0.97	0.93
E21	10.5	15.39	7.77	7.46	10.76	11.22	0.68	1.35	1.41	0.98	0.94
E22	13			7.28			0.84	1.67	1.79	1.21	1.16
E31	12	20.60	9.89	9.27	13.45	14.15	0.58	1.21	1.29	0.89	0.85
E32	12						0.58	1.21	1.29	0.89	0.85
E41	9.5	19.96	8.28	8.03	10.76	11.29	0.48	1.15	1.18	0.88	0.84
E42	10						0.50	1.21	1.25	0.93	0.89
E51	12	30.41	10.80	10.19	13.45	14.33	0.39	1.11	1.18	0.89	0.84
E52	11.83						0.39	1.10	1.16	0.88	0.83
E61	11.67	28.38	8.94	8.46	10.76	11.41	0.41	1.31	1.38	1.08	1.02
E62	13			8.32			0.46	1.45	1.56	1.21	1.14
Characteristic Ratios (X_k)							0.30	1.02	1.05	0.76	0.73

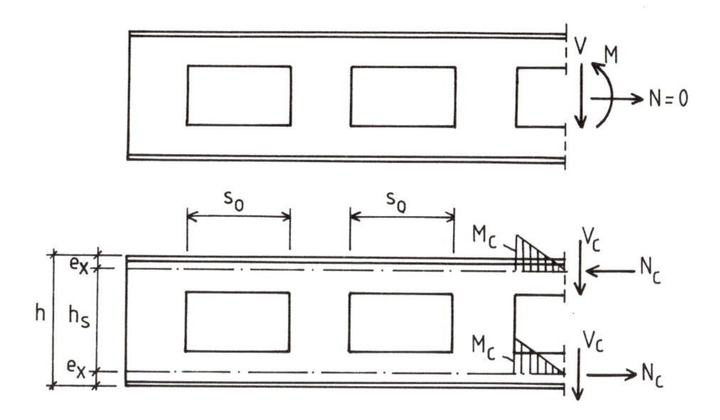

Figure 3 Internal forces in the beam chords

in chords are neglected and only the chord axial force, which is obtained by dividing the beam bending moment M to beam moment arm h_s, is taken into account. Using the simplified method, it is possible to calculate the ultimate load P_{1M} (see Table 1, column 6).

v) A Method Based on the Plastic Modulus of the Beam: In this method the first step is determination of the factored mean plastic modulus of the beam. This plastic modulus of the beam can be given as:

$$Z_m = \frac{l_p}{s} Z_1 + \frac{s_o}{s} Z_o$$

where Z_o is plastic modulus of a section with an opening, Z_1 is plastic modulus of its full section, l_p is the length of intermediate plate, s_o is the length of the opening and s is the opening step length. Using this method, it is possible to calculate the ultimate load P_{2M} (see Table 1, column 6).

To obtain the experimental/theoretical ultimate capacity ratios (P_e/P_t), the experimental ultimate loads of the twelve tested beams were divided by the theoretical ultimate loads which were determined by the five different methods mentioned above (Table 1). The characteristic ratios (X_k), with a 5% probability of error bound, were calculated for each theoretical method by statistical evaluation. These evaluations revealed that the best correlation between experimental and theoretical results was reached by the theoretical ultimate load P_{2T} (P_e/P_{2T} is 1.02, very close to unity).

This indicates that the theoretical results calculated by the panel mechanism including the $M_u + N_u$ interaction best reflects the experimental results.

The second best prediction was achieved by a characteristic ratio of 1.05, by P_{3T}, the method which additionally included the shearing force effect.

The traditional P_{1T} theoretical ultimate load which only take the bending moment into account is much below the acceptable range, with its characteristic ratio of 0.30. The other two methods which give P_{1M} and P_{2M} and which relate the theoretical ultimate load to beam bending in a simple way achieve relatively better results, with characteristic ratios of 0.76 and 0.73 respectively. This last comparison means that it is possible to obtain accurate results by multiplying the results of P_{1M} method, which is the simplest, by a factor of 0.76; at least for preliminary design calculations.

On the other hand, attempts to obtain the ultimate load of beams with openings by elastic methods were also investigated. Vierendeel beam and castellated beam elastic solutions were used for this purpose. Comparisons showed that these elastic methods can not be qualified as economical.

Serviceability Limit States

In addition to the ultimate limit states of the tested beams, their serviceability limit states were also investigated from the point of view of displacements. In this investigation, serviceability limit for the deflection has been taken equal to: L/300 in accordance with Turkish Standards(3). In the above ratio, L is the beam span.

The measurement related to the deflections made in the experiments have been compared with the values calculated theoretically by using various approaches:

i) The solution of Vierendeel beams with matrix-displacement method, with the aid of a computer,
ii) The simple beam simulation, assuming the beam with openings has only two chords as its cross section,
iii) A method which includes the shear deformations and uses the factored average of solid and hollow parts as the moment of inertia of the beam.

The experimental (P_{ef}) and theoretical (P_{tf}) loads that lead to the same serviceability limit state have been calculated with each of these three methods (Table 2). Then, the ratios of experimental and theoretical values (P_{ef}/P_{tf}) thus obtained for the twelve tested beams were examined and characteristic ratios (X_k) corresponding to a 5% probability of error were determined.

The ratios f_e/f_t, obtained by dividing the experimental values of deflection to the theoretical deflections calculated by the three methods mentioned above for each value of loading, have been examined. This examination, reveal the fact that the f_e/f_t ratio keeps on almost constant value from the beginning of the experiments till the approach of their end. Towards the end of the test, the ratio values are offset from the constant due to the shear panel mechanism involved. However, the cases when the ratios are offset from the constant value are beyond the serviceability limits of the beam. This enables the designer to use one of the theoretical results by multiplying it with a constant factor.

Table 2. Comparison between experimental (P_{ef}) and theoretical (P_{tf}) loads that leads to the same serviceability limit state

Name of the Beam Specimen	Simple Beam Allowance		Method Involving the Shear Deformations		Vierendeel Beam Assumption	
	L/300		L/300		L/300	
	P_{ef} x10³daN	P_{ef}/P_{tf}	P_{ef} x10³daN	P_{ef}/P_{tf}	P_{ef} x10³da N	P_{ef}/P_{tf}
E11	22.04	0.42	15.76	0.59	9.59	0.97
E12		0.44		0.61		1.00
E21	14.10	0.56	11.20	0.71	8.10	0.98
E22		0.51		0.64		0.88
E31	22.04	0.40	17.11	0.51	12.45	0.70
E32		0.41		0.53		0.73
E41	14.10	0.46	11.81	0.55	9.28	0.70
E42		0.50		0.60		0.76
E51	22.04	0.45	18.43	0.54	14.39	0.69
E52		0.43		0.51		0.66
E61	14.10	0.57	12.40	0.64	10.11	0.79
E62		0.58		0.66		0.80
X_k (L=4m)		0.39		0.48		0.54
X_k (L=5m)		0.45		0.55		0.66
X_k (Total)		0.37		0.49		0.60

CONCLUSIONS

The general evaluation of the comparison between the experimental and theoretical ultimate capacity indicates that the recommended procedure of calculation when accuracy is required is the usage of mechanism method involving $M_u + N_u$ interaction, i.e, P_{2T} method. The calculation method to be preferred when rapid results are needed, is the method resolving moment into force-couple, or P_{1M} method.

On the other hand, none of the methods of theoretical calculation has lead to the serviceability limits exactly. But the theoretical deflection values obtained by the use of these methods, can be multiplied by appropriate constant factors to reach the experimental deflection values.

REFERENCES

1. Bayramoğlu G. 'Yarım NPI Başlıklı, Dikdörtgen Boşluklu, Küt Kaynaklı, Yanal Burkulması Önlenmiş Çelik Kirişlerin Davranışı ve Taşıma Gücü', PhD Thesis, Istanbul Technical University, Institute of Science and Technology, (1992).
2. TS 4561, 'Turkish Standards', (1985).
3. TS 648, 'Turkish Standards', (1980).

40 DESIGN OF SEMI-RIGID COMPOSITE BEAM-COLUMN CONNECTIONS

Y. Xiao, B.S. Choo and D.A. Nethercot
University of Nottingham, UK

Results from the third phase of a series of tests on composite beam-column connections are reported. It is shown that the behaviour of the different types of composite connection tested is influenced by a wide variety of parameters. A unified design approach for both the moment resistance and the rotation capacity is proposed.

Keywords: Composite connections, Partial strength, Semi-rigidity, Beam-to-column joint.

INTRODUCTION

Composite frames are now quite commonly used both in North America and Europe. The reasons are simple to explain as:

- the two materials are combined in a complementary manner to best utilise their structural properties.
- properly organised the form of construction is fast and therefore cost-effective.

Guidance on the use of composite structures is available in relevant codes of practice such as EC4(1) and BS 5950:Part 3(2). Composite action of the concrete floor acting with the steel beam and metal decking has been widely researched and is utilised when designing both the composite slab and the composite beam. However, design of the beam-to-column joints in composite frames is largely based on 'pin joint' approach due to the shortage of fully understood test data and more importantly lack of appropriate design approaches to explain the composite action's contribution to the joint behaviour. Several previous studies(3)(4) have, however, revealed that the composite floor has an important influence on the main measures of joint behaviour such as moment resistance and rotational capacity. Neglecting this contribution to joint properties may result in an over conservative assessment of the actual behaviour of braced frames. There is thus a need to use the true strength and deformation characteristics of composite connections rather than relying on traditional design methods that treat the connections as bare steel joints. It is now becoming accepted that composite beam-column connections are capable of resisting appreciable bending moments.

Providing this is accompanied by sufficient rotation capacity to permit internal force redistribution, then quasi-plastic methods of frame design become attractive.

One part of an ongoing project supported by the Building Research Establishment is reported herein. A number of joint specimens were designed and tested to assess the behaviour of commonly used beam-column joints that were expected to function in a semi-rigid, partial strength composite fashion. Key measures of performance, including the moment resistance, rotational stiffness and rotation capacity of connections, have been examined when systematic changes have been made to the main variables. The experimental results have indicated that the behaviour of composite beam-column connections exhibited significant differences from that of similar bare steel details. These changes illustrate the illogicality of using current bare steel joint design approaches to deal with composite connections.

Following the first two phases of the experimental work, a further seven beam-column composite connections have been designed and tested. Several special features have been incorporated, resulting from earlier experimental findings. Based on an appraisal of all the test data obtained, a unified design approach to calculate the moment resistance and rotation capacity for different forms of composite connection is proposed. This aims to provide a more rational and more economical design for composite frames incorporating the true semi-rigid and partial strength nature of the connections.

TEST PROGRAMME

The usual form of metal deck flooring system comprising a concrete slab supported by profiled metal decking sitting on steel beams was chosen for all the connections. The three stages of the experimental programme were arranged to complement one another so that the principal variables of the next stage specimens could be adjusted according to the findings of the previous tests. A total of 19 composite connections were tested. The test results of the first and second phase work covering 12 specimens have been described in earlier papers[5][6]. Full details of the setting-up and casting procedure are given in ref.6, which contains a description of the test rig, instrumentation and test procedure. The seven test results from the third phase, which have not been reported previously, will be presented here. Both cruciform (internal) and cantilever (external) arrangements as shown in Figure 1 (a) and (c) are included. The basic components for the joint are summarized as: beams were 305 x 165 x 40 and 457 x 191 x 82 UBs, columns were 203 x 203 x 50 UC in grade 43 steel; reinforcement was A142 mesh, supplemented with T10 and T12 rebars in most cases; PMF CF46 metal decking was used as bottom shuttering with 19mm welded through shear studs connected to the steel beams; ready-mix normal weight concrete was used with a design strength of $30N/mm^2$.

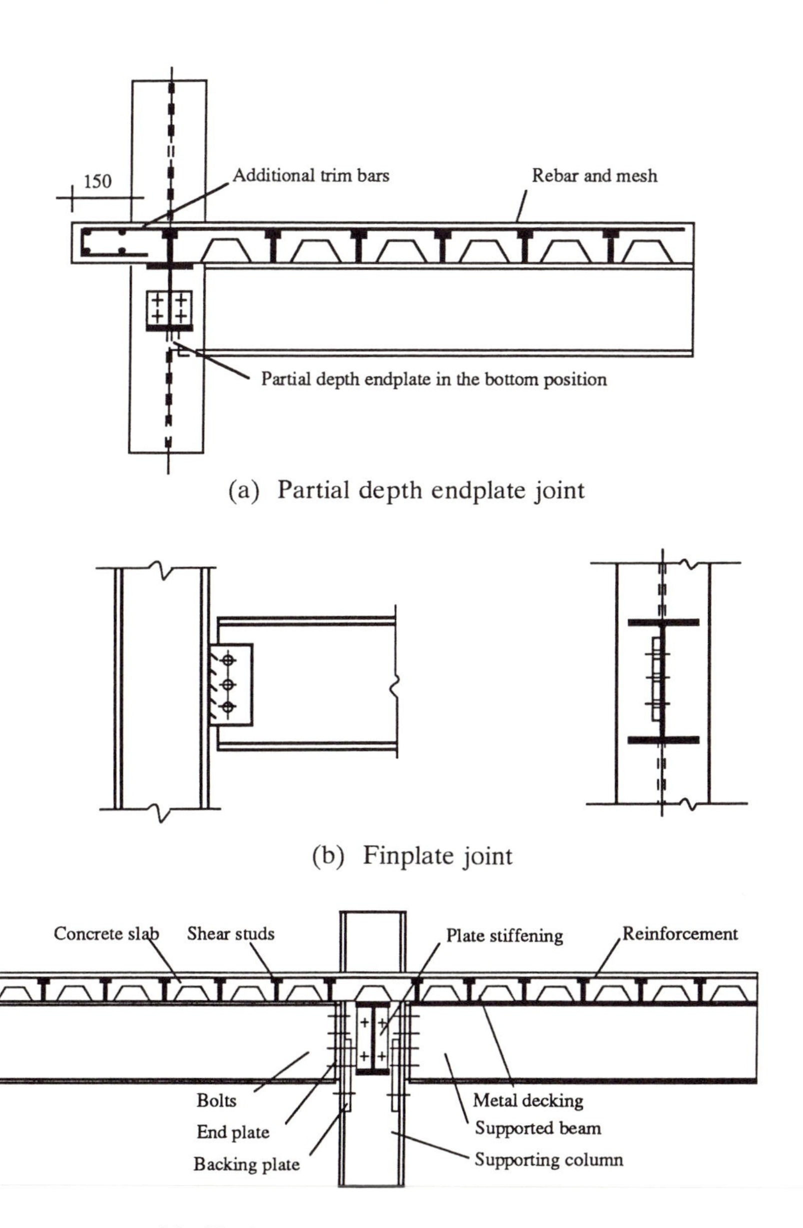

(a) Partial depth endplate joint

(b) Finplate joint

(c) Flush endplate joint with backing plate

Figure 1. Three types of joint and specimens and specimen construction

TEST RESULTS AND ANALYSIS

The seven connections tested included three types of steel detail as shown in Figure 1 (a), (b), (c), namely: partial depth end plate connected to the minor axis of the column; finplate incorporating a change of orientation of the metal decking and steel beam section size; flush end plate with backing plates on the flange of the column. A general description and list of the main test results for these seven specimens is given in Table 1.

Minor axis connections of the partial depth end plate
Four specimens have been tested with beams provided with partial depth endplates being connected to the minor axis of the supporting column. Specimens SCJ13 and SCJ14 were tested as cruciforms. Both connections had the same reinforcement ratio of A142 mesh plus 6 T10 and 2 T12 rebars (ratio of 0.8%). The results showed once again (as for previously reported tests SCJ8 and SCJ10) that suitable positioning of a partial depth endplate is crucial to the achievement of high stiffness and moment capacity of the connection. The most appropriate position is to locate the plate on the lower portion of the beam. Figure 2 shows that SCJ13 with the plate on the bottom of the beam has a larger initial stiffness and almost twice the resistance moment as that of SCJ14 with the partial depth plate at the top. SCJ13 exhibited no significant deformation in the joint area and failed due to excessive deformation of the slab and debonding of the concrete. Compared with the equivalent major axis connection (SCJ10), the failure mode of the minor axis connection was transferred from column web buckling to the slab. The stiffness and moment capacity was actually increased as the opportunity for compression zone deformation was effectively eliminated. SCJ14 failed due to buckling of the beam web (in the same way as SCJ8) as shown in Figure 3. The strength and stiffness of connections employing symmetrical partial depth end plates or flush end plates to the minor axis of the supporting column are usually increased as compared with equivalent major axis connections because the compressive force transfer effectively bypasses the column web. The connection will resist an appreciable moment and possess a high degree of rotation capacity if there is no significant moment transferred into the column. If, however, the joint must resist an unbalanced moment, then the minor axis connection is likely to be less efficient because of the low out-of-plane compressive strength of the column web. This point will be further discussed when presenting the test results for the edge joint specimens SCJ18 and SCJ19.

Composite edge joints
Two specimens SCJ18 and SCJ19 were designed as cantilevers with the object of investigating the anchorage situation of the reinforcement in the edge joint and the interaction with out-of-plane compressive strength of the column web. For SCJ18 the plate was placed at the bottom of the beam and for SCJ19 at the top. The experimental results of ref.7 have already indicated that poor anchorage

Table 1. Specimen Details and Test Results

Specimen	Joint Type	Specimen Shape	Reinforcement Ratio(%)	First Crack Moment (kN.m)	Ultimate Moment (kN.m)	Ultimate Rotation (mRad)	Maximum Rotation (mRad)	Failure mode*
SCJ13	Partial depth end plate (Minor Axis)	Cruciform	T10 Rebar & A142 Mesh (0.8%)	37.5	181.4	19.3	28.5	A
SCJ14	Partial depth end plate (Minor Axis)	Cruciform	T10 Rebar & A142 Mesh (1.0%)	22.5	89.8	34.4	39.4	F
SCJ15	Flush end plate (Major Axis)	Cruciform	T10 Rebar & A142 Mesh (1.0%)	37.5	185.5	23.2	27.7	A
SCJ16	Finplate Major Axis	Cruciform	T10 Rebar & A142 Mesh (1.2%)	30	224.7	27.1	43.5	H
SCJ17	Finplate Major Axis	Cruciform	T12 Rebar & A142 Mesh (0.5%)	15	97.4	44.1	48.2	H
SCJ18	Partial depth end plate (Minor Axis)	Cantilever	T10 Rebar & A142 Mesh (0.8%)	30	85.5	17.9	25.8	I
SCJ19	Partial depth end plate (Minor Axis)	Cantilever	T10 Rebar & A142 Mesh (0.8%)	10.5	60.8	41.6	46.7	I

A --- Excessive deflection of beams
F --- Buckling of beam web
H --- Web side plate twisting
I --- Anchorage failure of reinforcement

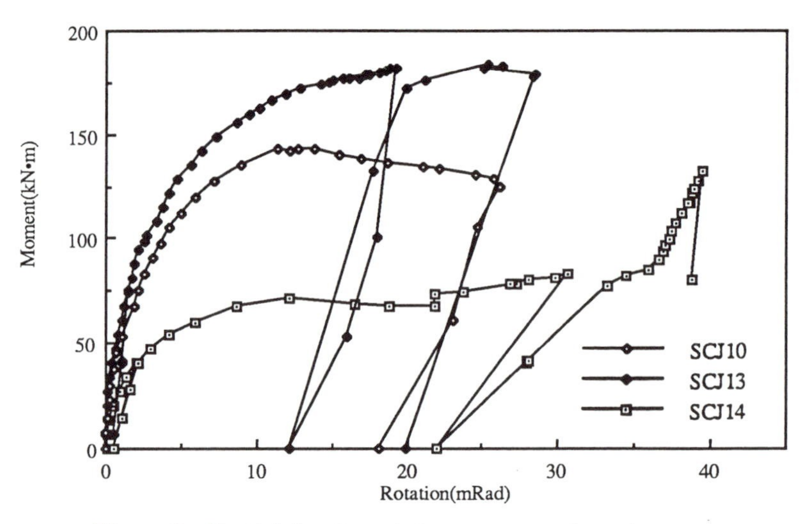

Figure 2. Partial depth endplate connected to the minor axis

Figure 3 Buckling of beam web for partial depth endplate

of the reinforcement was likely to prevent edge joints from developing high moment resistances. All four cantilever tests of ref.7 failed as a result of the concrete being pushed out from the slab behind the column thus causing loss of anchorage of the reinforcement. Much stronger details in terms of anchoring of the reinforcement in the column area were used this time. All the longitudinal bars were bent into the concrete and extra transverse reinforcement bars were added to the top and bottom of the slab at the back of the column. Trim bars (45°) were placed round the back of the column. This was found in ref.7 to give better control of the diagonal cracking. During the loading SCJ18 was very stiff until it exhibited a sudden failure. When the maximum load was reached, the specimen immediately lost stiffness and the load dropped down very fast. The failure initiated from cracking of the bottom of the slab at the back of the column and eventually tensile splitting of the concrete occurred. Anchorage of the reinforcement was lost due to failure of the slab and the failure was very brittle with a low residual strength. Yielding and large deformation of the column web was also detected although the load value achieved was not very high. The final moment strength was only 85.5 kN.m. A similar failure occurred for SCJ19 but with an even lower initial stiffness. The ultimate strength reached was 60.8 kN.m as a result of the failure of concrete slab. During loading the beam bottom flange of SCJ19 eventually came into contact with the web of the column. The strength of connection was increased and then failure of the slab occurred. Figure 4 presents the stiffness and strength characteristics for these two specimens. From these further tests on the edge connection, although suitable reinforcement design did increase the moment resistance somewhat, it was still difficult to prevent slab failure and loss of anchorage for the reinforcement as shown in Figure 5. The mode of failure was a very sudden and brittle one without prior warning and is clearly undesirable. The excessive deformation of the column web should also be noted. Although the failure load was just half that of the same joint detail when tested as cruciform type, early yielding occurred as unbalanced moment was transferred into the column web. Decreases of the moment resistance are inevitable under the influence of unbalanced moment to the minor axis connections.

It is theoretically possible to construct even stronger detailing of the reinforcement at the back of the column to strengthen the concrete slab and to take additional measures to stiffen the column web. The moment capacity will be increased and the brittle failure will possibly be eliminated. For simplicity, however, designing of edge joints and corner joints is more realistic if they are treated as simple connections due to the complexity of these special measures.

Flush endplate connection with backing plates

Usually, the tension zone of a bare steel joint design is very weak due to bending of the column flange. Its tensile strength then controls the design for the moment capacity of the joint. This is particularly true for the endplate joint when the beam is bolted to the flange of the column. Bolting plates to the back of the column flange in the tensile region can increase the stiffness and moment capacity of the joint[8]. When a concrete floor is present on top of the steel beam, the controlling

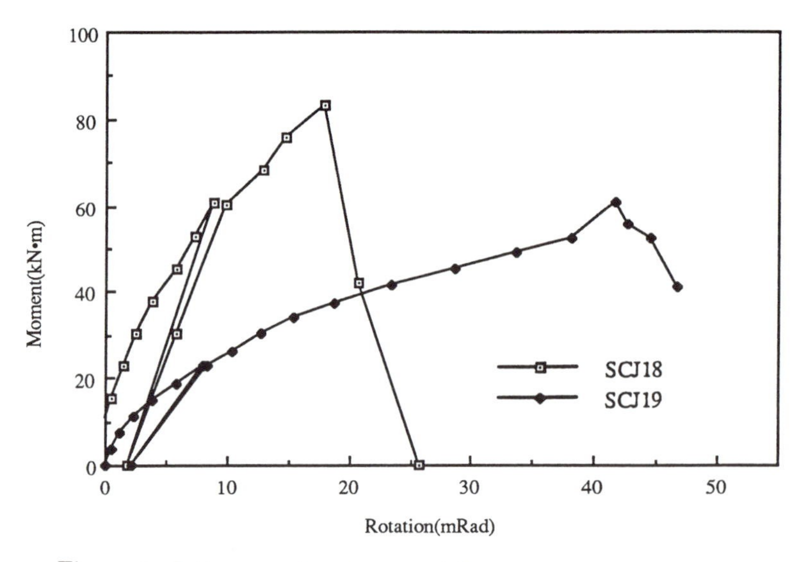

Figure 4. Influence of anchorage failure to moment-rotation of composite edge joint

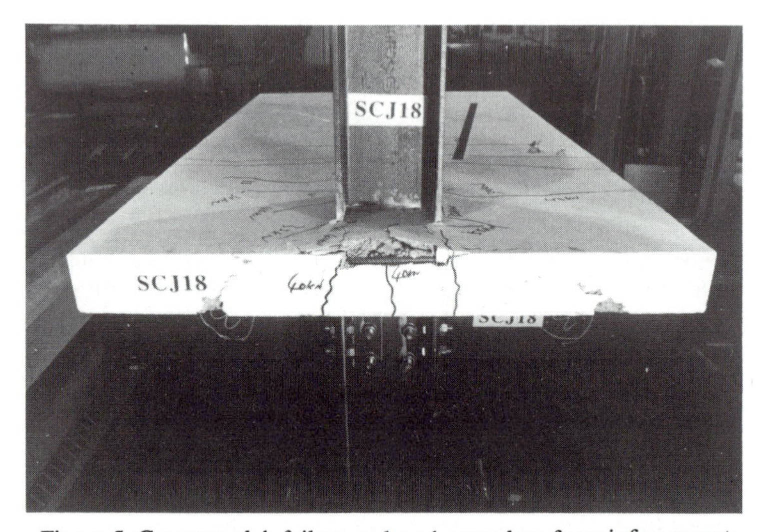

Figure 5 Concrete slab failure and anchorage loss for reinforcement

area of the composite connection is usually the compression zone of the column due to the restraint provided by the reinforcement to the tensile column area. Modes of failure for such composite connections will normally involve yielding of the reinforcement and compressive buckling of the lower portion of beam and/or column flange and web. The column web and flange can be stiffened by using welded web stiffeners. Flush endplate joints with orthogonal framed-in beams were also found to function as a kind of web stiffener in test SCJ7 in ref.6 when failure was caused by excessive compressive deformation of the column flange. One way of providing a stiffener to the column flange in the compressive zone is to use the concept of the backing plate to enhance the bending resistance so that the need for a welded web stiffener can be waived as indicated in Figure 1(c). Specimen SCJ15 was a flush endplate connection with backing plates on the column flanges. The test results indicate that this form of improvement for flush endplate joints is quite beneficial. Figure 6 shows an increased moment capacity, rotation capacity and ductility for SCJ15 when compared with specimen SCJ6 which had no web stiffening. Excessive deformation of the column flange was prevented as shown in Figure 7. When used with plate stiffening of the column web, backing plates can successfully improve the connection behaviour without the need for fitted and welded stiffening.

Finplate composite connection

A finplate connection with a much deeper steel beam section and increased slab depth was designed for test SCJ16. It was intended to increase the moment capacity for this type of very flexible connection. This type of joint is gaining popularity in this country due to its simple fabrication and ease of erection. Tests on bare steel finplates[9] have indicated a low stiffness and moment capacity with failure sometimes occurring prematurely by lateral torsional buckling. The shallow beam composite section with a light ratio of reinforcement tested as SCJ12 had already achieved an appreciable moment resistance; the composite slab for SCJ16 was then strongly reinforced with the aim of achieving higher stiffness and strength. The moment-rotation curve for the test is shown in Figure 8. Slip of the bolts in the horizontal direction at a fairly early stage is clear, causing loss of stiffness. The connection regained stiffness when the bolts came into bearing with the edges of holes and exhibited a larger ultimate rotation compared with equivalent end plate connections. The moment capacity was 224.7 kN.m, which is a high hogging region resistance. Debonding of the concrete from the metal decking was visible due to the large deflection of the beam. Eventual failure was by bearing of the bolts as the beam web moved towards the weld toe of the plate causing torsion of the finplate and web. The bottom flange of the beam finally touched the column flange.

SCJ17 was a finplate composite connection with the metal decking orientation the same as the beam direction. The aim was to check the contribution from the metal decking to the composite connection. Compared with the equivalent previous specimen SCJ12, the reinforcement ratio was decreased to 0.5% according to the pre-calculation of the moment capacity of the metal decking.

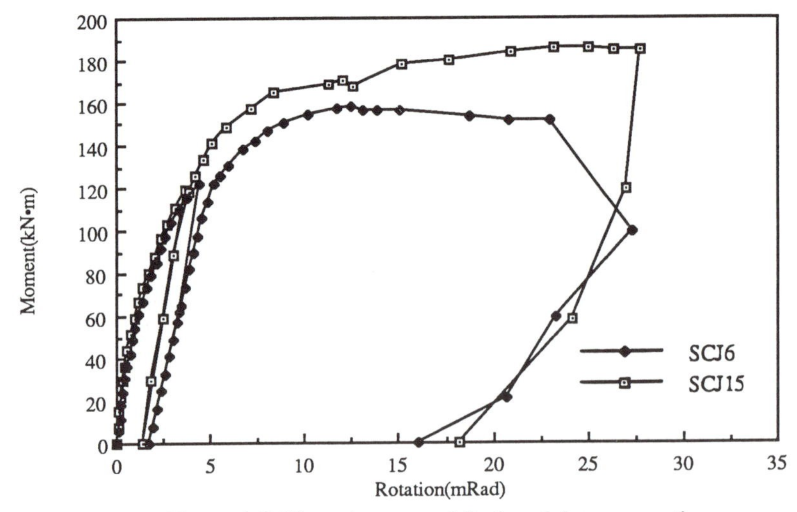

Figure 6 Stiffness increase of flush endplate connection with backing plate

Figure 7 Backing plates prevent excessive deformation of column flange

From the moment-rotation relation shown in Figure 8, it is clear that stiffness of the specimen was less due to the lower position of the metal decking compared with that of the mesh and rebar reinforcement. The final ultimate strength was similar to that of SCJ12 which means the metal decking can be regarded as equivalent reinforcement. Its influence on the initial stiffness of the connection should also be considered. The difference between a composite finplate connection and a bare steel finplate is that failure of the composite version starts from the bearing capacity of the last one or two bolt rows which must balance all the tensile forces. The premature horizontal torsional buckling caused by the eccentricity of the plate to the centreline of the beam web is avoided due to the restraint supplied by the concrete floor. Flexible joints like finplates can therefore produce high moment resistances when properly designed to function as composite connections. However, it should be noted that the contribution from the metal decking cannot always be relied upon, especially when it is arranged by butting sheets against a column rather than being continuous past the column.

UNIFIED DESIGN FOR THE MOMENT RESISTANCE AND ROTATION CAPACITY OF COMPOSITE CONNECTIONS

Calculation of moment resistance

Using basic concepts of the mechanics of force transfer within a composite connection, it is not difficult to construct a model for calculating moment resistance and full details for several different types can be found in ref.10. For the later derivation of a method to calculate rotation capacity, the main steps can be summarized with the aid of Figure 9 for a typical partial depth endplate connection as follows:

1. Determine F_s based on yield of the reinforcement
2. Calculate the available compressive capacity F_c
3. Determine the position of the neutral axis x_p
4. Decide whether any additional tensile resistance F_b is required from the bolts
5. Make any adjustments to the position of the neutral axis and the force components
6. Locate the centroid of either the total tensile force or compressive force, determine the lever arm and hence calculate the moment resistance M_u.

Different cases need slightly different treatment depending on the exact position of the neutral axis in the effective section and the different opportunities provided by different steel details.

The above approach assumes that the actual behaviour follows the assumptions of the stress block theory. In practice this may well require stiffening to prevent potentially weakening occurrences such as column web buckling, excessive column flange deflection in bending etc. Procedures to accommodate such complications into the basic approach have yet to be derived.

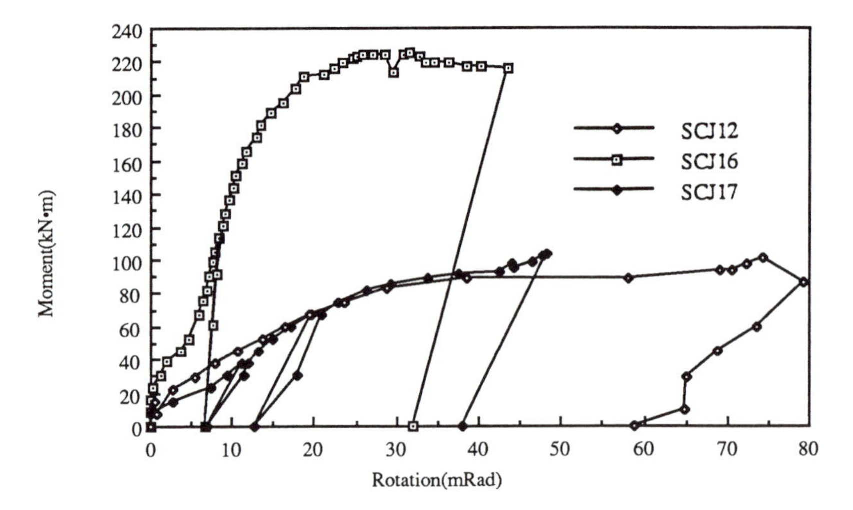

Figure 8. Influence of steel section and metal decking to composite finplate connection

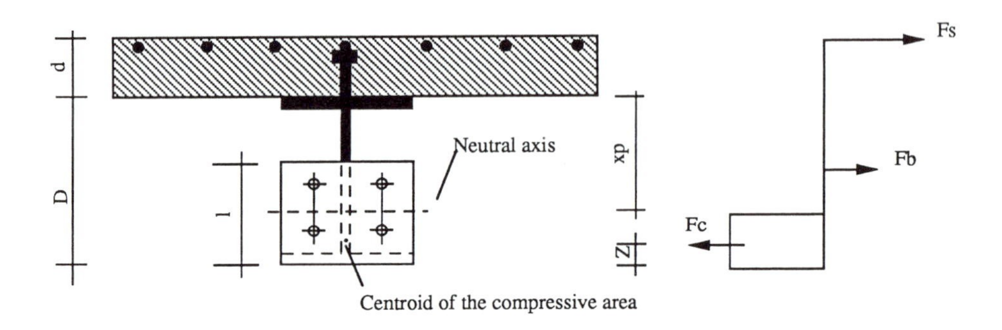

Figure 9. Partial depth endplate connection calculation section

Rotation capacity of composite connection

The numerous experimental results that are available indicate that composite connections can have appreciable moment resistances with various steel details, providing an appropriate reinforcement lay-out is selected for the column area. For frame design, it is also necessary for the composite connections to possess a certain degree of rotational capacity so as to permit internal force redistribution with the object of developing the full sagging moment capacity of the composite beams at mid-span. Examination of the test results for the composite connections has shown that the rotational capacities were significantly affected by changes to the variables considered in the various test series. The deformations of all the components in the joint will make contributions to the total value of the joint rotation. Major factors are plastic deformation of the reinforcement; elastic and plastic deformation of the column flange and web; elastic and plastic deformation of the steel beam section etc.. It is extremely difficult on the basis of the experimental and analytical work to justify a single deformation value for each component and then simply to add them together to get the general rotation capacity for the composite connection. Simplified calculation methods are needed to quantitatively assess the rotation capacity. A general procedure was proposed in ref.11 to calculate the joint rotation of endplate connections by considering mainly the elongation of the reinforcement and then deriving the rotation formulae based on full composite section depth. The elongation of reinforcement ΔL consists of two parts: deformation of the reinforcement in the plastic zone and deformation of the reinforcement in the remaining elastic region. The calculation formula can be written as:

$$\phi_{ult} = \frac{\Delta L}{D+D_r} \tag{1}$$

Where: D - depth of beam
D_r - distance from the top of the beam to the reinforcement

Both the concept and the calculation procedure is quite straightforward. But it requires some modifications after comparison against the experimental results. Some limitations are also present in the assumptions of the calculation. First, it neglects deformation of the compression zone which is quite substantial from many test results. Secondly, it assumes a fixed rotation point of the bottom flange of the beam which does not accord with the deformation behaviour of different steel details and is not consistent with the approach to calculate moment capacity. A modified calculation method based on this original procedure is therefore proposed to overcome the above shortcomings. Both assumptions and the calculation model will be unified with the resistance moment calculation. The following simplifying assumptions are made:

(a) Strain hardening of the reinforcement will be neglected
(b) Concrete tensile strain will be neglected

The elongation of the reinforcement in the plastic zone can be written as

$$L_1 = \alpha(p1 + p + \frac{D}{2}) \tag{2}$$

where p_1 = distance of the column flange to the first shear stud
p = pitch of shear studs
α is the reinforcement strain for the plastic zone

From test measurements of the strain distribution it is clear that the reinforcement is extensively yielded up to the second row of shear studs and so the plastic zone will be defined within the first two rows rather than to the first stud as in ref.11.

Elongation of reinforcement in the elastic zone is

$$L_2 = \beta\ (n-2)p_2 \tag{3}$$

where n is the number of shear studs needed to resist the longitudinal force
β is the average reinforcement strain in the elastic zone

The elastic zone will be considered as the remaining area up to the point of contraflexure of the composite beam. The tensile force in the reinforcement should be less than the total resistance force of n shear studs. Thus the number of shear studs will decide the length of the reinforcement.

Total elongation of the reinforcement is:

$$\begin{aligned} \Delta L &= \gamma(L_1 + L_2) \\ &= \gamma(\alpha(p_1 + p + \frac{D}{2}) + B(n-2)p_2) \end{aligned} \tag{4}$$

where γ is a steel detail factor which corrects for bolt slip.

The model covers not only the flush end plate(11) but also partial depth end plates, seating cleats and finplate joints. Suitable values for the above three parameters have yet to be derived for these details and will be proposed in a separate paper covering the stiffness and rotational capacity determination for composite connections.

As discussed before, the compressive deformation in the joint area is quite substantial and it is necessary to include this in the calculation. Location of the neutral axis position already forms part of the moment calculation of section 4.1. The rotation point is then assumed at the neutral axis for the calculation section so that the compressive deformation is naturally introduced. Calculation for the moment and rotation is then consistently unified together. From Figure 9, the rotational capacity can be written as:

$$\O_{ult} = \frac{\Delta L}{d_c + x_p} \tag{5}$$

Where dc is the effective depth of the concrete slab

x_p is the distance of neutral axis to the top of the beam; it can be found from section 4.1 moment calculation step 3 according to the different cases and different steel details.

CONCLUSIONS

Seven further composite joints were tested and results were analysed as a part of a programme of systematic study for composite connections in steel and concrete. Based on the experimental findings, a unified design approach for moment and rotation capacity was proposed. The following conclusions can be drawn from the work reported herein:

1) Minor axis connection of the partial depth end plate composite connection gives behaviour comparable to the major axis connections. Failure in the column area will be eliminated if there is no significant unbalanced moment transferred into the column.
2) Flexible finplate composite connections can produce appreciable resistance moment if properly design. The bearing capacity of the bolts in the lower portion of the finplate will control the final moment capacity of the connection.
3) Loose plates on the back of the compressive column flange zone of the flush end plate joint can effectively prevent excessive deformation of the column flange. Working together with the perpendicular plates connected to the column web, backing plates can stiffen the compressive zone of the connection which is usually very weak.
4) The provision of adequate anchorage to reinforcement in the edge connection is very difficult due to the slab around the column being pushed off. To avoid the need for complicated reinforcement detailing, a simple joint design is realistically suggested for the edge and corner joint.
5) The complex interplay of factors that control both moment and rotational capacities of composite connections can be simplified for calculation. A unified design approach for the resistance moment and rotational capacity is proposed.

ACKNOWLEDGEMENT

The financial support from the Building Research Establishment and assistance from Dr D.B. Moore throughout the research work are greatly acknowledged.

REFERENCES

1. Commission of the European Communities, *Eurocode No.4, Design of Composite Steel and Concrete Structures*, Draft. (October 1990)
2. BSI BS 5950: *Structural Use of Steelwork in Building Part 3:Design in Composite Construction*. (1990)
3. Zandonini R. *Semi-Rigid Composite Joints*. Structural Connections-Stability and Strength ed. R. Narayanan, Elsevier Applied Science, pp63-120. (1989)
4. Nethercot D.A. *Tests on Composite Connections*. Second International Workshop: Connections in Steel Structures, Behaviour, Strength and Design, Pittsburgh, Pennsylvania, U.S.A, (April 1991)
5. Xiao Y. Nethercot D.A. and Choo B.S. *Moment Resistance of Composite Connections in Steel and Concrete*. Proceedings of the First World Conference on Constructional Steel Design, Acapulco, Mexico. (November 1992)
6. Xiao Y. Nethercot D.A. and Choo B.S. *The Influence of Composite Metal Deck Floorings on Beam-Column Connection Performance*. Proceedings of 11th International Speciality Conference on Cold Formed Steel Structures, Missouri-Rolla, U.S.A. (October 1992)
7. Lam D. Davison J.B. and Nethercot D.A. *Semi-Rigid Action of Composite Joints*. The Structural Engineer, Vol. 68, No.24, pp489-99. (December 1990)
8. SCI report RT-216. *Rules for the design of backing plate connections-calibrations with additional experimental studies*. (1992)
9. Owens G.W. and Moore D.B. *Steelwork Connections*. The Structural Engineer, Vol. 70, No.3/4, pp37-46. (February 1992)
10. Xiao Y. Nethercot D.A. and Choo B.S. *Composite Connections in Steel and Concrete*. Progress report to Building Research Establishment, No.2 SR91024, Department of Civil Engineering, University of Nottingham, (July 1991)
11. Gibbons C. *Partial Strength Moment Resistance Connections In Composite Frames*. SCI report, No. SCI-RT-257, (April 1992)

41 OPTIMAL CONFIGURATION OF STEEL-SKINNED SANDWICH PANELS

C.M. Davies and A.J. Stevens
British Steel Technical, Rotherham, UK

This paper describes the development and use of a software package which can rapidly produce information on weight-optimised and cost-optimised configurations of steel based sandwich panels for particular applications. This fulfils a requirement for a preliminary design tool which can assess the viability of a sandwich solution, select the most appropriate skin and core materials and provide a first approximation to the optimum configuration.

INTRODUCTION

In a wide variety of applications there is a drive to reduce structural weight. In steel structures the use of higher strength steels would initially appear to be an effective method of achieving this goal. However there is a limit to the extent to which this can be done, since considerations of deflection and elastic stability are frequently the governing criteria rather than strength. Designers are therefore turning to lower density materials, such as aluminium alloys and fibre reinforced polymers.

Although steels have the highest Young's modulus of any commonly available engineering metal, they also have a relatively high density. In order to achieve weight reduction in steel structures it is necessary to employ forms of construction which take maximum advantage of the stiffness of steel.

Sandwich construction, in which stiff skins are attached to a low density core, has been used in the aerospace industry for 50 years and is increasingly being employed in other (principally transport) applications. The potential applications for steel-skinned sandwich panels have been studied and a number of promising possibilities identified. Such steel-based composites could take several forms;

- honeycomb core
- corrugated/dimpled sheet core
- polymeric foam core

For each of these types of sandwich construction there are a number of design variables (such as cell size, cell wall thickness, cell wall material, pitch of corrugations/dimples, foam material and foam density) in addition to the obvious design variables of skin steel grade, skin thickness and core depth. Once these variables have been fixed, techniques such as finite element analysis(1) can be used to validate the proposed design. However, there is a need for a preliminary design tool which can be used to:

(a) assess the viability of a steel-based composite compared with a traditional 'monolithic' solution,
(b) select the most appropriate skin and core materials,
(c) produce a first approximation to the optimum skin/core configuration.

Such a design tool would need to consider economic factors in addition to weight, strength and stiffness criteria.

This paper outlines part of a BS Technical Strategic Research Programme which is aimed at developing steel-based composites, which offer a range of attractive properties.

REVIEW OF EXISTING THEORY

Considerable effort has been devoted to the analysis of beams, panels and struts of sandwich construction, and the results have been summarised in a number of publications(1)(2)(3). For the present purposes it has been deemed satisfactory to model a sandwich panel as a beam, with the simplifying assumptions that the skins are thin relative to the core and that the core material is homogeneous and much less stiff than the skin material. The approach is based upon that presented by Allen(2) and developed by Gibson and Ashby(3).

The deflection of such a panel under load consists of a contribution governed by the flexural rigidity of the faces and a contribution governed by the shear rigidity of the core. In contrast to most traditional forms of construction the shear deflection of a sandwich beam is not negligible in comparison with bending deflection, and might even be the dominant component of total deflection.

A sandwich beam or panel could collapse or become unfit for its purpose by several mechanisms, namely:-

(a) deflections might exceed tolerable limits,
(b) the skin might suffer gross plastic deformation,
(c) the skin might suffer local buckling ('wrinkling'),
(d) the core might fail to withstand the stress imposed,
(e) the bond between the core and the skin might fail,
(f) the skin and core might fail locally under a concentrated load.

The design could be subject to a number of constraints:-

(a) the panel or beam should not collapse under less than the fully factored load (ultimate limit state),
(b) the deflection under working load should be within allowable limits (serviceability limit state),
(c) application of the working load should not produce a permanent deflection,
(d) the weight of the panel/beam should be minimised or kept within a specified limit,
(e) the cost of the panel/beam should be minimised or kept within a specified limit,
(f) the skin thickness could be subject to a practical minimum and/or restricted to certain preferred values,
(g) the core thickness might be restricted to certain preferred values,
(h) the total thickness of the sandwich might need to be kept lower than a specified limit.

Gibson and Ashby[3] have developed a procedure for optimising sandwich beams with respect to weight. The procedure can be applied treating the density of the (foam) core as a variable and using certain empirical relationships between foam density and mechanical properties. For the present purposes the core density is regarded as fixed in any given case. It can be shown that the compliance, δ/P, of the beam is given by

$$\frac{\delta}{P} = \frac{2\,\ell^3}{B_1 E_f b t c^2} + \frac{\ell}{B_2 b c G_c} \qquad (1)$$

where:- δ = maximum deflection
P = total applied load
ℓ = span
B_1, B_2 are constants relating to the mode of loading
E_f = Young's modulus of skin material
G_c = shear modulus of core material
t = skin thickness
c = core thickness
b = width of beam

The weight W, of the beam is given by

$$W = 2\rho_f g b \ell t + \rho_c g b \ell c \qquad (2)$$

where ρ_f = density of skin material
ρc = density of core material
g = acceleration due to gravity

Equations (1) and (2) can be re-written in terms of the ratios (t/ℓ) and (c/ℓ);

$$\frac{t}{\ell} = \frac{W}{2b\ell^2\rho_f g} - \frac{\rho_c}{2\rho_f}\left(\frac{c}{\ell}\right) \tag{3}$$

$$\frac{t}{\ell} = \frac{2B_2}{B_1}\frac{G_c}{E_f(^c/_\ell)}\left\{\frac{1}{B_2(^\delta/_P)bG_c(^c/_\ell)-1}\right\} \tag{4}$$

By plotting these equations as graphs of $(^t/_\ell)$ versus $(^c/_\ell)$ a chart of the form shown in Figure 1 is obtained. Equation (4) produces a curve which is the locus of all the sandwich configurations which just satisfy the stiffness constraint. Equation (3) produces a series of lines, all of the same slope, each of which is the locus of sandwich configurations of a given weight. This chart can be used either to determine the range of configurations which satisfy the stiffness constraint whilst meeting a given weight limit, or to determine the optimum configuration (which satisfies the stiffness constraint at minimum weight). Equation (2) can be replaced with a similar expression for cost, as the sum of the costs of skin and face material, in order to produce a chart of the range of configurations which meet the stiffness constraint within a given cost limit, or the configuration which satisfies the stiffness constraint at minimum cost.

The optimisation approach outlined above can be extended to take account of strength criteria as well as stiffness. The equations describing various failure

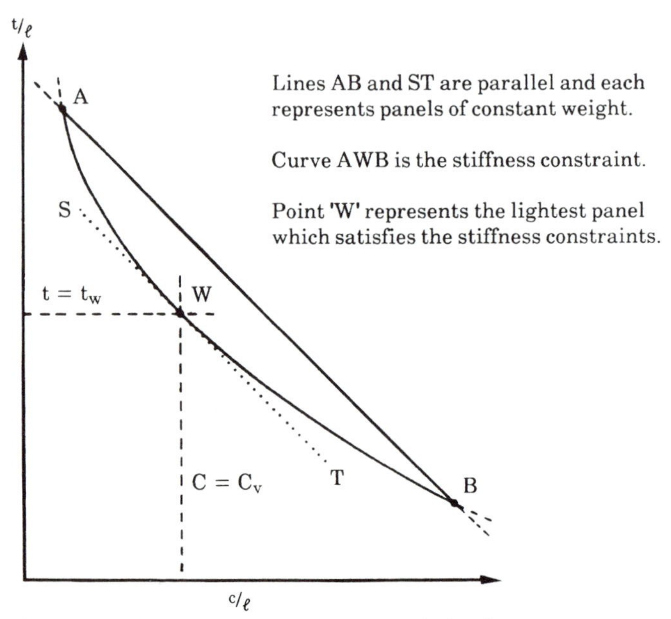

Figure 1. Principle of sandwich optimisation

modes can be presented in a form which allows them to be plotted on the $(^t/_\ell)$ vs $(^c/_\ell)$ chart. Gibson and Ashby[3] argued that, of the failure modes, bond failure and failure under concentrated load can be so readily avoided that they need not be considered. Their equations for face yield, face wrinkling and core failure can be rearranged to give;

$$\frac{t}{\ell} \geq \frac{P\ell}{B_3 b c^2 \sigma_y}\left(\frac{c}{\ell}\right) \tag{5}$$

where B_3 is a constant relating to the mode of loading
σy = yield strength of face material,

as the condition for avoiding face yielding,

$$\frac{t}{\ell} \geq \frac{P\ell}{0.57\ B_3 b E_f^{1/3} E_c^{2/3} c^2}\left(\frac{c}{\ell}\right) \tag{6}$$

where E_c is the Young's modulus of the core material

as the condition for avoiding face wrinkling, and

$$\frac{c}{\ell} \geq \frac{P}{B_4 b \ell \tau_c} \tag{7}$$

where τ_c = yield strength of foam in shear
B4 is a constant relating to the mode of loading
as the condition for avoiding failure of the core in shear.

Using equations (3), (4), (5), (6) and (7) together with other limiting conditions such as maximum allowable core thickness and minimum practical skin thickness, it is possible to define an envelope such as that shown in Figure 2. All combinations of skin thickness and core thickness lying within the shaded area satisfy the design criteria. This allows the design to be optimised readily, with respect to either weight or cost.

COMPUTER BASED OPTIMISATION OF STEEL FOAM COMPOSITE PANELS

The equations presented above describing the limits of each of the potential modes of failure can be calculated using a computer program and the results presented graphically. This enables the designer to specify the optimum skin/core configuration. The program developed is, currently, limited to the analysis of a single mid-span point load or uniformly distributed load with cantilevered, simply supported or built in end conditions.

The software package used to calculate the limits of each mode of failure is TK Solver[4]. TK solver is a proprietary programming environment in which

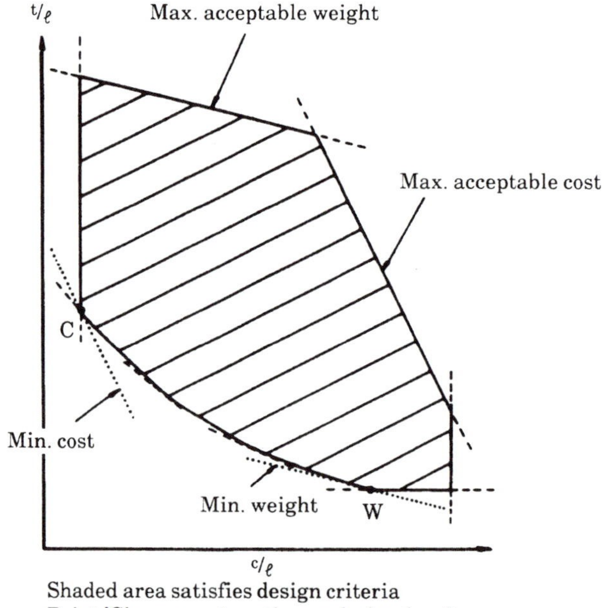

Figure 2. Design envelope

equations can be input in any order and are rearranged internally by the software. In addition the user can use expressions on both sides of the equality and the unknown variable can appear anywhere in the expression. This enables the end user to use whatever information is available when specifying the problem. For example a uniformly distributed load can be specified as a total load or a load/unit area. A table of conversion factors can be stored within the program such that the end user can use a wide range of units to describe the problem, i.e. the span could be specified in metres, millimetres, feet or inches. This provides a useful facility to the user and reduces the opportunities for error.

TK Solver provides a pre-formatted input and output screen. The input screen for the calculation of the limits of failure is shown in Figure 3. Constants B1-B4 describe the boundary conditions. 'Span' and 'width' are self explanatory. 'Deflection' is the maximum allowable deflection of the panel at mid span. 'Load' is input either as a total load or as a load per unit area. The 'load factor' is used to factor the loads for limit state design. Factored loads are applied when calculating the limits of failure such as face yield, wrinkling and core shear but unfactored loads are used for calculating deflection, which is a serviceability limit state. The weight constraint and target cost are used to generate lines of constant weight and cost used in the optimisation process. The skin properties are mostly self explanatory although there is a facility to specify a minimum skin thickness

St. Input	Name	Output	Unit	Comment
				Boundary conditions
384	B1			Boundary Constant(=384/5 SS,=384 BI)
8	B2			Boundary Constant(=8 SS,=8 BI)
12	B3			Boundary Constant(=8 SS,=12 BI)
2	B4			Boundary Constant(2 SS,=2 BI)
				Loading/Geometry Inputs
2500	ℓ		mm	Span
1000	b		mm	Width
30	y		mm	Deflection
10	PRESS		kN/m^2	Uniformly distributed load)One
	ρ	25000	N	Total load)only
1.6	LF			Load factor
30	W		kg	Weight constraint (total)
100	COST			Target cost
				Skin properties
0.2	tmin		mm	Minimum skin thickness
500	Yf		MPa	Yield stress of face density of skin material
7900	Rst		kg/m^3	Density of skin material
2E5	Ef		MPa	Skin modulus
1.5	SCt		/kg	Price of skin material/kg
				Core Properties
90	Cmax		mm	Maximum core thickness
40	Efoam		MPa	Stiffness of foam)One
	Es		MPa	Stiffness of solid core)only
45	RoC		kg/m^3	Density of foam core
	RoS		kg/m^3	Density of solid core (Optional
L50	c		mm	Core thickness
18	Gc		MPa	Core shear modulus
0.9	Yfoam		MPa	Core shear limit (foam))One
	Yc		MPa	Core shear limit (solid))only
0.3	C11			Foam material constant
12	CCt		/kg	Price of core material/kg

Figure 3. Sample input sheet

to ensure that the design generated is practical. A maximum core thickness can be input as actual foam properties or as the properties of the solid material and the density of the foam enabling the software to determine the actual foam properties using equations developed by Gibson and Ashby(3).

The results are plotted for limiting skin thickness against core thickness, using TK solver. Core (c) and skin thickness (t), rather than the non-dimensional (c/ℓ) and (t/ℓ), are used to make the format more readily applicable to a specific design.

Figure 4 shows a sample output from the model. The line plots represent the combination of skin and core thicknesses which satisfy specific failure criteria considered. The line A-E represents the locus of the skin and core thickness configurations which satisfy all the design requirements. The panels meeting the design requirements at minimum cost and minimum weight are determined by the points at which the lines of equal weight and equal cost are tangential to the locus of the design limit (X and Y). A further plot is produced by the model showing the effect of core thickness on panel weight and cost. This is produced by calculating the weight and cost of the panel configurations which lie on the line A-E. A sample plot is shown in Figure 5. This enables the user to make an informed compromise between weight and cost depending on the relative importance of weight saving. It can be seen in the example shown that cost is very sensitive to a change in core thickness whereas weight is relatively insensitive. In this example the 'optimum' panel would be biased towards the minimum cost configuration.

FURTHER DEVELOPMENT OF THE COMPUTER MODEL

The example of the use of the computer based optimisation shown above relied upon the user to select the most suitable skin and core materials. Ideally the user should only have to eliminate materials known to be unsuitable for other reasons such as lack of fire resistance or undesirable thermal properties and the computer model should consider all other suitable candidates and determine the most suitable combination of skin and core materials. This could be achieved with the use of a database of properties of potential skin and core materials that could be edited by the user either directly or by answering relevant questions about the proposed application. It is proposed that the primary database be of general nature and not, for example, contain all the possible densities of polyurethane but would have information about three, high, medium and low density. If after the first run of the program polyurethane foams appeared to be suitable a further run could be performed using a more detailed polyurethane database.

The results of the analysis would be available as a plot of weight versus cost for minimum weight and minimum cost for each skin/core combination. This would be extended by plotting the cost and weight of intermediate configurations of panel which meet the specification, to enable the sensitivity of these parameters to be assessed. The most promising would then be selected by the

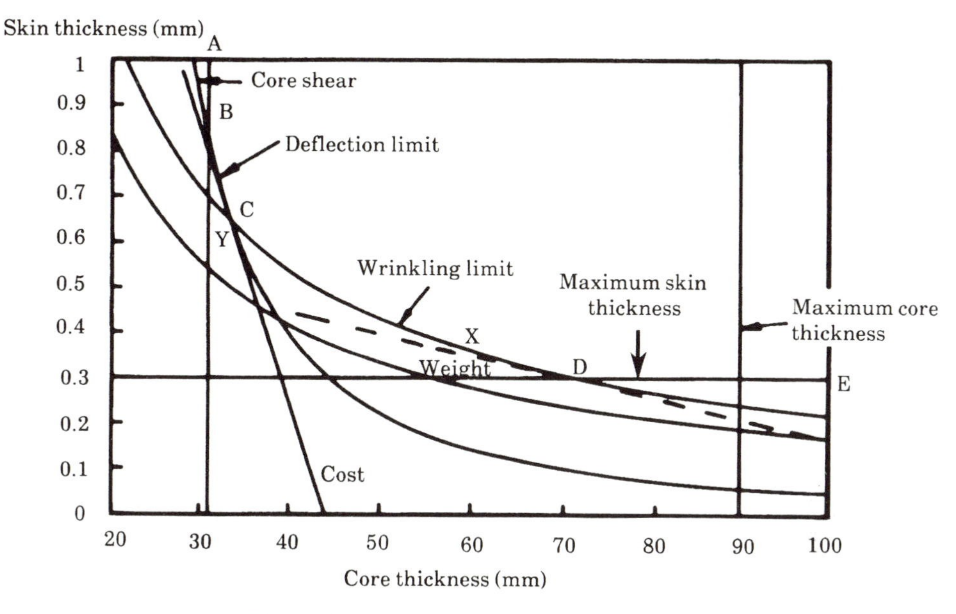

Figure 4. Locus of design limit

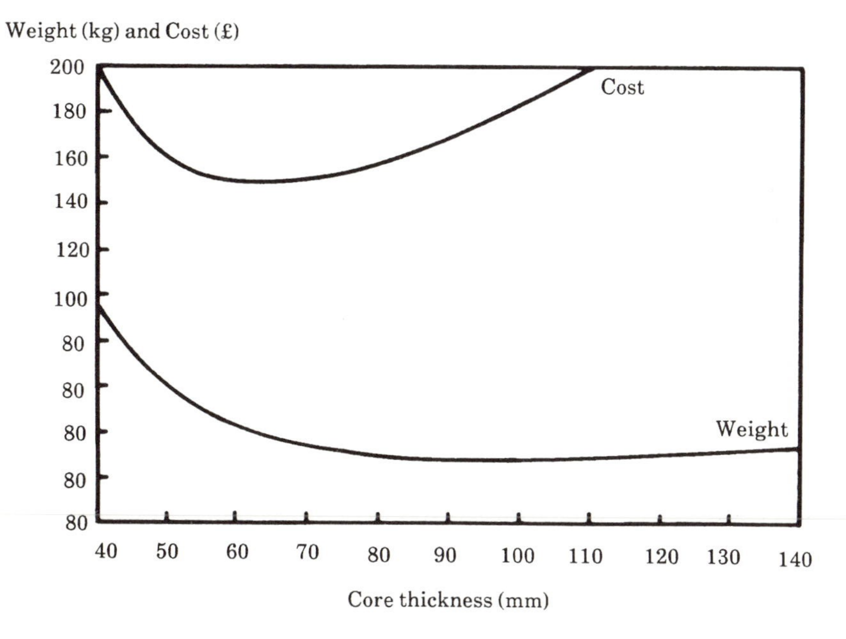

Figure 5. Weight optimisation of simply supported steel/pvc/steel panel

user and plots of skin versus core thickness and cost and weight versus core thickness examined to determine the configuration to be used. A flow diagram of the proposed program is shown in Figure 6.

The proposed system could be further extended, if required to include:-

(a) Automatic editing of potential skin and core materials based on the proposed span, load and environment to eliminate materials that have been shown from experience or previous analysis to be unsuitable.
(b) Minimum requirements for other performance parameters such as thermal and sound insulation.
(c) Other types of core such as metal honeycombs or trapezoidal sections.
(d) Analysis of plates with various boundary conditions.
(e) Analysis of complex loading conditions.

THE ROLE OF PHYSICAL TESTING

Experimental work is necessary at several points in the design process;

- to verify the description of beam behaviour,
- to evaluate the mechanical properties of the constituent materials,
- to justify the application of equations derived for a beam to the design of a wide panel,
- to confirm that the prototype panel fully meets the specified performance levels.

British Steel Technical is engaged in the testing of scale models and full size prototypes in relation to a number of applications.

ACKNOWLEDGEMENTS

The authors wish to thank Dr M.J. May, Manager Swinden Laboratories and Product Technology and Dr R. Baker, Director of Research and Development, British Steel plc, for permission to publish this paper.

REFERENCES

1. Anon: *Divinycell Technical Manual*, Barracuda Technologies AB, Laholm, Sweden (October, 1991).
2. Allen H.G. *Analysis and Design of Structural Sandwich Panels*, Pergamon Press, Oxford (1969).
3. Gibson L.J. and Ashby M.F. *Cellular Solids - Structure and Properties*, Pergamon Press, Oxford (1988).
4. Anon: *TK Solver Reference Manual*, Universal Technical Systems Inc., Rockford, Illinois, (1989).

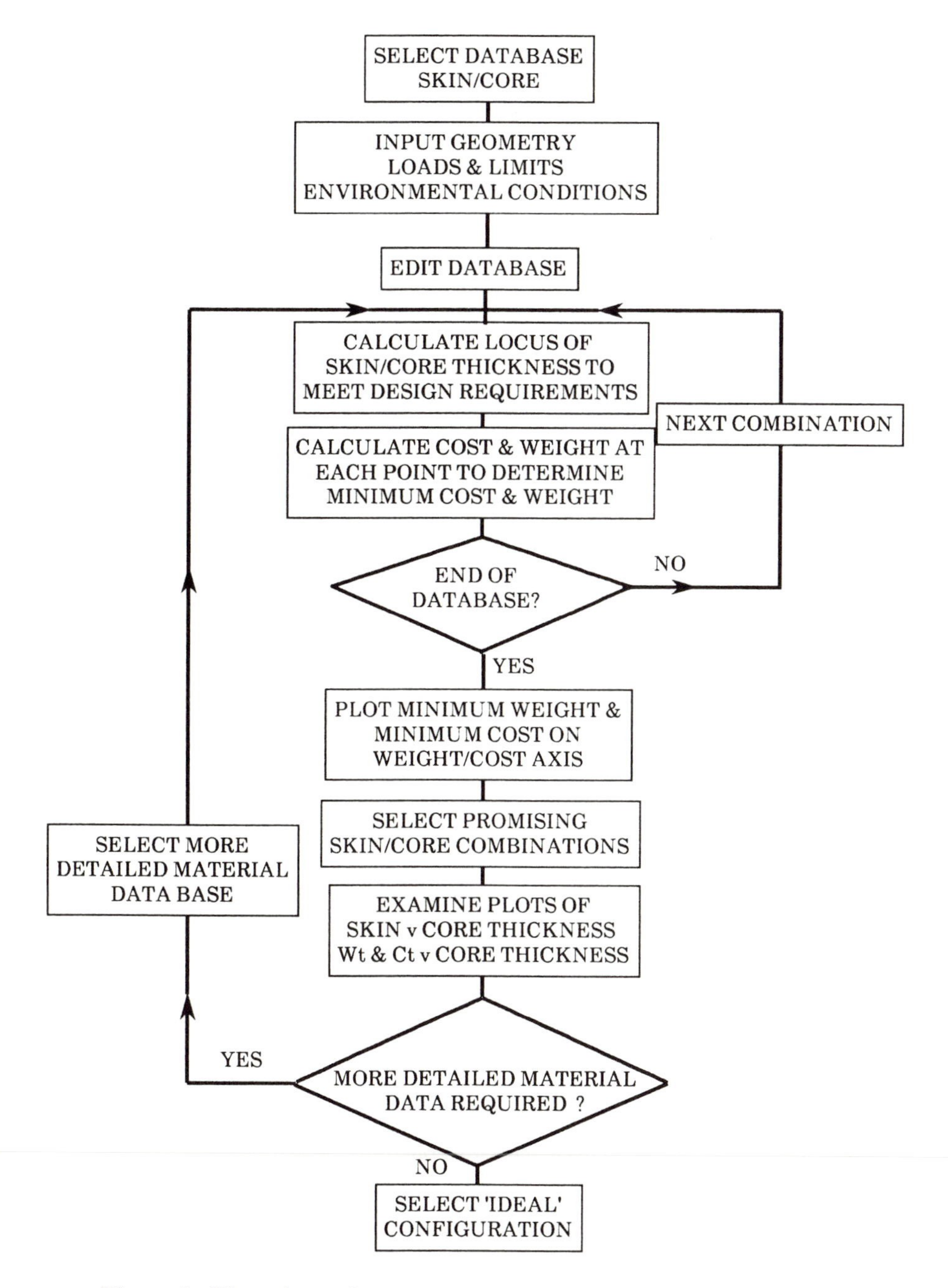

Figure 6. Flow chart of proposed composite panel design program

42 SLIM FLOOR CONSTRUCTION

D.L. Mullett and R.M. Lawson
The Steel Construction Institute

The design of slim floor beams, where the floor is contained within the slab depth, is reviewed. Slim floor construction can incorporate either precast concrete units or profiled steel decking (210mm deep decking). The paper reports on research that is being carried out at the Steel Construction Institute into aspects of both these types of Slim Floor buildings. The research and development for this project was funded by British Steel, General Steels and British Steel, Strip Products.

INTRODUCTION

In the last decade the Nordic countries, Sweden in particular, have developed a method for constructing floors which is colloquially known as slim flooring. This form of construction has dramatically increased the market share for steel framed multi-storey buildings. For example, in the early part of the 1980's, the market share for steel buildings in the Stockholm region was approximately 5%. In 1989 this figure had risen to about 80%. In view of this success, British Steel plc decided to send a team of structural engineers to Sweden to investigate the merits of this system. The findings of that investigation plus further research into slim floor construction has been used to develop the 'Slimflor'* beam, specifically aimed at UK Construction. Figure 1 shows typical fabricated slim floor beams, but the simplest system is shown in Figure 1a. This section is known as the 'Slimflor' beam which comprises a single horizontal plate welded to the bottom flange of the universal column (UC). The plate extends beyond the width of the bottom flange (UC) thus providing a platform for the precast units to rest on. As with all slim floor systems the beam is contained within the slab depth, hence providing a flat soffit throughout the floor area. This approach provides unhindered passage for the services, i.e. no internal downstands.

* Registered trademark of British Steel; further information and advice is available from the SCI and British Steel General Steels on all aspects of slim floor construction.

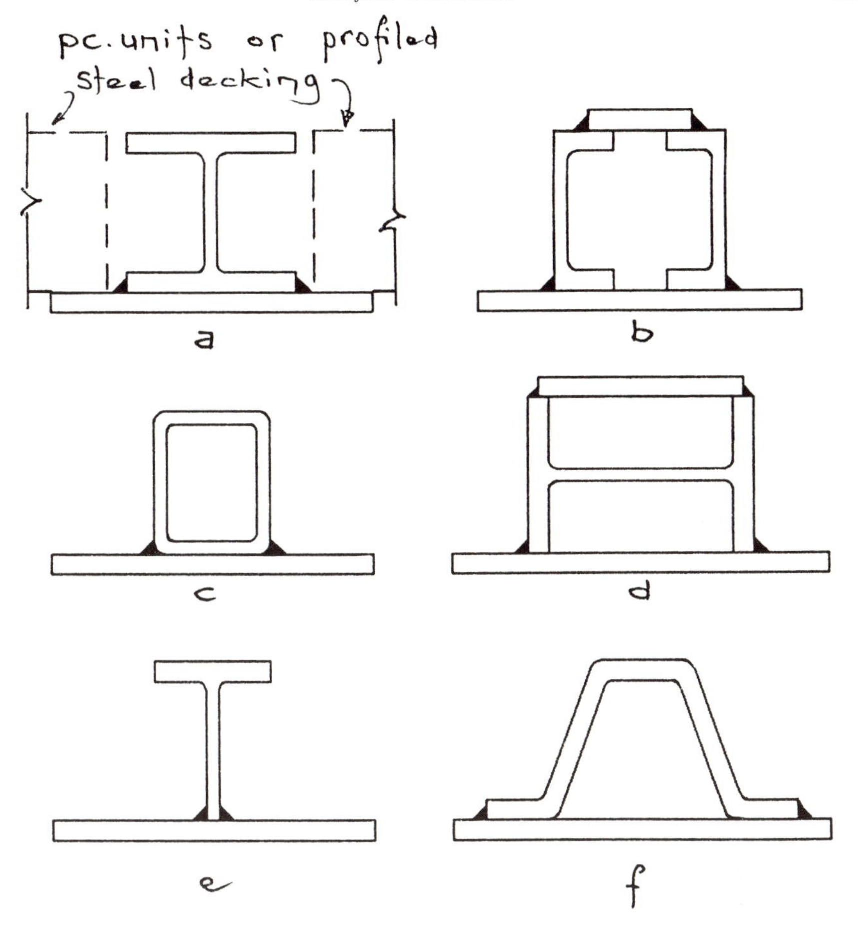

Figure 1. Types of slim floor beam

An SCI design guide entitled 'Slim floor design and construction' has been produced which covers all aspects of slim floors - see Reference 3. A computer program complementary to the publication is also available for purchase.

British Steel (General Steels) publication 'Design in Steel 2, Slimfloor Construction' gives information on the advantages and scheme design of the slimflor beam. An estimating computer program is also available from the Advisory Service (see Reference 4).

The 'Slimflor' beam concept was introduced to the UK in November 1991. Since this introduction a high level of enthusiasm has been shown by specifiers. In addition, the Slim floor concept is being considered for use in the rest of Europe.

The 'Slimflor' beam was initially developed for use with precast units. Precision Metal Forming (PMF), a UK manufacturer of profiled steel decking has recently developed a deck which can span 6m unpropped and has a depth of 210mm and is trapezoidal in cross section. The steel deck is an ideal replacement for precast units as it will reduce the dead weight of the floor. Also it lends itself for passing minor services within the slab depth between the ribs of the deck. Holes constructed through the web of the beam enable these services to pass through the floor.

In November 1992, full scale tests were carried out on the above form of construction at City University. These tests determined the serviceability and ultimate load carrying capabilities. Also, in November, this method of construction was subject to a fire resistance test at Warrington Fire Research Centre.

The first part of this paper deals with the 'Slimflor' beam and precast concrete units.

Three forms of construction are recognised.

Type 1: Dry construction with grout only for sound insulation purposes (Figure 2(a)).

Type 2: Encased construction, offering better fire resistance and stiffness properties (Figure 2(b)).

Type 3: Composite construction with an in-situ concrete slab and welded shear connectors (Figure 2(c)).

The slim floor beams and columns form a 2-dimensional frame. In British practice it would be normal to 'tie' the columns together in the third dimension using a tie member. This tie can be contained within the slab depth (see Figure 3).

Slim floor construction provides a number of benefits:

- The floors have a flat soffit.
- The overall floor construction depth is reduced. This can reduce cladding costs.
- It improves the fire resistance of the section. The concrete that surrounds the beam partially insulates the section. This can lead to the elimination of the fire protection giving up to 60 minutes fire resistance.
- The concrete that surrounds the beam produces an increase in the second moment of area of the section. This enhancement is helpful in reducing deflections.
- It offers unhindered passage for services.
- In the case of local element instability the concrete will improve the load carrying characteristics of the beam. For the future this could prove an asset for continuous construction.
- In certain circumstances, 'dry construction' can be employed, thus saving time before the building is occupied.

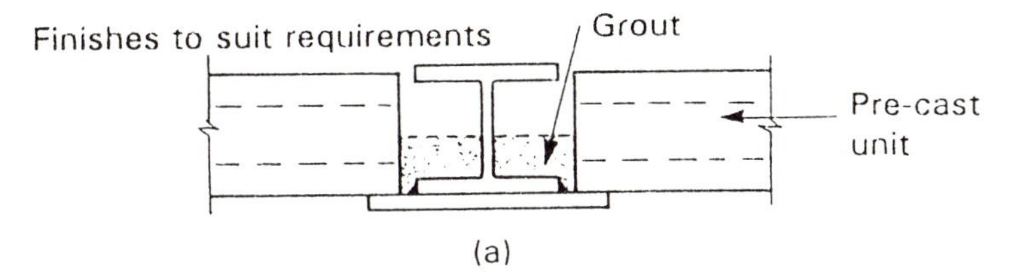

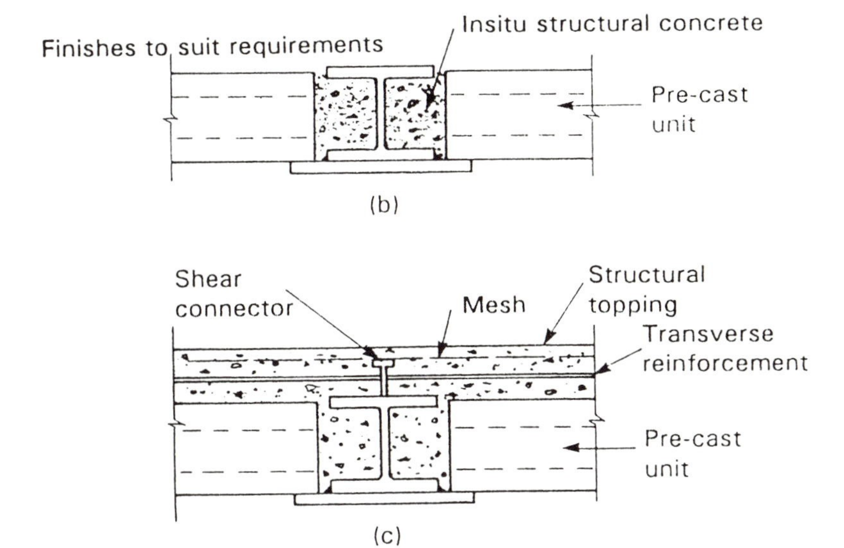

Figure 2. Illustration of three methods of slim floor construction

The modified slim floor beam shown in Figure 2 has the additional benefits:

- It uses standard steel sections.
- The beam is easy to fabricate with full depth end plate connections. Only two fillet welds are required to attach the longitudinal plate which can be automatically welded without turning the section.
- The system provides relatively long spans with minimum construction depths. This will have the effect of reducing cladding costs.
- No internal voids for sound or heat transfer in fire are created. This reduces the amount of fire protection.
- The system is inherently versatile to suit the requirements of a given building. This is emphasized by the three forms of construction shown in Figure 2 which range from dry to composite construction.

  Note: Torsional effects in the construction stage and possibly at the edges of the building may occur under eccentric loading. This can be reduced by design, or by propping during construction.

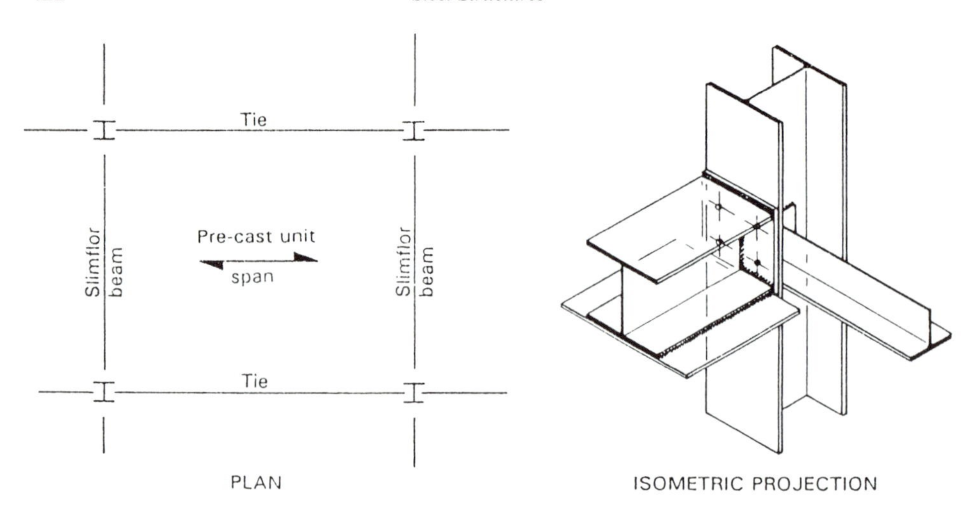

Figure 3. Typical framing arrangement

DESIGN

Forms of construction

The design procedures are based on the use of BS5950: Parts 1[1] and 3[2] (or to the forthcoming Eurocodes 3 and 4). The cross-sections will be limited to plastic (Class 1) or compact (Class 2) sections. Semi-compact (Class 3) sections limit the design of the section to the elastic moment capacity. This complicates the design procedure but, more importantly, is an uneconomical use of steel.

The basic design assumptions are summarised as follows:

a. Unpropped simply supported beams subject to uniformly distributed loading.
b. Use of plastic or compact cross-sections.
c. Plastic analysis of the cross-section based on rectangular stress blocks.
d. Serviceability checks use elastic analysis with unfactored loads. To ensure that irreversible deformation (under normal service loads) does not occur in the steel, the extreme fibre stress is limited to p_y, and in situ concrete stress is likewise limited to $0.5f_{cu}$.
e. Deflections of beams are limited to span/360 under imposed loads, which applies to buildings of general usage; allowances should be made where deflections under serviceability loads could cause damage to the finishings. The total deflection of the beam is limited to span/200.

Only some of the salient aspects of slim floor design can be discussed here; full guidance is given in the SCI guide[3]. Design assumptions for assessing the moment capacity and torsional effects for the three forms of construction are as follows:

Type 1: Non-composite

a. Slim floor beam is unrestrained in the construction stage.
b. Slim floor beam is restrained under imposed load.
c. Out-of-balance loads to be considered for construction and imposed loads.

Type 2: Semi-composite

a. Slim floor beam is unrestrained in the construction stage.
b. Slim floor beam is restrained under imposed load.
c. Out-of-balance loads to be considered for the construction stage only.

Type 3: Composite. As for Type 2, and in addition

a. Self weight and construction loads resisted by steel section (if unpropped).
b. Imposed loads resisted by composite section.

Construction stage
In the construction stage it is not always possible to ensure that the precast units are placed in such a way as to eliminate out-of-balance loads. The complexities introduced by torsion are best avoided in building structures. However, in this instance this is not easily achieved unless strict erection procedures are adhered to. Out-of-balance loads will also have an undesirable influence on the lateral torsional buckling (LTB) of the section.

The combination of these two effects is highly complex and requires simplification for general use.(5) This has been achieved by developing an expression similar in form to that used in BS5950: Part 1 for the combination of stresses.

$$\frac{M_x}{M_b} + \frac{M_y}{M_{cy}} \leq 1.0 \quad (1)$$

where M_x is the applied moment about the x-x axis; M_b is the buckling resistance moment about the x-x axis, determined using BS5950: Part 1 ($M_b \leq M_c = S_x p_y$). M_y is the applied moment to the top flange of the steel section about the y-y axis, which is obtained by considering the torsional moment as two opposing forces in the flanges as shown in Figure 4. M_{cy} is the plastic moment capacity of the top flange of the steel section about the y-y axis.

An alternative method of eliminating torsion is to prop the beam during construction. This can have an adverse influence on construction times, and generally has a nuisance factor in site operations. However it may be appropriate on small projects, or for long span beams.

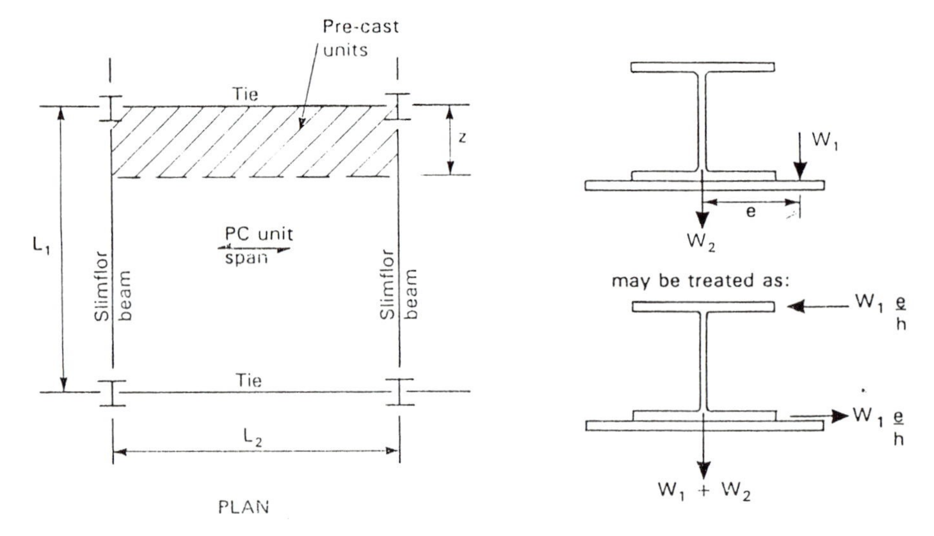

Figure 4. Simple treatment of torsion on slim floor beams during construction

Moment capacity (non- and semi-composite sections)
In the construction stage, LTB and pure torsion in slim floor beams are treated in the same manner as in a non-composite beam. The concrete that surrounds the beam is used for stiffness purposes only and is assumed to provide adequate lateral restraint to the member for ultimate load conditions. It is difficult to show that the moment capacity of the steel beam is enhanced by composite action unless additional shear connection is provided. It is for these reasons that the composite action is neglected at the ultimate limit state. Figure 5 shows the method of deriving the equation for moment capacity (steel member only) using plastic analysis for the design of the cross-section at ultimate loading.

Figure 5(a) shows the position of the plastic neutral axis (PNA) in the web, below the centreline of the steel section at a distance y. In order to simplify calculation of the moment capacity, M_c, of the section, Figures 5(b) and (c) show a standard method of rearranging the rectangular stress blocks.

If moments are taken about the centreline of the section:

$$M_c = M_s + R_p\left(\frac{D}{2} + \frac{t_p}{2}\right) - R_p\frac{y}{2}$$

where

$$y = \frac{R_p}{2p_y t_w}$$

$$M_c = M_s + \frac{R_p}{2}(D + t_p) - \frac{R_p^2}{4p_y t_w} \tag{2}$$

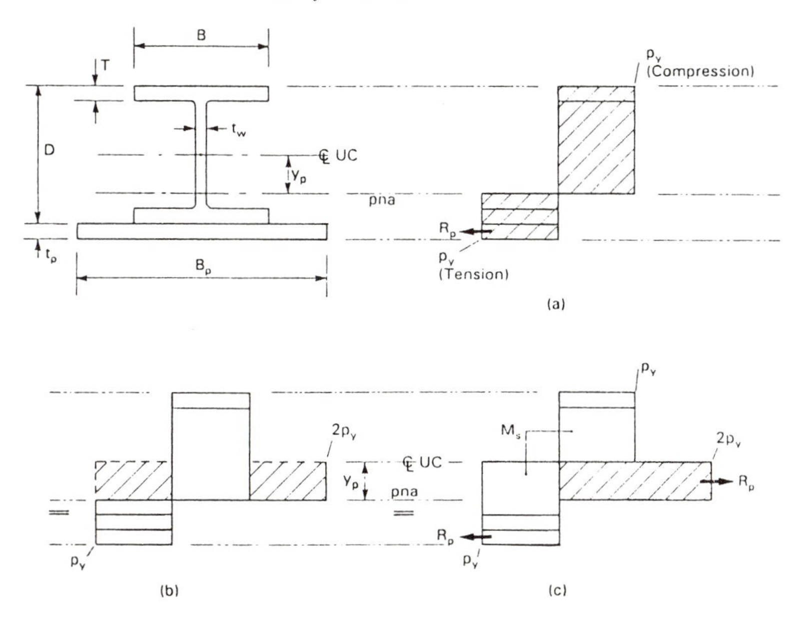

Figure 5. Plastic analysis of non-symmetric steel section

where M_s is the moment capacity of the steel section (ignoring the plate) = $S_x p_y$
R_p is the tensile resistance of the plate of thickness, t_p
D is the depth of the steel section
p_y is the design strength of steel
t_w is the web thickness.

Moment capacity (composite section)
The moment capacity of composite slim floor beams is dependent on the area of concrete in compression and on the degree of shear connection between the concrete and steel beam. The effective breadth of the slab, B_e, is taken as beam span/4, as for a normal composite beam.[2]

Full shear connection Figure 6 shows a typical cross-section through a composite beam. Also shown is the plastic stress distribution across the section. Full shear connection occurs where the number of shear connectors provided is sufficient so that the force they transfer is greater than the resistance of the concrete or steel member. This will generate the maximum moment capacity of the cross-section. In the majority of cases, the concrete resistance R_c is less than the resistance of the steel member, ie.

$$R_c < (R_s + R_p)$$

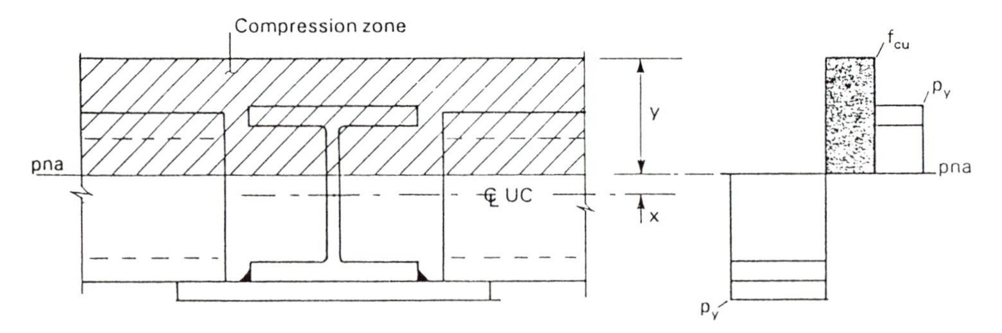

Figure 6. Typical cross-section through composite slim floor beam with full shear connection

where $R_c = 0.45f_{cu}B_eD_s$ and $R_s = Ap_y$.

and D_{su} the slab depth above the section (assuming $D_{pc} < D$)
f_{cu} the cube strength of concrete ($\approx$ 1.2 x cylinder strength).

In cases where the plastic neutral axis (PNA) for full shear connection lies within the precast units, higher forces may be generated in the shear connectors, potentially leading to an overload in longitudinal shear. Therefore, to account for this effect the number of shear connectors is determined for the maximum possible force developed in the concrete (including the precast units), or in the steel section. Conversely, when considering the moment capacity it would appear prudent to assume that this bonding of the concrete no longer exists. In these circumstances a conservative approach has to be adopted, otherwise, the degree of shear connection could be lower than the minimum required value. The moment capacity calculation is best illustrated by referring to the following case:

Partial shear connection Partial shear connection design is attractive where the moment capacity is much greater than the applied factored moment. When this situation occurs BS5950: Part 3 and Eurocode 4 permit a reduction (to a maximum of 40%) in the number of shear connectors relative to full shear connection. The principle is that the number of shear connectors is reduced so that the longitudinal force, R_q, transferred by the shear connectors is sufficient to provide an adequate moment capacity.

The moment capacity, M_c, is given by the following expression based on the stress block approach as illustrated in Figure 7:

$$M_c = M_s + R_q\left[D_s + \frac{D}{2} - \frac{R_q}{0.9f_{cu}B_e}\right] + \frac{R_p}{2}(D + t_p) - \frac{(R_q - R_p)^2}{4p_yt_w} \qquad (3)$$

where: $M_s = S_x\, p_y$

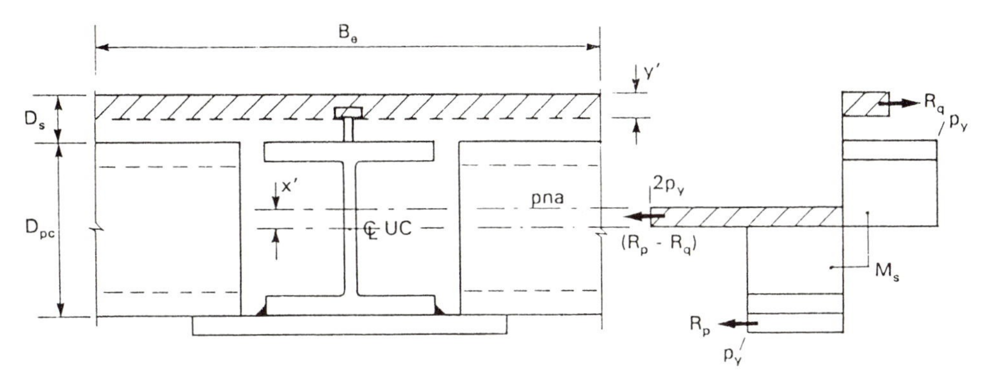

Figure 7. Typical cross-section through composite slim floor beam for partial shear connection

In all cases M_c is less than the value for full shear connection. The degree of shear connection is defined as R_q/R_c for $R_c < R_s + R_p$, which is usually the relevant case.

For full shear connection $R_q = R_c$ in equation (3).

FURTHER DESIGN CONSIDERATIONS

Biaxial stress effects in the flange plate

Biaxial stresses in the flange plate have to be considered as a direct result of the way the loads are applied to the flange plate. The plate is subject to longitudinal and transverse effects as shown in Figure 8. The longitudinal stress due to overall bending of the section, σ_1, has an influence in reducing the resistance of the plate when also subject to a transverse bending stress, σ_2. This is irrespective of whether the stresses are plastic or elastic. Plastic analysis of the section may be continued but using a reduced strength of the plate taking account of σ_2.

Design for torsion

It is usually not possible to eliminate out of balance loads which cause torsion of the section. A simple way of taking this into account in design is to treat the beam as subject to transverse forces in the flanges.(3) Longitudinal and transverse bending stresses are then linearly combined using equation (1). Torsional moments are resisted at the ends of the beams by four bolted end plate connections (see Figure 3).

FIRE RESISTANCE

The fire resistance of slim floor beams has been investigated both experimentally, by a number of fire resistance tests, and analytically using computer modelling

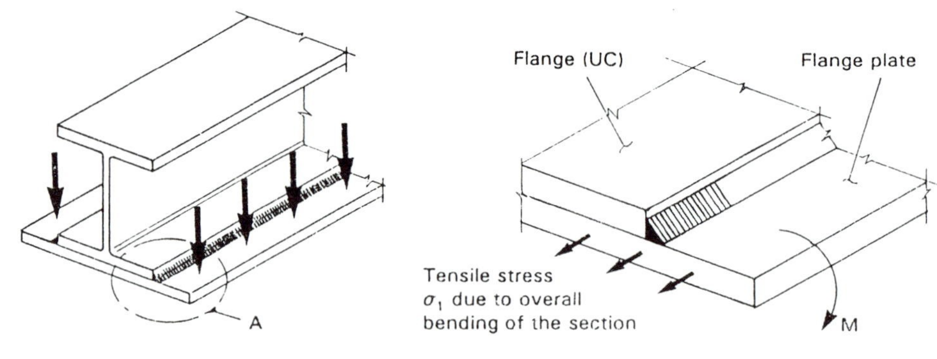

Figure 8. Biaxial stresses in flange plate

techniques. As a result of these studies recommendations on the use of unprotected sections have been prepared for up to 60 minutes fire resistance[6]. For more than 60 minutes fire resistance additional fire protection is generally recommended. This may take the form of a board pinned to the underside of the flange plate.

The fire resistance of a slim floor beam is inherently good because most of the section is shielded from the fire by the floor units. For 60 minutes fire resistance, with minor exceptions, Type 1 and Type 2 systems can be used unprotected. However, the composite system, (Type 3), or those using the deep deck option may require additional protection or may have restrictions on design loading.

Summary of fire tests

The recommendations are based on the analysis of 8 fire resistance tests on loaded beams[3]. These are summarised in Table 1. In tests 1 and 2 the floor units were built directly into the beam web and no supporting plate was used. In tests 3 to 8, 15 mm bottom plates were used together with 8 mm fillet welds.

For tests 7 and 8 it was decided to test a Type 1 design with the section only half filled with concrete as temperatures recorded would then be conservative when applied to a Type 2 or 3 design. In tests 3 and 5 sand infill was used which resulted in a lower fire resistance compared to concrete infill.

It can be seen from Table 1 that there is always an appreciable temperature difference between the plate and the bottom flange. This difference is very important and is due to an interface resistance between the two surfaces. However, in test 7 the interface resistance between the 152 × 152 × 30 UC and the plate was appreciably lower than in the other tests.

The behaviour of the beams can be modelled mathematically using the measured and predicted temperatures and correlation with the test results was good.

Table 1. Summary of fire resistance tests

Test	Section UC size x kg/m	Type	Fire Resistance (mins)	Load Ratio	Plate (†) Temperature (°C)	Flange (†) Temperature (°C)
1	254 x 254 x 73	-	44	0.56	-	746
2	254 x 254 x 89	-	93	0.42	-	783
3	254 x 254 x 107	1	60	0.55	799	661
4	203 x 203 x 86	3	68	0.44	727	558
5	203 x 203 x 60	1	82*	0.51	812	691
6	254 x 254 x 73	2	78	0.47	778	578
7	152 x 152 x 30	1	75*	0.48	788	731
8	305 x 305 x 283	1	115*	0.17	728	411

Note: * Test discontinued before failure
† at 60 minutes

USE OF DEEP PROFILED STEEL DECKING

Precision Metal Forming (PMF), a UK manufacturer of profiled sheeting has recently developed a steel deck which can span 6m unpropped and has a depth of 210mm. The deck when acting compositely with an in-situ concrete slab is an ideal replacement for the pc units as it will reduce the dead weight of the floor. This has the effect of reducing steel weight and the size of foundations. Also, it lends itself for passing minor services through the slab depth between the ribs of the deck (Figures 9 and 10).

A special end diaphragm (stop end) to the decking is used so that the concrete fully encases the steel section except for the bottom surface of the plate. The insulating effect of the concrete provides the beam with 60 minutes fire resistance. Higher periods of fire resistance may be achieved with additional fire protection to the plate. The soffit of the deck is left exposed and fire resistance of the slab is achieved by provision of additional reinforcing bars in the ribs.

The decking is designed to be *unpropped* during construction for beam spacings up to 6 m, and may achieve reasonable imposed load capacities, if propped, for beam spacings up to 7.5 m. The slimflor beams may be designed economically within the constraints of the slab depth for spans of 6 to 9 m.

Two forms of construction are considered:

- Type A - Non-composite beam (utilising the concrete encasement only for increased stiffness and fire resistance).
- Type B - Composite beam with additional shear connectors and achieving composite action with the slab.

These two forms of construction are illustrated in Figure 10. In principle, the beams are designed to be *unpropped* during construction, although there may be circumstances where propping is used to provide the minimum steel beam size for

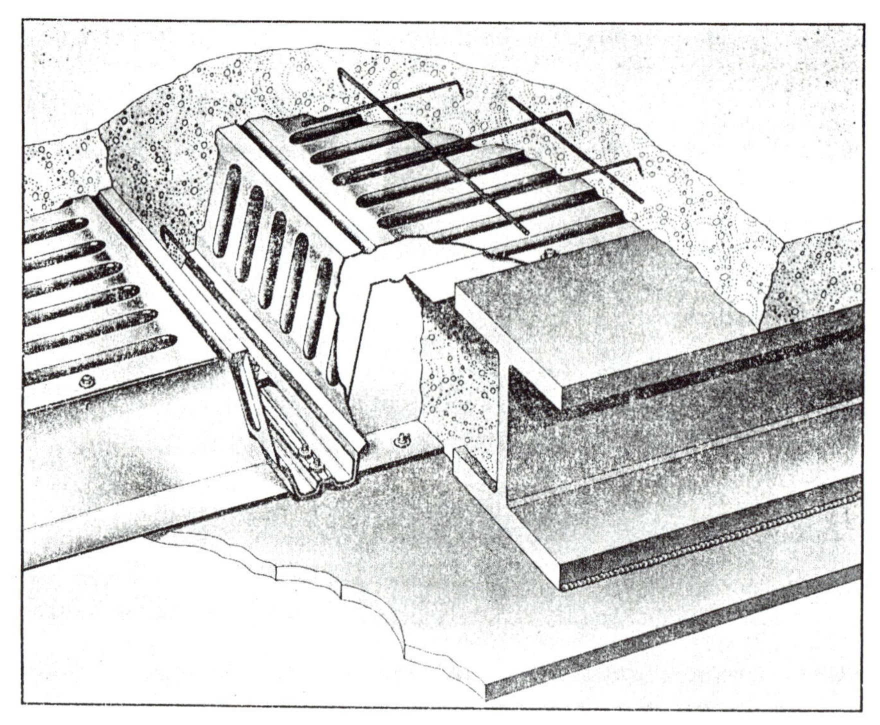

Figure 9. Basic components of the 'Slimflor' system using deep decking

a given span. Grade 50 steel and *lightweight* concrete are the preferred materials. However, where serviceability criteria control the design, use of grade 43 steel is more cost effective.

Interim design guidance(7) for this method of construction has been published by the SCI. This will be followed by a formal SCI design guide, similar in form to the guide for precast concrete units. Revised software will be issued to cover both forms of construction.

ADVANTAGES OF SLIM FLOORS USING DEEP DECK COMPOSITE SLABS

- Comparable to other 'fasttrack' methods for speed of construction.
- Savings in cladding cost and service distribution.
- Additional zones are available between the ribs for services (holes can be cut in the steel web of the UC section for service runs (see Figure 10).
- Lightweight slab relative to p.c. units or reinforced concrete alternatives.
- Easy erection (bundles of decking are delivered cut to length, and may be lifted into place, and deck man-handled as required).
- Acts as a safe working platform during construction.

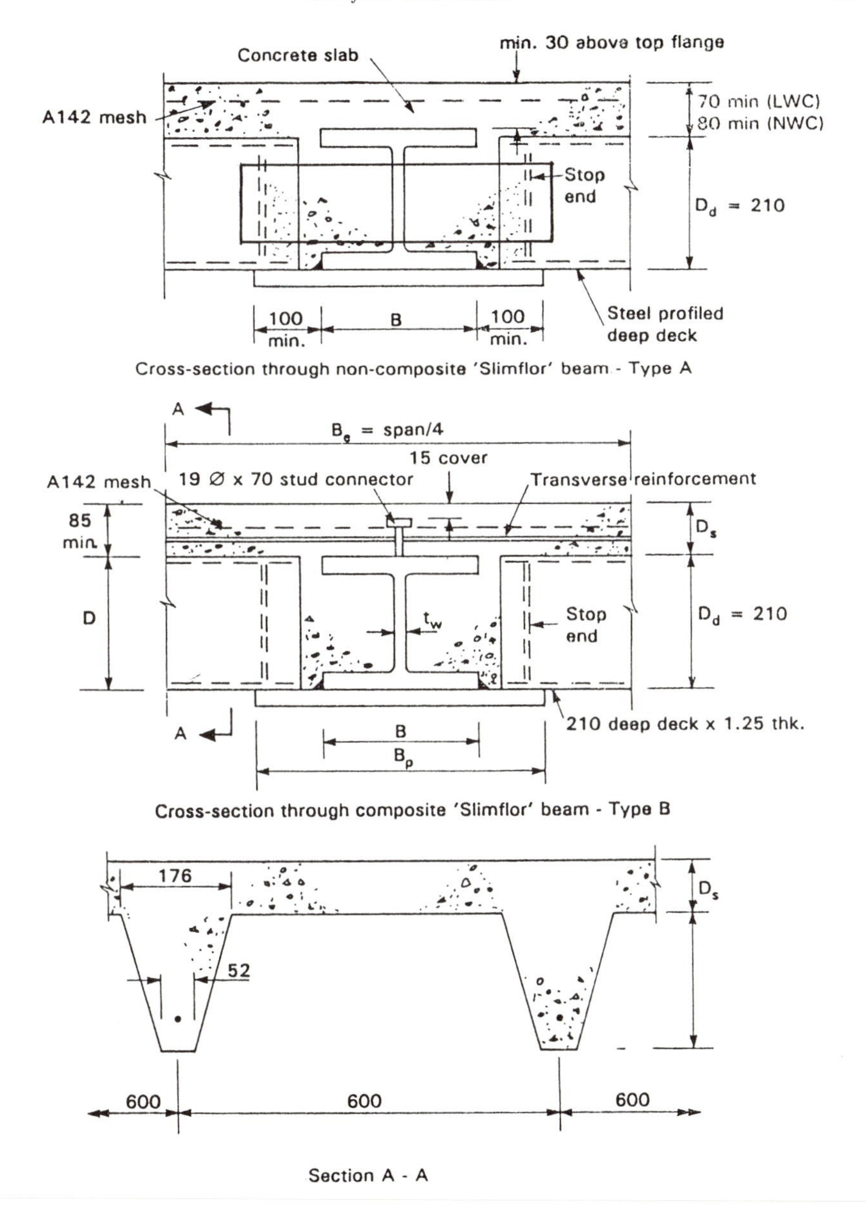

Figure 10. Construction details for Type A and B construction

- Easy fixing (by pins or screws) and reinforcing detailing.
- Pumped concrete saves cranage relative to pc units.
- Flexibility of detailing, provision of openings such as stairwells etc.
- Acts as a 'diaphragm' to resist in-plane forces (no wind bracing needed).
- Good serviceability performance due to the monolithic nature of the floor slab.
- Robust from the point of view of explosions, fire etc.
- Inherent fire resistance (60 minutes fire resistance can be achieved with no further protection, and higher periods of fire resistance require only additional protection to the soffit of the beam).
- Service holes through the floor slab are easily achieved as opposed to p.c. units which are difficult to cut due to the prestressing wires.

Design assumptions for Type A and B forms of construction

a) The Slimflor beam is considered to be laterally unrestrained for the construction stage.
b) The Slimflor beam is laterally restrained for the imposed loading condition.
c) Out-of-balance loads do not need to be considered for the imposed loading condition because of the increased torsional strength of the 'composite' section.

The basis of design is similar to the precast concrete method of construction.

FIRE RESISTANCE OF SLIM FLOORS WITH COMPOSITE SLABS

Composite Slab

The fire resistance of the composite slab is achieved by provision of reinforcing bars in the ribs. No further fire protection is required and the deck is assumed to provide no tensile resistance. The rate at which the bars heat up is dependent upon their 'cover' and the period of fire exposure. The 'fire engineering' method of analysis[8] may be used to determine the size and cover of the bars needed for a particular application. The mesh reinforcement over the beams is not included in this calculation.

Sufficient concrete topping for composite slabs is to be provided over the decking to satisfy insulation requirements in fire. Data for composite slabs is given in BS 5950: Part 8. The minimum depths may be taken as 80 mm for normal weight and 70 mm for lightweight concrete for 60 minutes fire resistance. This represents a slight (10 mm) increase over the data in BS 5950: Part 8 because of the greater spacing of the troughs in the deep deck profile.

Further fire resistance tests are to be carried out to provide improved guidance.

Slimfloor beam

The steel section is partially insulated from the fire by the concrete. The only part of the section that is exposed is the bottom plate (note: special end diaphragms ensure that concrete covers the upper part of the bottom plate and flange; see Figure 9 and 10).

A series of fire tests has been carried out on 'Slimflor' beams using p.c. units. A computer program has been especially written by the SCI to analyse the thermal and structural behaviour of the beams. This program has been adapted to analyse slim floors for use with deep decks.

An important parameter is the load ratio, or load applied at the fire limit state relative to the load capacity of the beam. A load ratio of 0.6 is the maximum normally considered. 60 minutes fire resistance can be readily achieved with no further calculation required.

Board-type fire protection can be easily attached to the soffit of the beam to achieve longer periods of fire resistance. Refer to the ASFPCM/SCI/FTSG publication *Fire protection for structural steel in buildings*(9).

The composite 'Slimflor' option may be analysed in a similar manner, but because of the different structural behaviour, slightly different fire resistance periods result from the analyses.

TEST PROGRAMME FOR THE DEEP DECK METHOD OF SLIM FLOOR CONSTRUCTION

With such a novel method of construction it was decided that an extensive test programme would be required to justify the proposed methods of design.

Table 2 gives the overall test programme

Date of Test	Test Centre	Component to be tested
21.7.92	British Steel Welsh Labs	PMF deck
15.1.92	Warrington Fire Research Station	Deck/slab (light weight concrete)
16.10.92	Warrington Fire Research Station	Deck/slab (normal weight concrete)
21.10.92	Salford University	PMF deck + in situ concrete slab
4.11.92	Warrington Fire Research Station	Deck/slab + slimfloor beam
20.11.92	City University	Slimfloor beam + deck/slab

The following is a brief description of each test:-

Construction Stage Test
PMF CF210 Deck -
In the construction stage the deck supports the wet weight of the concrete plus a construction load which represents the weight of operatives, tools and materials etc.

To simulate this condition the deck was tested at British Steel's (Strip Products) Welsh Laboratory at Port Talbot. This laboratory has the facility for vacuum testing which is an ideal representation of the true conditions found in practice. The deck was tested over a 6m span with end diaphragms at the support position. The deck achieved an ultimate test load of 4.8k kN/m^2 which compared favourably with the hand method of calculation.

Figure 11 shows the deck mode of failure which occurred at mid-span.

PMF 210 Deck plus in situ concrete slab

In this test the deck and in situ concrete are intended to act compositely together. BS 5950: Part 4, testing, requires a series of tests to be carried out in order to establish the shear-bond characteristics of the deck. These tests were carried out at Salford University. The loading frame comprised single jack load applied to the slab via four spreader beams placed at intervals to simulate a uniformly distributed load. It has been observed, in all cases, that the failure mode was flexural with a typical failure load of 13.0 kN/m^2, well in excess of design loads.

The results of these two tests show that in the majority of cases the construction stage will dominate the design procedures.

FIRE RESISTANCE

The three fire tests which were carried out at WFRS have been used to determine the temperature and thermal distribution through the floor components. This data is then used to develop a computer model which enables the user to predict whether the floor components require fire protection for the period of time under consideration.

Figure 11. Construction load tests, showing mode of failure for the unpropped deck which occurred at mid-span

Composite Slab

The first two fire tests shown in Table 2 considered the composite (steel deck and in situ concrete) only. The specimens were 1.0m square and the slab was formed using light and normal weight concrete with reinforcing bars placed in the ribs of the deck.

'Slimflor' beam

The third test shown in Table 2 was a full scale test which combined the slab and the 'Slimflor' beam. The method of construction is shown in figure 10. This test achieved 62 minutes without the use of fire protection.

CITY UNIVERSITY TEST

This test investigated a 'Slimflor' beam comprising a composite slab formed using deep (210mm) steel deck and acting as formwork to in situ concrete. The deck rested on a steel plate welded to the bottom flange of a Universal Column section (see Figure 10). The test specimen spanned 7.5m with an overall construction depth of 300 mm. Four jacks placed on top of the beam simulated a uniformly distributed load.

In addition there were holes in the web of the beam which permitted short lengths of cylinders to pass through the beam and these continued to the stop ends. This type of deep deck 'Slimflor' beam enables minor services to be passed through the cylinder and within the decking profile and thus to be incorporated within the depth of the floor.

The test was discontinued when the total jack load had reached 1016 kN (100 tonnes) and the central deflection was 150 mm (span/50). Results showed that the test specimen had achieved approximately twice the ultimate design load. This was due to the 'Slimflor' beam acting compositely with the concrete. It is too early to recommend partial composite action without the use of shear connectors but with further research, it may be possible to take advantage in this increase in strength. The load/deflection graph for the cycles of loading is shown in Figure 12.

REFERENCES

1. British Standards Institution BS 5950: Structural use of steelwork in building Part 1: Code of practice for design in simple and continuous construction: hot rolled sections. BSI (1990)
2. British Standards Institution BS 5950: Structural use of steelwork in building Part 3: Codes of practice for design in composite construction. Section 3.1: Design of simple and continuous composite beams. BSI (1990)
3. Mullett D.L. Slim floor design and construction. The Steel Construction Institute, (1991)
4. British Steel Advisory Service, Steel House, Redcar, Cleveland TS10 5QL. Telephone (Hotline) 0642 474242

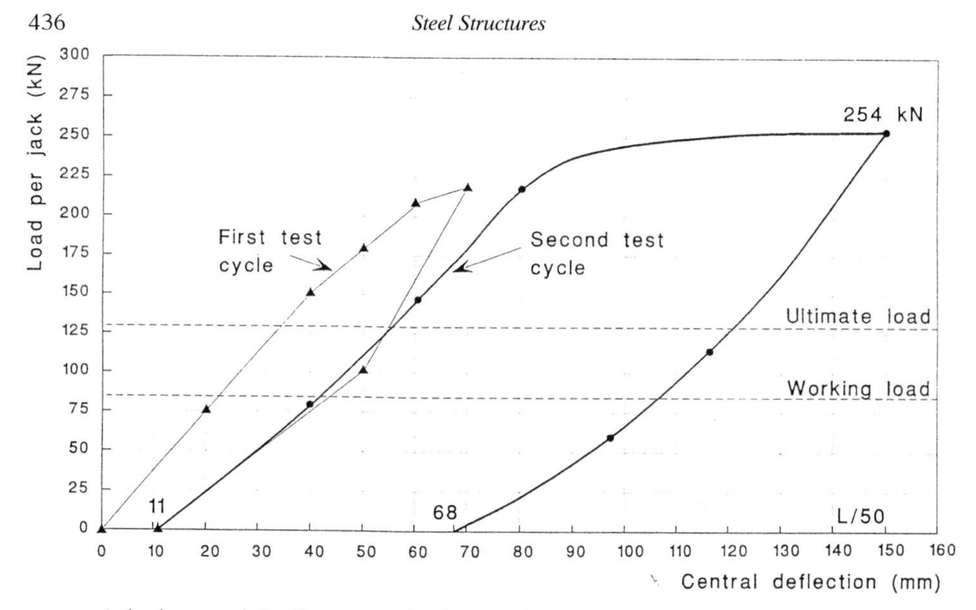

Figure 12. Load/deflection graph for the full scale test at City University

5. Nethercot D.A., Salter P.R. and Malik A.S. Design of members subject to combined bending and torsion. The Steel Construction Institute, (1989)
6. British Standards Institution BS 5950: The structural use of steel in building Part 8: Code of practice for fire resistant design. BSI (1990)
7. Mullett D.L. and Lawson R.M. Slim floor construction using deep decking. Interim design guidance. The Steel Construction Institute (1992).
8. Newman G.M. The Fire Resistance of Composite Floors with Steel Decking (2nd Edition) The Steel Construction Institute (1991).
9. ASFPCM/SCI/FTSG. Fire protection for structural steel in buildings. Revised Second Edition (1990)

PART EIGHT

NEW CONSTRUCTION TECHNIQUES

43 A ROBOT FOR MASONRY CONSTRUCTION

D.A. Chamberlain
City University, UK

Masonry production in the form of brickwork and blockwork is a high value element in an active construction industry, and thus a natural target for automation research. However, the automation established in the manufacturing industry cannot simply be re-applied, due to difficulties arising in the areas of part tolerances, material supply, work repetition, access, mobility, physical environment and integration with other processes. The automation of masonry production cannot be interpreted as the classical 'pick and place task', so amenable to automation. A comprehensive production cell will be expected to work with standard production masonry units, whilst ensuring the quality of materials used. It will need to be easy to operate, self navigating, and perform progress surveys of constructions. Substantial machine intelligence is implied, with comprehensive sensing for reliable and safe operation. Machine compatible work descriptions will need to be derived automatically from the CAD definition of a project. Some progress has been achieved in the enabling technology for such a device.

INTRODUCTION

In an active construction industry, the total value of labour in masonry production is substantial. As other repetitive and labour intensive activities, it is thus a natural target for automation development. However, apart from some mechanical handling devices[1][2] associated with the use of large masonry blocks, only experimental automation, and robotic devices have been produced so far. Here we differentiate between automation which implies fixed cycles of operation, and truly robotic devices which can intelligently modify their operation in response to sensed states.

Broadly speaking, the application of automation technology to masonry production will tend to lead to requirements for precisely prepared masonry units, bonding by thin layer adhesives, and accurate performance and location of the plant. In general, the simpler the machine, the more precise and structured the task needs to be. On the other hand, the use of advanced robotics technology has the potential to relax this regime, with regular production masonry units employed, together with mortar type bonding. Furthermore, this plant could be expected to need only rough positioning, being self navigating and locating.

However, accepting the current form of masonry, plant for producing it can be expected to be highly sophisticated. Having explained the important distinction between 'automation' and 'robotization', the term 'automation' will be taken to cover either in the remainder of this paper.

On reviewing progress in masonry automation, it is apparent that the various national characteristics of masonry have a substantial influence on the methods. In Israel, where a large, fixed location, industrial robot has been used(3), commonly available interlocking gypsum blocks have been dry bonded with posterior reinforcement applied. This work has been complemented by computer simulated studies of the productivity of fully mobile robot concepts. In former Russia , where brickwork tends to be massive, a prototype machine(4) has been developed which applies mortar to a single brick face, and then delivers this to the point of lay. Units are delivered at a high rate to the manual workers (about one every three seconds) who place the units and fill spaces with pumped mortar. This system, from which a full solution is being pursued, would be suitable for large scale projects, though plan access is necessary from all brickwork areas. Whereas both of these approaches accept standard production masonry units, work in Finland and elsewhere(5) has been based on the use of masonry units which are prepared to fine dimensional tolerances and bonded using adhesives. An interesting mobile access system, based on a modified scaffolding system has also been built for manual bricklaying to the exterior of buildings(6). In the UK, research has proceeded on the lines of standard production bricks and blocks used together with standard and modified mortar mixes(7).

The objective of the research described in this paper is to further the enabling technology for an advanced masonry tasking robot. To this end, a robot cell has been devised for the purpose of investigating fully automatic masonry constructions. Excluding mobility issues, which are are receiving extensive consideration elsewhere, all aspects of masonry production are represented in the cell, including material supply, quality checks, bond mix delivery and dispensing, manipulation and placement, survey and remedial action. In this, spin-offs are apparent which might aid existing manual process.

A description of the robot cell elements follows this, after which the materials employed and operation of the cell are discussed.

DESCRIPTION OF ROBOT CELL

Supply Conveyer

Material supply is a key issue in any form of automation. For the experimental cell, an industrial conveyor (Figure 1) of the type used for rubble clearance has been suitably adapted. The three main functions of this are dimensioning units, assessing surface and edge damage, and presenting satisfactory units for the robot to pick.

On account of the way in which brick tolerances are specified in the UK, individual bricks are occasionally rejected purely on dimensions, these determined by a group of distance measuring sensors which scan passing units. More likely, in the case of facing bricks, surface and edge damage will lead to rejection. This decision is based on assessments derived from digital image processing. Edge and line finding algorithms have been applied to detect departures from the perfect shape. The surface roughness occurring in broken areas has also been usefully exploited in this, with some measure of correlation to human assessment achieved. A typical part processed image is shown in Figure. 2.

Masonry units which are determined as satisfactory for use are moved on to the pick location point. The location of units on the belt need not be precisely

Figure 1. Material Conveyor

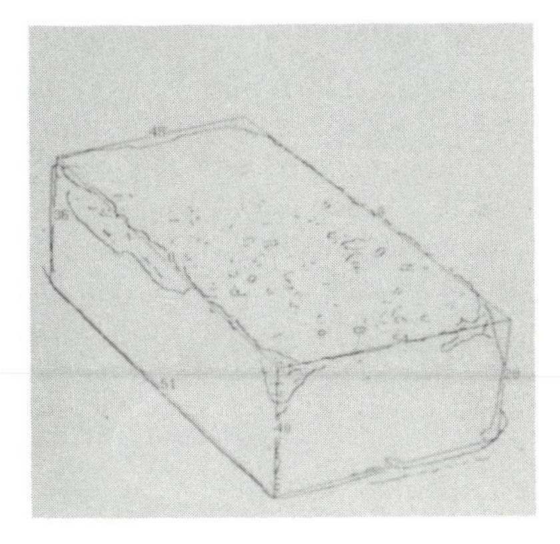

Figure 2. Damage Detection By Image Processing

known, as the robot is capable of locating units at the pick up stage. Furthermore, the conveyor can be relocated during the process of a project, a realistic requirement in a situation where there may be competition for working space. This provision marks a distinct departure from the conventional manufacturing environment where supply lines are generally maintained in precise locations.

Robot and Gripper

The robot, which is shown in Figure. 3, is a large gantry type robot with X,Y,Z translational and revolute axes. Its working envelop allows wall constructions within a plan area of 4.0m x 2.5m to 2.0m height. Motion control for the robot is implemented on a series of SMCC's (Smart Motion Control Cards) which are slaved to a master PC. This PC also commands other slave processors located on the conveyor, brick/block gripper and the mix dispenser and navigation beacon described later.

A clamp type gripper is fitted to the base of the robot arm which enables it to pick a range of units from the conveyor, manipulate them for mortar application and locate them in the wall project. Near the gripper is mounted a combined PC (credit card size) and A/D (Analogue to Digital) signal converter which services the sensors located on the gripper. Sensors are provided for (i) locating masonry units prior to picking them from the conveyor, (ii) confirming the grip action, (iii) determining the position of a picked unit relative to the gripper, (iv) survey during and after placement of unit and (v) general collision avoidance. Highly focused ultrasonic ranging sensors are used for (i) and (iv).

Figure 3. Gantry Robot: Dry Wall Construction

Strain gauges are used for (ii), a pair of plunger type displacement transducers for (iii), and wide angle ultrasonic ranging sensors for (v). A companion eye for the navigation beacon is also located near the gripper.

The amount of sensor data is potentially enormous. However, sensors can be addressed selectively according to the current activity. A library of rule sets have been developed for this purpose, these implemented in 'C' code on the slave processor. These rules have a high level form, an example being as follows:

```
Unit_in_hand:rule
    If grip_action is firm
    and position_hand is clear_belt
    and Sen_A > A_okay
    and Sen_B > B_okay
    then Angle_Make = (Sen_A - Sen_B)/AB_Length
```

For this rule to fire, the unit must be gripped satisfactorily, lifted clear of the conveyor belt, and both grip locating sensors operational. Angle_Make,the angle correction necessary to allow for a unit to be placed level, is returned to the master PC.

Navigation Beacon

A laser beacon has been included in the cell to enable navigation experiments, an important issue for mobile robots. This comprises a Spectral Physics rotating laser, mounted on a vertical drive axis for which motion control is achieved by a microprocessor slaved to the master PC. Levels can be requested or set by the robot, these detected by the companion eye located on the gripper. Whilst developed specifically for the robot cell, the rising laser beacon could be used to aid manual activity, a hand held command unit replacing the master PC.

Bond Mix Dispensing

Three possible arrangements for mortar application have been considered: (i) insitu, (ii) on conveyor and (iii) at dispensing station.

Whilst they are all technically possible, (i) is the least satisfactory, mostly because it requires the robot to carry a feed line or reservoir. Experiments have shown that a feed line, which has substantial weight and rigidity when full, tends to deflect the robot's end effector in an uncontrollable manner. A large volume of mix is necessary in order to prime the feed line, blockages are frequent and substantial pumping power is demanded. Although the use of a reservoir eliminates the need for a feed line, the combined weight of its case, motor and mix content would be unacceptably large (estimated at 30kg-40kg).

Option (ii) is the most attractive because it would make the greatest contribution to productivity. However, to prepare a unit with mortar on two faces, it is necessary to either rotate it under a stationery nozzle, or move the nozzle around it. Either process could be built onto the convey, though reliability and costs may be questionable, particularly as the robot is readily able to provide the necessary manipulation. For this and other reasons, option (iii) has been adopted in the research.

Figure 4 shows the mix dispensing station which comprises 'SmartPump', a sophisticated peristaltic pump, a feed line and extruding nozzle. Whilst the pump is a general purpose device suitable for applications, such as grouting and pressure pointing under manual control, fully automatic operation is possible using its serial communications port. By this means, the flow can be stopped and the pump speed and direction controlled from a remote PC. Other pumps have been tested[8] including an air powered 'Metrix' pressure pointing pump.

Using the arrangement shown is Figure. 4, the robot provides the motion and manipulation necessary to apply mortar, visiting the station between picking the unit from the belt and laying it in the wall. Research is continuing in this area in order to determine the optimum combination of flow rate, passing speed and nozzle inclination. Sensing methods are also being investigated, which will enable the quantity and shape of the extrusion to be gauged and controlled on a real time basis. In current experiments the pump is preset to provide an output corresponding to an average joint thickness of 10mm.

Various mixes have been used in pumping experiments including the use of 'Pozament.' For convenience, the polymer 'Polyox 301' has also been used to bind soft sand to a realistic consistency. Experiments are continuing to determine the best materials for the application and also a reliable means for predicting their

Figure 4. Mix Dispensing Station

pumping and bedding down properties. To this end, the bedding table shown in Figure. 5 has been developed to directly estimate the amount of immediate settlement occurring on placement of a unit. In use, the tray is filled with mortar and a steel box released onto it using electro magnets. Weight can be added to match an average brick or block using the appropriate box. Whilst this is a useful test, no noticeable correlation has been found with standard tests such as the drop ball and flow table. Work is also being carried out on electrical conditioning of mortar as a means of improving pumpability.

PROJECT DESIGN, SIMULATION AND CONSTRUCTION

A task specification for the robot is derived from the output of a CAD facility which has been programmed in AutoLisp, the programming language of AutoCad. Figure 6 shows a simple wall project which has been designed using this. The raw output from this utility is a descriptive list having one element per masonry unit as follows:

...),(Xc, Yc, Zc, Orn, Typ, Spl, RV),(....

where the atoms 'Xc', 'Yc' and 'Zc' are the centroid coordinates, 'Orn' the plan orientation indicator, 'Typ' the type indicator, 'Spl' the length in the case of a special block and 'Rv' the existence indicator (means of distinguishing an opening).

Figure 5. Initial Settlement Device

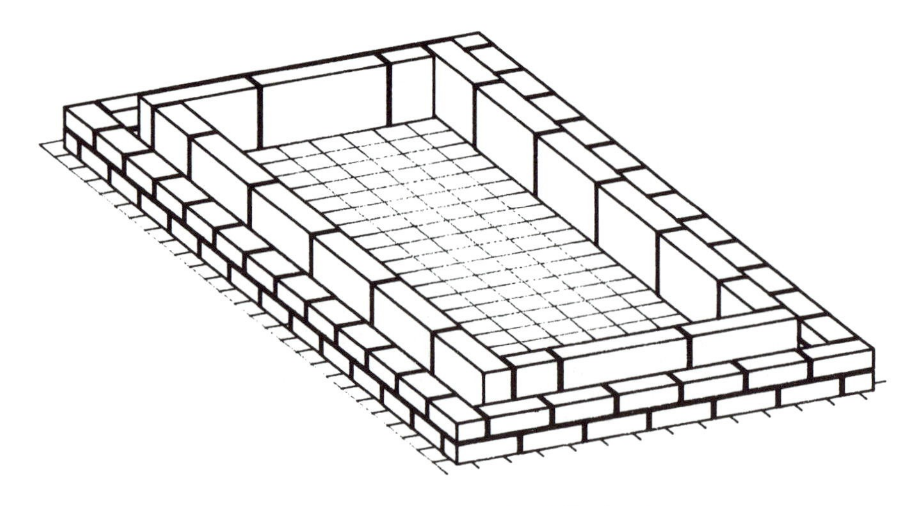

Figure 6. Cad Output of Wall Project

Single skin brickwork or blockwork can be specified and cavity wall construction where the cavity width can be varied. Openings corresponding to standard window and door openings can be introduced, with the logic ensuring that vertical joints do not align in blockwork. Whilst the list is built up by in a continuous spiral order starting form the lowest level upwards, this is not necessarily the order of laying. In fact, the robots work specification file is the outcome of rule-based processing of the descriptive list. On completion, this comprises not only the construction order and location data for masonry units, but also path parameters for collision free, optimised, robot motion. To avoid collision, it is necessary to consider the position and orientation of the end effector in relation to the partly completed wall. In the optimisation, we seek the least time move, the effective move time being the sum of the travel and end of move settling times. Generally speaking, the faster the robot travels between two points, the longer it will vibrate at the end of its journey. Furthermore, it cannot perform useful work until this vibration has reduced to an acceptable level.

The work specification file is compatible with a computer based robot simulation facility in which the robot cell is modelled, a frame from a typical project being shown in Figure. 7. Whilst this activity is not essential to the performance of the actual robot, it does enable the risk of collision and project completion time to be assessed.

Once transmitted to the robot, the work specification file is operated with runtime adjustments according to sensor information. Relocation of the conveyor or mixing station during the process of work leads to the work specification file being reprocessed. To date only dry wall projects similar to that shown in Figure. 3 have been completed, bonded wall construction to commence in the near future. Automation using other brick and block systems is also to be investigated, the 'TriBrick' shown in Figure. 8 being an example of these.

Figure 7. Computer Simulation

Figure 8. TriBrick System

CONCLUSIONS

The background to the automation of masonry production has been presented, and the bases for the solutions being research. In the project reported, all aspects of the automation, excluding overall mobility, are represented. On account of the complexity of the task when accommodating standard production material with mortar, and the need for the automation cell to respond to sensor information at runtime, there is a need for sophisticated machine intelligence. This is being achieved with the help of rule based logic of the form used in so called 'Expert Systems'. In the quality assessment of masonry units by the cell there appears to be a useful role for computer based image processing.

Whilst the cell has so far only performed dry wall constructions, the progress achieved in controlled dispensing of mortar is paving the way to the automation of bonded wall construction. To this end, a prototype peristaltic pump has proved effective. However, further work is necessary in the areas of mix design and the dispensing arrangements. A CAD facility has been developed with generates the starting data for determination of the robot's work specification. In order to arrive at this however, it is necessary to consider the path requirements for collision free optimum motion.

In common with the automation of other activities, progress masonry automation is dependent on progress in the areas of (i) overall mobility, (ii) effective sensing, (iii) machine intelligence and (iv) combating motion induced vibration.

ACKNOWLEDGEMENTS

The author wishes to acknowledge the support of the Science and Engineering Research Council and Wiltshire PLC without who's help the research could not continue. Thanks is also due to Spectra Physics for supplying laser equipment, Metrix engineering for the use of their pump, and Pozzament their cooperation in the development of mixes. Special thanks are due to my friend and colleague Mr Peter Thomas, for making his 'SmartPump' available.

REFERENCES

1. Bohm D. 'The Mason's Elevator-Handling-Machine', Proc. 8th Int. Symp. on Automation and Robotics in Construction, Stuttgart, (June 1991).
2. Lohja Oy Ab, 'EU 542 Mechanization of Bricklaying Technology on the Building Site', Finland, Eureka Project (current).
3. Rosenfeld Y. and Warszawski A. 'Robotic Performance of Interior Finishing Works: Development of Full-Size Applications', Proc. 7th. Int. Symp. on Automation and Robotics in Construction, Bristol, (June 1990).
4. Malinovsky E., Borschevsky E.A., Eler, E. and Pogodin V.M. 'A Robotic Complex for Brick-Laying Applications', Proc. 7th. Int. Symp. on Automation and Robotics in Construction, Bristol, (June 1990).

5. Lehtinen H., Salo E. and Aalto H. 'Outlines of Two Masonry Robot Systems', Proc. 6th. Int. Symp. on Automation and Robotics in Construction, San Francisco. (June 1989).
6. Paul Wurth S.A. 'EU 377-FAMOS BRICK, Highly Flexible Automated and Integrated Brick Laying System', Luxemborg, Eureka Project (current).
7. Chamberlain D.A., Speare P.R.S. and West G.A.A. 'A Masonry Tasking Robot', Mechatronics Systems Engineering, Vol. 1, pp 139- 147, Kluwer Academic Pubs. (1990).
8. Chamberlain D.A., Speare P.R.S. and Ala S.R. ' Progress in A Masonry Tasking Robot', Proc. 8th. Int. Symp. on Automation and Robotics in Construction, Stuttgart, (June 1991).

44 INTEGRATED AUTOMATION FOR DESIGN AND PRODUCTION

M.M. Cusack
University of the West of England, UK

The potential for robotic assembly in construction is examined with particular reference to the evaluation of the building design. The broad issues of design for construction assembly are explored and the relation to manufacturing assemblies identified. The paper presents the idea of representing the building design in a generative way determined by a rule system derived from the spatial relations among building components. It is argued that this provides the means to evaluate designs based on(1) component features appropriate for automatic mating and assembly,(2) component and sub-assembly delivery, and(3) construction planning for effective use of assembly resources.

INTRODUCTION

Design for assembly has become a standard phrase in Manufacturing Automation and forms part of a general theme which attempts to create the relationship between design and production in manufacturing. This concept is equally applicable to construction although traditionally design and production are divorced.

There are many difficulties with implementing design for assembly which manufacturing research has addressed[1][2]. The main approaches have emphasised the rationalisation of assembly moves, particularly directions of assembly moves, the types of fixing employed after mating and the detailed design of the mating components to guide assembly by contact forces.

A separation of the global and local aspects of the problem can be observed in this approach. The design of component mating emphasises the local aspect with automatic assembly devices and their supporting sensing systems, guiding components through a sequence of locally constrained moves to achieve final placement. The constraints are essentially kinematic in nature and the problems centre around the ability of the assembly device and associated sensing systems to respond to spatial constraints and to recognise when goal spatial relations have been achieved.

DESIGN FOR ASSEMBLY

To improve design for assembly it is necessary to be able to analyse the consequences of design decisions at both levels. It is proposed here that appropriate methods of constructing the design be adopted which either guarantee the requirements for automatic assembly or can be guided by the requirements for automatic assembly. The construction of designs according to rule based generative schemes offers the potential to realise this aim[3,4,5].

The recognition that component features and partially completed designs are central to a design description forms the basis for current approaches to CAD. The attention to features will lead to understanding during the design process of the complexity and difficulty of mating. However, it will not necessarily contribute to understanding the aggregation of these features which form the spatial context for planning the assembly sequence and the types of approach move required[6]. Features based CAD requires augmentation by rule based approaches to the aggregation of features to create the final design. In the case of assemblies which are aggregates of features across several components, the developing relationships among features across many components becomes central to the task of planning assembly moves.

Design for assembly must thus be based not only on the local mating of features but also on the relations of features across the design. In this way the spatial context of assembly is determined and the information required for assessment of motion planning made available. The nature of assembly design rules should thus encapsulate the requirements for rationalised assembly operations. A route to this goal is to consider the design rules as mirroring assembly actions. Designs will be created by sequences of constructive rules which act at the component level to bring together features and then act at the sub-assembly level to bring together aggregates of features. The design is thus described as a sequence of rule applications based on the spatial relations between features and components. However, there still remains the central problem of inferring the features of components and sub-assemblies which emerge from the rule applications but are not specified explicitly in the rules. It is these emergent spatial relations which provide the context for assembly operations.

The scope for systematic and rule based design systems which can encapsulate assembly knowledge, is considerable. The opportunities provided by the construction industry are particularly significant in this area as it is largely an unautomated activity of considerable size, exhibiting complex material and component delivery problems as well as the mating and fixing problems associated with a wide range of components. A major lesson from manufacturing assembly is that without fundamental attention to design for automatic assembly there is a tendency to move towards reduced cost or easily manufactured components at the expense of being unable to assemble automatically. The building design is a complex spatial assembly characterised by its static, evolving nature. Assembly operations take place inside and around the current state of the building structure. Assembly 'stations' are moved around the partially completed structure.

Access and emerging features are critical in building design. The robot assembly device will be intimately linked with the building structure. This emphasises the need to examine design for assembly in parallel with the development of assembly automation. The design and construction sequence will determine the possibility for automatic assembly to a greater extent than component design for successful parts mating. It is argued that the design of the building must be understood in terms of a developing assembly of components which form the spatial environment for these assembly operations. The building design description required to plan and assess automatic assembly is thus not static but phased and sequential. The rule based descriptions indicated above for manufacturing assembly appear to have particular relevance for building design(7,8).

The planning of the construction process requires the transport and fixing of large numbers of parts. The design process has tended to emphasise the compositions of these parts in terms of functional relationships to satisfy functional specifications such as support, weather protection, lighting, heating and ventilation. Construction planning emphasises the sequence and spatial relationships of these components as they are brought into place on site. The ability of design systems to exhibit knowledge of these construction sequences would be a great advantage in planning for automatic or robotic assembly. The designer should be aware of the spatial relations required between features and the spatial context in which they are to be realised. If it were possible to make these special relationships an integral part of the means of design, then rules could be constructed based on these spatial relations. Constructive rules to implement defined spatial relations then form the basis for creating building designs and would open the way for a systematic link between building design and construction planning. The spatial relations between components are now the central units of the design. The developing building during construction then corresponds to the developing design as rules of construction are applied.

MODULARITY

A criterion often applied to design for assembly especially in flexible manufacturing assembly is modularity. This may refer to the use of similar components, components within a modular dimensional system, or the use of sub-assemblies common to different final assemblies. The complex spatial nature of the developing building can be considerably simplified if the component assemblies obey a system of dimensional co-ordination. Not only are the local operations of handling, mating and fixing simplified, but also the determination and updating of the spatial properties of the developing building.

The concepts of modularity can be effectively put into practice using rule based generative design methods. The selection of design rules based on the spatial relations between a vocabulary of modular components will ensure resulting modularity in the developing and final design. The modularity may thus be incorporated into design generation rather than made an imposed constraint on the design. This can avoid a cascading process by which small local changes, made to ensure modularity, have an effect on the whole design in potentially drastic and unforeseen ways. Traditional modular schemes are often considered to impose undue constraint on design. This is caused by the concentration on component modularisation, without the formal representation of the possible ways that the components can be assembled. Modular ways of relating components contained in constructive design rules will ensure the dimensional coherence of the whole design and provide the freedom from the apparent constraint imposed by modular components. Modularity and dimensional co-ordination across disparate elements of the building is essential for simplifying assembly and for planning the sequence and hierarchy of assembly operations. Further, effective planning for robotic assembly across building projects will be facilitated by the adoption of agreed systems of dimensional co-ordination and component tolerances.

COMPONENT DELIVERY

Planning robotic assembly deals not only with the assembly itself but also with the presentation and delivery of components to the assembly system. In manufacturing this aspect of automatic assembly, it is not directly concerned with product design. However, for construction assembly this becomes a critical area of the design. Components need to be delivered to locations within the building. The geometry of constraints and supports afforded by the current building state needs to be understood at each stage. The building structure itself may be used as the basis for component transport and delivery. Building design must consider how developing geometry affects material transport. These considerations range from the needs for additional delivery structures to the evaluation of component access to the site of assembly. Component routing and access will be dynamic as construction proceeds and effective construction planning needs knowledge of how the building geometry will evolve on site.

The concepts of rule based design provide the basis for considering the dynamic geometry for component access. Explicit requirements can be placed on the generated designs to ensure adequate access for both components and assembly requirements. These constraints will affect the type and sequence of rule applications and provide the means to control design generation as well as forming the basis for creating material delivery plans. The major difficulty encountered at this stage is that complex spatial conditions need to be recognised in the developing design which are the consequence of composite rule applications in different areas and at different levels of detail in the design. The inference of emerging spatial properties is a critical problem.

The design of assemblies has often attempted to group components into functional sub-assemblies. The design principle has application to building assembly by the use of sub-assemblies prefabricated on or off site. The potential advantages in factory based prefabrication resulting from a simplified working environment needs to be set against the spatial access and fixing problems for the complex sub-assemblies. The evaluation of relative merits poses a significant problem for the construction planner and for the methods of rule based design proposed. Effectively there are two assembled systems, both generated by sequences of rule applications and both with complex aggregates of spatial properties. Ensuring access, transport and handling requires mutual interaction between the two generative schemes and the emergent spatial relations they create.

BUILDING ASSEMBLY DESIGNS

The use of rule based design in building is not confined to rules corresponding to physical operations. The arrangement of spaces within the building can also be generated by rule based systems[7,8]. The rule applications are not then explicitly concerned with the building construction, but rather its spatial structure and architectural style. The rule sequences take no account of construction principles and serve to create spatial aggregates which meet functional and aesthetic specifications. The architectural and construction modes of considering building design thus have a common formal base when considered in terms of rule based design systems. This should provide the opportunity to integrate the two approaches so as to provide the architect with the formal tools needed for spatial design of the building for site assembly. Both methods describe the same spatial composition, but in different ways. The link between the two descriptions is needed to effect integration of design and construction. Expressed in a different way, the translation is required between the formal languages derived from separate rule systems to provide the interpretation of architectural design as construction procedures.

CONSTRUCTION PLANNING

The main thrust of the paper has concentrated on the need for means to represent the developing building so that construction operations, particularly assembly, can be planned effectively. The use of such methods may only have a manual effect on the functional and aesthetic features of the building, but there will be a significant impact on the nature of the building structure and the design of components to facilitate automatic assembly. Design for assembly in construction must consider not only potential construction plans but also the precise details of those plans in order to make adequate evaluation of the overall use of construction resources, whether robotic, machine or manual. To this end, attempts are being made to link computer based production systems[9] with CAD systems. These integrated models should allow problems to be formulated in a more rigorous manner than hitherto and provide solutions that have not previously yielded to manual methods. Although the design may be suitable for robotic assembly, it is possible that further evaluation shows that time and cost far outweighs any advantage in labour saving or quality. It is important, therefore, that effective methods evaluating generated construction plans and making iterative improvements are available. Time and cost implications must be fully explored for any construction plan and every effort made to optimise the relationship of these two parameters[10].

More significantly, a major problem relates to the variability of construction sites and site layout case studies are being analysed. The main planning features identified are movement, storage, activities, access and control with particular attention given to movement and action density in the various activity flows. It is important that on all sites a central focus or series of central focuses, about which all activities will revolve, is determined. In other words, the "centre of gravity" of each particular structure is determined around this focal point. All other operations in that area can be coordinated geometrically for the site to produce a common centre of gravity which becomes a focus for all operations. In particular, the point of access will influence the focal point. It must be remembered once again that this will not be static but will change with the dynamic nature of the activities involved.

ROBOT PLANNING

The creation of an evaluated construction plan identifies robotic requirements where appropriate and necessary. The broad feasibility based on design geometry of robotic assembly will be established. However, the problem still remains of planning robot actions to realise the spatial relations between components as specified in the design rules. The power of the rule based approach to building design is significant at this stage. Motion plans are constructed within the current building geometry for each stage for construction. The spatial environment for the robot, when moving and handling components, is derived from the corresponding design description for that stage of construction. The detailed programmes of assembly moves are now constructed and requirements for sensor guidance and navigation specified. At this stage, it may be appropriate to leave the local planning of the assembly moves to the execution phase of the robot task. The nature of the construction site may demand this separation of planning and execution since it may be difficult to foresee all contingencies in constructing the assembly plan. Given an inherent uncertainty in the construction environment it would be misplaced effort to attempt detailed planning of robot moves before the corresponding stage of construction is reached. The requirements of rule based design which have been proposed as appropriate for creating building design in such a way that can capitalise on the advantages of robotic assembly and for which effective construction plans can be generated to make optimal use of construction resources, will impose particular needs on the nature of CAD systems used in architectural design and construction planning.

Architectural CAD has generally used a formal modelling framework based on geometric elements for entering, recording and displaying the spatial features and characteristics of the final design or significant sub-assemblies. The process of creating such models involves the informal application of design rules structured according to levels of detail and types of building service.

CONCLUSIONS

Design evaluation for construction planning and assembly methods should be accommodated within the design process and guide design generation at each stage. The argument is that a new approach to design is required. Autonomous robotic machines require appropriate design descriptions of the building to make available the necessary information about the developing geometry of the building as work progresses on site. It is proposed that rule based methods, based on assembly operations of components on site, provide the foundations for this new approach.

REFERENCES

1. Boothroyd G. Dewhurst P. *Design for Assembly Handbook.* University of Massachusetts Press (1983)
2. Redford A. Lo E. *Robots in Assembly*, OU Press (1986)
3. Medland A. Mullineux G. 'The investigation of a rule based spatial assembly procedure', Proc CAD-86 (1986)
4. Earl C. 'Creating Design Worlds', Planning and Design, 13 (1986)
5. Earl C. Cusack M. 'Building Design for Robotic Assembly', Proc 5 ISRC (1988)
6. Slocum A.H. 'Development of the integrated construction automation methodology', Proc 3 ISRC Marseilles (1986)
7. Stiny G. 'Introduction to shape and shape grammars', Environment and Planning B,7 (1980)
8. Flemming U. et al. 'A generative system for building design layouts', Department of Architecture, Carnegie Mellon University (1988)
9. Cusack M.M. 'A simplified approach to the planning and control of cost and project duration', Construction Management and Economics, 3, pp. 183-198 (1985)
10. Cusack M.M. 'Optimisation of Time and Cost', Project Management 3 (no.1), pp. 50-54 (1985)

DISCUSSION PAPER 44

David R Moore, De Montfort University, Leicester

a) The robot has been used to construct 'dry-wall' blockwork. There is mention of a possible approach to providing a means for the robot to construct a wall with mortar joints. The mortar in this case seems to be specifically designed for ease of use by the robot. Given that the process of bricklaying (and site organisation in general) is in effect being changed to suit the capabilities of the robot, is there any intention of testing the strength of the resultant walls? It would seem important to be sure of the performance of the product in comparison to a 'traditionally' produced equivalent.

(b) Assuming that the robot is not intended to be used within a covered, factory style environment and also considering that 24 hour employment is one of the potential benefits of using robots, is it intended to supply the robot with a facility for the measurement of moisture content within the bricks/blocks being used? In order to maintain the quality of the product in a site environment the measurement of a possibly wide range of moisture contents could be argued to be vital.

(c) With regard to the quality of the product resulting from the use of a robot, in particular the quality of 'wet-wall' products, is it planned to match output rate to the rate at which the product would gain sufficient strength to carry its self-load? This may not be a problem on brickwork but the greater output rate on blockwork may cause stability problems.

45 MULTI-STOREY MODULAR BUILDINGS

R.P.L. McAnoy and J.W. Richards
Taywood Engineering Limited

This paper outlines the work underway by a leading European team of industrial and academic partners to establish design guidance and standards for fully fitted modules for multi-storey buildings. It provides an indication of the project's technical objectives, describes some observations from an initial structural analysis and the planned programme of testing which will be used to verify the theoretical results.

INTRODUCTION

The European construction industry is always searching for potential improvements in construction practice to increase speed, quality or cost effectiveness. Many organisations are considering an increase in the use of off-site prefabrication. The use of fully fitted modules for buildings has already been recognised as an important development of prefabrication meriting further attention. This technique is currently used by certain hotel groups for bedroom/ bathroom construction. It was selected also for some of the prestigious commercial buildings constructed in London in the late 1980's, where complete factory finished toilet and washroom modules were produced to a very high quality specification (Plate 1). There are many other building forms where complex repetitive fitting out could be suited to modular construction.

The successful uses of modules for buildings can lead to significant client benefits through construction management improvements associated with quality control (for example by reducing the demand for some skills traditionally in short supply), planning complexities (as a result of modern finishes and services requiring a large number of different individual specialist subcontractors) and speed restraints (due to sequential operations, space limitations or drying times for wet trades).

In adopting such a technique, the UK building industry's nature is quite correctly cautious and the potential benefits noted above need to be considered in individual cases against the architectural and functional requirements of the building, the construction demands and the benefits of modules against the possible initial cost disadvantage that is commonly accepted as being in the range of 5-25% (due to transportation costs and the additional material and labour involved in forming the box of the module). The potential may be limited at

Plate 1. Recent module technology to be extended to multi-storey modular construction

present by the lack of any rational means to compare producers' claims on functionality, technical specification and quality and also the engineering and design capability of the module producers.

In recognition of the potential for extending the scope of modular techniques for permanent buildings, a group of interested organisations applied to the Commission of the European Communities (CEC) for financial support to investigate design criteria and production techniques and to establish a technical base from which standards and codes of practice could be developed. A major effort was essential to achieve the strategic objective of the project to dramatically improve the efficient use and competitiveness of multi-storey building modules.

In December 1991, a three year project was started with partial funding under the CEC Brite-Euram programme. The contributing organisations in the project team are :

Taywood Engineering Ltd (project leader)
Dragados Y Construcciones
R B Farquhar Ltd
Torvale International Ltd
Haden Young Ltd
Styling International Ltd
A I Systems
University of Liege

PROJECT DETAILS

The project has been divided into 8 separate and distinct technical areas. Each of these addresses a fundamental aspect. However, the work in each technical area involves completing several tasks, often interdependent upon those in other areas.

The technical areas are :-

- definition phase
- development of the module box structure technology
- development of building materials technologies
- development of services technologies
- research into production technologies
- development of computer software
- development of inter-connection technologies
- prototype construction and technical documentation

The project will integrate existing technologies from various different industry sectors, to modify and develop these to provide the appropriate solutions to global specifications set up for the module as a whole. As part of their role within the project, various partners within the team will bring in background technologies and experiences.

The primary focus of the project will be on specifications for volumetric modules, as they have the greatest potential for maximising the use of factory production methodology. However, the specifications will be appropriate to most systemised elements produced in a pre-formed manner for delivery and fabrication on site.

The main design elements under consideration are:

- structural
- fire
- thermal and acoustic insulation
- internal environment
- architectural detailing
- power, light and mechanical demands
- weather and external environmental protection
- maintenance
- production and installation quality control

PRESENT MODULAR CONSTRUCTION

The supply of modular units for developments ranging from single to three storey heights is adequately covered. Such modules are usually built in a timber framework, structural steelwork, or a mixture of these materials.

The leading producer of timber framed structures in the UK does not use timber framing for building developments over three storeys. Another company building modular system units manufactures large single storey modules using trusses to deal with the larger spans involved, and considers this an economical method of providing clear working areas in a short time.

A few specialists in modular construction produce separate units which can be built to 4 storeys in height, typically using structural steelwork elements.

Buildings of five storeys and above in height in the UK need to be designed to meet the requirements of the disproportionate collapse clauses in the Building Regulations, and 5 storey construction of this form is rare.

The modular market lacks suppliers of modular units for heights over 4 storeys. However, a research and development company licensing a technique to building companies on a project by project basis, claims that their system could be built to 15 storeys. The system involves pouring concrete between the modules on site to create insitu reinforced concrete walls and floors. No positive details of the use of the system have been identified at this time.

Some manufacturers using steelwork consider building with modules up to five storeys to be outside the limits of their systems, and state that extra, expensive, steel would need to be introduced. However, lift shafts and service modules have been installed by a few suppliers up to 8 storeys.

There is the absence of a system covering the high rise market (including hotels and offices) having a factory produced 'dry system' approach.

Small lightweight hotel bedroom modules at the lower price end of the market measuring 6m by 3.5m weigh about 3 tonnes, whilst an 8m by 4m unit built using structural steelwork framing can weigh as much as 8 tonnes. These bedroom units would all be moved in the UK by road to site.

Generally suppliers of modular units are able to meet the requirements of the UK Building Regulations for sound reduction, fire resistance periods, and insulation. Many suppliers are able to improve on the minimum 'U' values required by the current Building Regulations.

The policy of most manufacturers is to build a module and to deliver it to site immediately, without further storage.

A leading timber frame producer can erect 8 bedroom hotel modules per day, and build a 40 bed motel in less than 20 weeks from the start of foundation work.

STRUCTURAL CONSIDERATIONS

There are fundamental differences between the structural behaviour of a traditional multistorey frame and that comprising interconnected stacked modules. These differences have not been subjected fully to a rational study in which data from analytical and physical modelling have provided confidence in the design assumptions.

Firstly, the traditional frame and floor slabs are designed to transmit the dead and imposed loads to the foundations without taking account of the possible additional stiffness and carrying capacity of secondary elements, such as wall and floor panelling, external cladding etc. Secondly, the insitu connections at node points have a fixity which has little possibility of major variance from the design standard. Thirdly, primary loads are transmitted through structural members whose performance is well understood i.e. columns, beams and floor slabs. The designer can leave the means of connecting wall panels and other finishes to the structure to the detailer. Any uncertainty in the load carrying capacity of these connections due to site practices, or indeed any uncompleted details at the start of the frame's construction, does not compromise its structural integrity.

In contrast, a structure developed by volumetric modules has very different characteristics. For low rise construction, the main structural demand is due to the temporary displacements imposed by racking, occurring as a module is handled during fabrication, transportation and installation. It is only as the number of storeys increase above three that the permanent loads dominate. Above four storeys an additional load case of 'disproportionate collapse' needs to be considered. Column interconnections are possible weak points, as they need to take up any tolerances between modules and also provide site connections as the structure is formed. There are many more structural members, as each column taking the permanent loads comprises those from a set of the four corner posts of the original modules and nodes for different load cases are not coincidental, possibly adding torsional factors. Finally, the confidence in factory-made connections, with far greater accuracy and reliability and fixing mechanism of the finishing panels to the frame, provides an opportunity to consider the added stiffness which this provides. In analysing module structures, this final point was found to be particularly important in assessing structural performance.

It should be noted that many parts of Europe have codes of practice which require earthquake loads to be taken into account. In these cases, the critical aspect is normally the lateral stiffness under the appropriate design accelerations.

PRELIMINARY ANALYSIS

For the purposes of an initial detailed study within this project, a module sized at approximately 8m long x 2.8m wide x 2.8m high was selected having 4 corner posts. Modules were designed as 'stackable', freestanding units of up to 6 storeys in height. The size selected provides for use as a 'typical' hotel bedroom/bathroom unit, with adjacent corridor and integral services provision. Such units will accommodate a variety of finishes and are transportable by road in the UK and elsewhere. A sectional elevation of one of a variety of modular structures and sub elements under investigation is shown in Figure 1.

A structural design analysis is being undertaken of various alternative structural box configurations against designs generated by the team. A number of options have been considered to-date.

Initially a design was considered which comprised two dimensional-frames, made of welded cold formed galvanised profiles, bolted together on the factory floor into a box-like three dimensional frame.

An initial finite element model, as illustrated in Figure 2, was analysed under the simplest handling situation of a point load lifting one corner of the box structure, to compare the numerical results later with test data from a prototype frame.

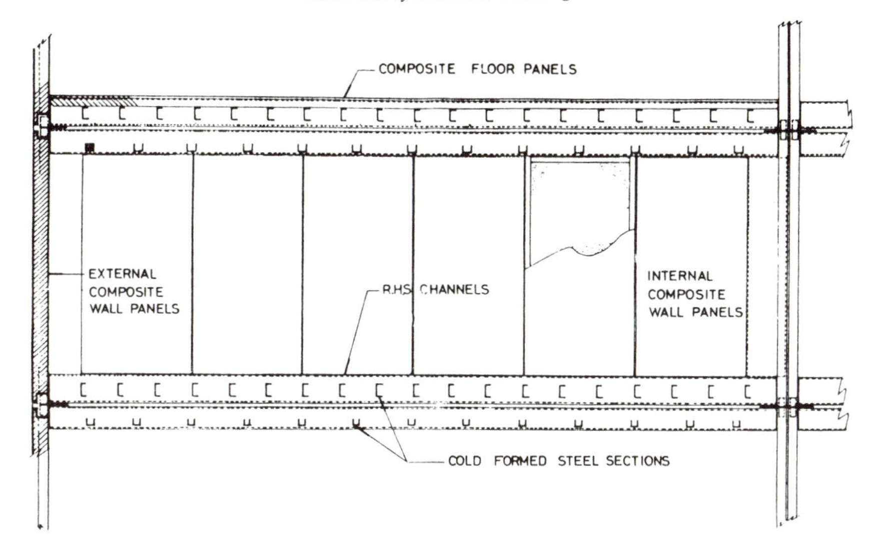

Figure 1. Section through panelled modular unit

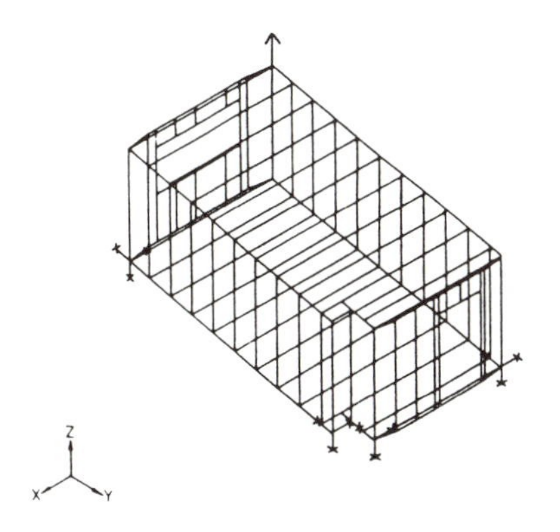

Figure 2. F.E. Model-preliminary prototype
Simple displacement under single point lift from factory floor

Analysis was then undertaken on a block of 60 stacked modules shown on Figure 3a. The intention was to obtain a first appraisal of the stability of the structure. Many simplifying assumptions were made concerning the loading and the behaviour in order to speed the computation. Several models of decreasing complexity were considered, the purpose being to reduce computer analysis time.

A numerical model of the whole building was planned and the finite element discretization of one pair of modules with the corridor is shown on Figure 3b.

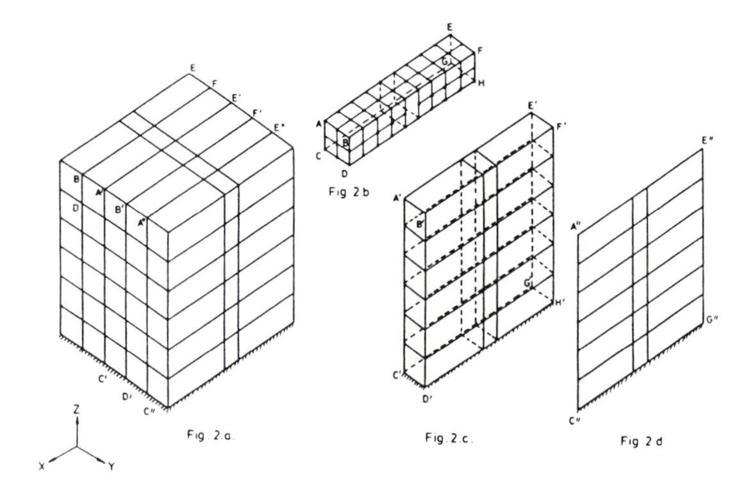

Figure 3. F.E. Model-stack of module pairs, plus corridors without panels. Self weight and wind. Out of plane bending

To obtain a reasonable accuracy, 8 shell (plate and membrane) elements were used for each of the floor, roof and wall panels, 4 elements were used for the end wall and corridor wall.

The most severe loading condition is likely to occur when the building is subjected to selfweight, surcharge and wind or earthquake applied along the X or Y direction. Because of the module arrangement, the structure is stiffer when loaded in the Y direction - short spans, with closely spaced columns to resist the overturning moment.

Hence it was decided rather to analyse the structure loaded in the X direction. One stack of modules such as A'B'C'D'E'F'G'H' shown in Figure 3b was considered to be representative. The same finite element mesh was used as that explained for a single module.

As a first approximation, it was assumed that all panels were the same and made of two layers of plywood, 12mm thick, separated by 100mm of insulation material, and acting as a sandwich plate. Significant conclusions were obtained from the analysis:

1. The panels have a very significant effect on the overall behaviour of the structure.
   - When the stiffness of the panels is taken into account, the lateral deflection at roof level is 70 times smaller than the deflection obtained without panels.
   - The bending moments in the beam elements and columns are noticeably reduced when the panels are taken into account.
2 The columns of the first storey are the most heavily loaded.

3. The combined load case of self-weight plus surcharge plus wind, is the most severe.
4. Bending in the Y direction and the accompanying torsion of the frame elements in the XZ plane are insignificant.

The results led to the conclusion that a two- dimensional analysis was sufficient, for a preliminary design. An example of bending moments obtained in such a model without panels is given in Figure 4 and with panels is given in Figure 5.

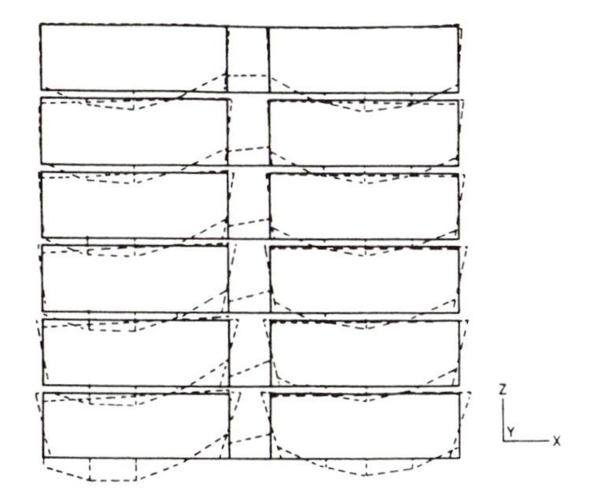

Figure 4. F.E. Model-stack of 6 module pairs, plus corridors without panels. Self weight and wind. Out of plane bending

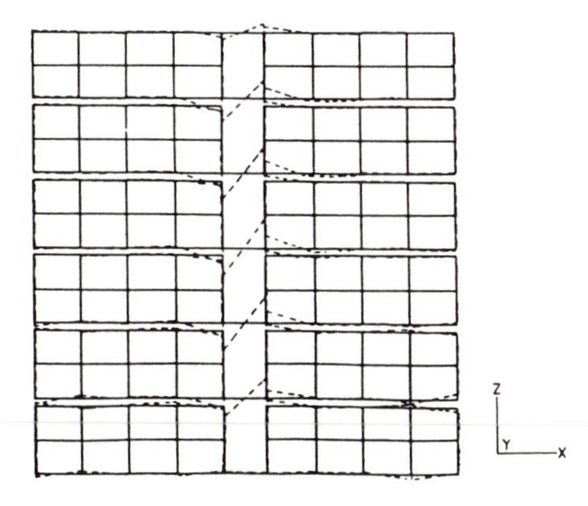

Figure 5. F.E. Model-stack of 6 module pairs, plus corridors with panels. Wind load. Out of plane bending

The same model was also used to investigate the replacement of the wall panels by a numerically equivalent diagonal bracing. It was found that the stiffness brought by the bracing increases much faster than the cross section of the braces. This means that relatively weak panels already play a significant role. Detailed analysis of a variety of other box configurations and designs and partial components will continue to be analysed and methods devised for incorporating stiffnesses of the sub modular elements within the structural analysis of the box module and evaluating the stiffnesses of different jointing mechanisms.

TEST PROGRAMME

A number of tests will be undertaken to examine the physical performance of modules and sub elements at varying stages of fitting out. Two stages of testing are planned, where fabrication of modules or modular elements will be undertaken by the Team in Scotland and in Spain and transported by sea and road for testing.

Stage 1
It is initially necessary to test the effectiveness of factory fabricated wall and floor panels and assess their stiffness and actual contribution into the structure. Tests on these sub modular elements will include the application of static vertical and horizontal loads and observations and measurements made of deflections under a range of values.

Stage 2
The object of the second stage is to validate the prototype designs both against detailed performance specifications.

Full scale testing of two fully fitted out prototype modules, complete with structure, finishes and services will be undertaken. These modules will be constructed to designs and details prepared by the Team during the preceding period. The modules will be 'stackable', self supporting elements of a multistorey structure.

The range of tests for Stage 2 will include the following:-

- Visual inspection for damage caused by transportation, especially of the prototype module shipped from Spain. A full photographic record will be made prior to transportation.
- Subject the modules to a full range of water penetration tests, including dynamic wind loading.
- Application of physical loading to simulate stacking of modules.
- Measurement of the acoustic performance of the modular wall panels and linings.
- Measurement of thermal properties of the modular wall panels.
- Physical interconnection of the two modules and laterally loading them to test the performance of the interconnection.

- Connection of the services outlets to the mains systems and testing at each service element. Monitoring of noise and vibration levels.
- Inter-connection of the services outlets and monitoring the ease with which they can be made.
- Comparison of the test results from physically and environmental loading the module prototypes with the expected behaviour of these from computer modelling.

It is the Team's intention to invite various outside building developers, specifiers and architects to inspect the prototype modules and tests.

CONCLUDING COMMENTS

On completion of the project in 1994, a technical document will be prepared for publication, aiming to assist the development of a European technical guide for building modules, with guidelines setting out construction industry recommendations for module use and production.

The development of a rational design method for modules is complex as the traditional design process of a building has to be reconsidered from first principles to take into account factory production, transportation and installation methods. The fitting of services and finishes in a factory provides advantages but great care must be taken to ensure that the selected materials, connections and detailing are suitable for the extra load cases imposed by this method of construction.

The final deliverables from the project need to be as broad as possible to encompass the variety of designs on offer from European manufacturers. The first half of the project has included the preparation of global and particular specifications so that each of the specialist areas have common input information, followed by an analysis of the design options available and current practice. The design assessment stage is clearly showing the importance of the structural base. This is the critical component governing factory production methods and its accuracy and rigidity has a major influence on the installation procedure. There is great potential for manufacturers to improve quality and reduce costs and there is a scope for innovation.

The test programme is due to start in mid 1993 and the details are presently being discussed and agreed by the team. The main purpose will be to demonstrate both innovative cost effective features and the capacity of modules to accommodate structural and environmental loadings without damage to the services and finishes.

ACKNOWLEDGEMENTS

The authors wish to thank the directors of Taywood Engineering Ltd and their partners in the Brite Euram project for permission to publish this paper. They also thank the Commission of the European Communities for their financial contribution to the project.

46 FREESPAN CORRECTION OF SUBMARINE PIPELINES USING DIVERLESS TECHNIQUES

John Anderson
International Composites Ltd

INTRODUCTION

Submarine Pipelines are used extensively to transport hydrocarbon products from offshore fields to shore based processing facilities and also for interfield connections between platforms, floating bulk storage and tanker loading terminals. Several of the main arterial submarine pipelines in the North Sea conveying crude oil and gas from offshore fields to land based processing facilities are large diameter (32, 36 inch) and several hundred kilometres long.

The stability and structural intergrity of these submarine pipelines can be seriously affected by strong currents which erode the seabed from underneath the pipe causing freespans to develop. If the freespan length and/or height continue to develop without corrective action then damage to the pipeline can result. This means significant potential loss of revenue to the pipeline operator, a potential safety risk for personnel and equipment near the pipeline and potential damage to the environment.

Companies operating submarine pipelines in the North Sea are required by government legislation to inspect their pipelines annually and to carry out any intervention work which may be necessary to keep the pipelines within standard minimum operating criteria set by the Health and Safety Executive (HSE).

Remotely operated vehicles have been used for carrying out pipeline survey and inspection work for the past 10 year. When the survey and inspection of the pipeline is complete the pipeline operator then must decide what corrective action(s) must be taken to rectify any problems that exist. Any work associated with corrective action to the pipeline has traditionally been carried out by diving operations which typically require a team of hyperbaric divers and a dedicated diving support vessel (DSV).

Freespan problems are usually corrected either by rock dumping, for large areas of scour, and sand bag or grout bags for smaller areas. Grout bags are fabric forms made from PVC or polypropylene canvas which the diver places below a pipeline in a predetermined position. The fabric form is then injected with grout from the surface. After curing the grout forms a structural support below the pipe.

INNOVATION SPONSORED BY SHELL EXPRO

New techniques have now been developed which allow remotely operated vehicles to carry out the corrective action as well as the survey and inspection work necessary on submarine pipelines.

In 1991 and 1992 Shell Expro, a joint venture between Shell UK Exploration and Production (the operating partner) and Esso UK Exploration and Production, sponsored the development of grout bag technology which could be deployed by diverless techniques.

International Composites Ltd (ICL) were responsible for developing the grouting services which were required to interface with the Remotely Operated Vehicle (ROV), provided by Subsea Offshore Ltd (SSOL), a major international diving company who operate a fleet of 26 ROV's.

SPECIFICATION FOR THE SUPPORT VESSEL AND ROV

The Kommandor Subsea, a dedicated ROV support vessel operated by SSOL was selected for carrying out the given scope of work. The ROV is launched and recovered via a moon-pool through the centre of the vessel. The main lifting point for the machine passes through the centre of mass of the vessel minimising the effects of pitch, roll and heave. All ancilliary equipment is launched from as near to this point as possible to give the vessel a greater weather capability, and therefore maximise operating time.

The launch area and workshops are contained within a hangar where all launch and maintenance activities can be carried out in a controlled environment. From the moment the ROV enters the water, until it is on deck again, the parent vessel maintains contact through the umbilical which supplies electrical power, receives video pictures, other data and feedback from the ROV control system.

The position of the ROV is calculated from through-water acoustic signals and the parent vessel maintains position on the ROV by feeding this positional information to a computer controlled engine management system. A surface positioning system which uses satellite information to plot the vessel position is incorporated in the underwater navigation system.

The reliability and confidence levels associated with these integrated sub-systems has increased during the past ten years to the point where oil companies are "relaxed" about using ROV's about their live transmission pipelines, which transfer hydrocarbons worth millions of pounds each day.

Pioneer 11, an ROV system designed, developed and manufactured by SSOL was used for the grout bag intervention work. The ROV was fitted with two manipulators which attempt to duplicate the human arm. The manipulators are not particularly sophisticated but are very strong, robust in construction and easy to maintain.

A seven function manipulator was used on the front right side and five function on the front left side of the ROV which provided the following capabilities for use in the deployment of the grout bags, grout hose release and collecting reusable items:

Right Side	**Left Side**
shoulder up or down	shoulder up or down
arm left or right	arm left or right
elbow in or out	elbow in or out
elbow rotate	no rotation
wrist rotate	no rotation
wrist left or right	wrist left or right
jaws open or close	jaws open or close

GROUTING SERVICES

Onshore Trials

The specification for the equipment and consumable items to be used offshore were tested in full scale trials at an on-shore test tank facility in Findon.

For the trials, Shell Expro provided several 3metre long sections of 36" diameter pipeline which had 3" concrete weight coat to simulate the Offshore pipelines. The sections of pipeline were supported at each end to provide freespan heights corresponding to Offshore requirements.

Grout

A special hydrostable grout with mix design minimum 30 N/mm2, made from Ordinary Portland Cement, seawater and Sika admixtures was developed and tested in laboratory trials before the test tank grouting trials. The grout mix was designed to have high fluidity to assist pumping and filling of the grout bag, and also antiwashout properties which allow good underwater visibility at all times.

The quality of the grout was maintained by checking the specific gravity on each alternative batch. Cube moulds (3 no. x 75 mm moulds per batch) were made from specific batches of grout and cured for 28 days to check compressive strength.

Grout bags

Several grout bag designs were manufactured and tested in the tank trials under controlled conditions before selecting the final grout bag design to be used

Offshore. The grout bags were made from a high visibility yellow polypropylene canvas. The main features of the grout bags will be demonstrated.

Grout Equipment Spread
A grout equipment spread was selected for the onshore trials to demonstrate that it was "fit for purpose". The equipment spread consisted of a high shear grout mixer, grout pump and an air powered hose winch complete with grout hose.

Bulk handling of cement is well proven technology therefore ICL considered that setting up a large capacity grout handling system was not necessary as the quantity of grout required for the trials was relatively small. Bagged cement was therefore used to make up the grout for the trials.

The grout equipment package that was used Offshore is shown in the deck lay-out of the Kommandor Subsea.

METHOD OF DEPLOYMENT

Once the support vessel was on-station at a location where freespan correction was required the ROV was launched and carried out a preliminary survey of the freespan area. The position for the freespan support(s) was then marked using a high visibility marker system and a transponder.

A purpose built support/launching frame fitted on the port side access platform at mid-ship was used to handle the work basket for transferring grout bag(s), with grout hose attached, from the support vessel to the seabed.

The work basket was placed on the seabed approximately 4 metres from the pipeline at the freespan location. The ROV then deployed the grout bag from the work basket into position below the pipeline using a purpose built bridle system which was an integral part of the grout bag construction.

After surveying the grout bag on both sides of the pipeline to confirm correct alignment and position the hydrostable grout was injected into the grout bag until it was full.

A final survey of the filled grout bag was then carried out to confirm that the grouting operation was complete. The ROV then released the grout hose from the grout bag using an autodisconnect unit developed for this purpose. The grout hose was then recovered to the support vessel after purging clean with water.

CONCLUSION

The diverless freespan correction operations proved to be extremely efficient and cost effective compared with conventional diving operations under the conditions experienced in the middle and northern sectors of the North Sea ie. relatively good visibility and low currents.

The following objectives were proven as a result of the project:

a) Major savings in operational costs can be achieved using diverless techniques as day rates to hire non-diving support vessels are significantly less than DSVs.

b) The operational time available was higher with ROV operations than for diving (based on previous project information) as the ROV was able to operate for longer periods in the water and also under conditions where normal diving operations would not have been possible.
c) The rate of grout bag deployment and installation was significantly faster with the ROV than diving operations.
d) Through the use of ROV technology the level of Safety in Offshore operations is significantly improved as it is not necessary to have a "man in the water".
e) Time spent onshore in preplanning each stage of the project and testing unproven equipment, consumable materials and operating techniques allows team members to familiarise themselves with the technology and to develop good operating procedures for Offshore. Fostering good working relations between the "project team members" resulted in good teamwork and efficient organisation throughout the Offshore operations which was the key to the success of the project.

Looking to the future, we would predict that further development of the diverless technology will allow similar work to be carried out in conditions of high currents and reduced underwater visibility. The robotic system will be used to carry out more complex operations and for working in deep water where normal diving operations are not possible.

The use of underwater robotic technology should also prove to be interesting for many marine Civil Engineering applications where project restrictions and delays created as a result of diver involvement can be minimised. This will lead to improved control of underwater project operations.

ACKNOWLEDGEMENT

The author would like to express appreciation to the following personnel who contributed to the overall success of the project:

Shell
- Mr John Newton (Operations Manager-Northern Pipelines Group)
- Mr Ken Vaughan (Offshore Representative)
- Mr George Hogg (Manager ROV Operations)

Subsea Offshore Ltd
- Mr Paul Brain (Manager ROV Operations)
- Mr Douglas Bathgate (Project Manager-Kommandor Subsea)
- Mr Trevor Archer (Kommandor Subsea - Operations Supervisor)
- The crew on board the "Kommandor Subsea"

International Composites Ltd
- ICL Project team members

AUTHOR INDEX

SUBJECT INDEX

This index has been compiled from the keywords assigned to the individual papers, edited and extended as appropriate. The numbers refer to the first page number of the relevant paper.